T0302020

Series on Analysis, Applications and Computation – Vol. 10

ISAAC

Fractional Differential Equations and Inclusions

Classical and Advanced Topics

Series on Analysis, Applications and Computation

More information on this series can be found at http://www.worldscientific.com/series/saac

Series on Analysis, Applications and Computation – Vol. 10

Fractional Differential Equations and Inclusions

Classical and Advanced Topics

∘ Saïd Abbas
Tahar Moulay University of Saida, Algeria

∘ Mouffak Benchohra
Djillali Liabes University of Sidi Bel-Abbes, Algeria

∘ Jamal Eddine Lazreg
Djillali Liabes University of Sidi Bel-Abbes, Algeria

∘ Juan J Nieto
Universidade de Santiago de Compostela, Spain

∘ Yong Zhou
Xiangtan University, China
& Macau University of Science and Technology, China

World Scientific

NEW JERSEY · LONDON · SINGAPORE · BEIJING · SHANGHAI · HONG KONG · TAIPEI · CHENNAI · TOKYO

Published by

World Scientific Publishing Co. Pte. Ltd.

5 Toh Tuck Link, Singapore 596224

USA office: 27 Warren Street, Suite 401-402, Hackensack, NJ 07601

UK office: 57 Shelton Street, Covent Garden, London WC2H 9HE

Library of Congress Cataloging-in-Publication Data
Names: Abbas, Saïd, author.
Title: Fractional differential equations and inclusions : classical and advanced topics /
  Saïd Abbas, Tahar Moulay University of Saida, Algeria [and 4 others].
Description: New Jersey : World Scientific Publishing Co. Pte. Ltd., [2023] |
  Series: Series on analysis, applications and computation, 1793-4702 ; vol. 10 |
  Includes bibliographical references and index.
Identifiers: LCCN 2022033519 | ISBN 9789811261251 (hardcover) |
  ISBN 9789811261268 (ebook for institutions) | ISBN 9789811261275 (ebook for individuals)
Subjects: LCSH: Fractional differential equations.
Classification: LCC QA372 .A227 2023 | DDC 515/.35--dc23/eng20221102
LC record available at https://lccn.loc.gov/2022033519

British Library Cataloguing-in-Publication Data
A catalogue record for this book is available from the British Library.

For any available supplementary material, please visit
https://www.worldscientific.com/worldscibooks/10.1142/12993#t=suppl

Desk Editors: Soundararajan Raghuraman/Lai Fun Kwong

Typeset by Stallion Press
Email: enquiries@stallionpress.com

Printed in Singapore

We dedicate this book to our family members. In particular, Saïd Abbas dedicates to the memory of his father Abdelkader Abbas (1926–2008), to his mother, his wife and his children Mourad, Amina, and Ilyes; Mouffak Benchohra makes his dedication to the memory of his father Yahia Benchohra and his wife Kheira Bencherif; Jamal E. Lazreg dedicates it to the memory of his father Mohammed Lazreg; Juan J. Nieto dedicates it to the memory of his father Fidel Nieto and Yong Zhou makes his dedication to the memory of his father Shaoji Zhou (1930–2005) and his mother Yaoqing Chen (1927–2018).

Preface

The term fractional calculus (FC) is more than 300 years old. It is generally believed to have stemmed from a question raised in the year 1695 by L'Hôpital and Leibniz. Frequently, FC is called fractional-order calculus: including fractional-order derivatives and fractional-order integrals. It is a branch of mathematical analysis which deals with the generalization of operations of differentiation and integration to fractional order. FC have recently been applied in various areas of engineering, mathematics, physics and bioengineering, and other applied sciences. The subject has witnessed volcanic growth in the hands of so many professionals of mathematics, in the form of research papers and books.

This book is devoted to the existence and stability (Ulam–Hyers–Rassias stability and asymptotic stability) of solutions for various classes of functional differential equations or inclusions involving the Hadamard or Hilfer fractional derivative. Some equations present delay which may be finite, infinite, or state-dependent. Others are subject to impulsive effect which may be fixed or noninstantaneous. The tools used include some fixed-point theorems, as well as some notions of Ulam stability, attractivity and the measure of noncompactness as well as the measure of weak noncompactness. Each chapter concludes with a section devoted to notes and bibliographical remarks and all abstract results are illustrated by examples.

The content of this book is completely new and complements the existing literature in fractional calculus. It is useful for researchers and graduate students for research, seminars, and advanced graduate courses, in pure and applied mathematics, engineering, biology, and all other applied sciences.

We owe a great deal to R.P. Agarwal, W. Albarakati, A. Alsaedi, M. Darwish, J.R. Graef, J. Henderson, G.M. N'Guérékata, A. Petruşel, and S. Sivasundaram for their collaboration in research related to the problems considered in this book.

S. Abbas
M. Benchohra
J.E. Lazreg
J.J. Nieto
Yong Zhou
October 2021

About the Authors

Saïd Abbas is a Full Professor at the department of mathematics at Tahar Moulay University of Saida since October 2006. Abbas received his master's degree in Functional Analysis from Mostaganem University, Algeria, 2006, and his doctorate's degree in Differential Equations from Djillali Liabes University of Sidi Bel Abbes, Algeria, 2011. His research fields include fractional differential equations and inclusions, evolution equations and inclusions, control theory and applications, etc. Abbas has published four monographs and more than 200 papers.

Mouffak Benchohra (born in 1964, Algeria) is a Full Professor at the department of mathematics of Djillali Liabes University of Sidi Bel Abbes since October 1994. Benchohra received his Ph.D. degree in mathematics from Djillali Liabes University, Sidi Bel Abbes, Algeria, 1999. His research fields include fractional differential equations, evolution equations and inclusions, control theory and applications, etc. Benchohra has published more than 500 papers and six monographs. He is a Highly Cited Researcher in Mathematics from Thompson Reuters (2014) and Clarivate Analytics (2017 and 2018) and word's top 2% researcher from Stanford University (2020 and 2021). Benchohra has also occupied the position of head of department of mathematics at Djillali Liabes University, Sidi Bel Abbes. He is in the Editorial Board of 12 international journals.

Jamal Eddine Lazreg is a Full Professor at the department of mathematics, Djillali Liabes University of Sidi Bel Abbes since 2016. Lazreg received his Phd in mathematics from Djillali Liabes University of Sidi Bel Abbes, Algeria, 2014. His research fields include fractional differential equations and inclusions. Lazreg has published more than 40 papers in international journals.

Juan José Nieto is a Full Professor of Mathematical Analysis at the University of Santiago de Compostela, Spain, and a Fulbright Fellow at the University of Texas, USA. He has carried out an intense research activity that is responsible for numerous projects and directing many PhD dissertations. He was the Editor-in-Chief of *Nonlinear Analysis: Real World Applications* and is currently Editor-in-Chief of *Fixed Point Theory and Algorithms for Sciences and Engineering*. His research has a great relevancy and impact not only in mathematics: in 2010, he was listed in the exclusive list of the 13 scientists with most "Hot Papers" among all scientific fields in the world. He has been uninterruptedly on the list of *Highly Cited Researchers* since 2014.

Yong Zhou is a Full Professor at the School of Mathematics and Computational Science at Xiangtan University since 2000. He is also a Distinguished Guest Professor at Macau University of Science and Technology since 2018. His research fields include fractional differential equations, functional differential equations, evolution equations and inclusions, and control theory. Zhou has published five monographs in Springer, Elsevier, De Gruyter, World Scientific, and Science Press, respectively, and more than three hundred research papers. He was included in the *Highly Cited Researchers* list from Thompson Reuters (2014) and

Clarivate Analytics (2015–2021). Zhou has undertaken five projects from the National Natural Science Foundation of China. He was the Editor-in-Chief of *International Journal of Dynamical Systems and Differential Equations* from 2007 to 2011. In addition, he had worked as an Associate Editor for *IEEE Transactions on Fuzzy Systems* and an Editorial Board Member of *Fractional Calculus and Applied Analysis*.

Contents

Introduction

The fractional calculus may be considered an old and yet novel topic. It is an old topic because, starting from some speculations of G.W. Leibniz (1695, 1697) and L. Euler (1730), it has been developed progressively up to now. A list of mathematicians, who have provided important contributions up to the middle of the 20th century, includes P.S. Laplace (1812), S.F. Lacroix (1819), J.B.J. Fourier (1822), N.H. Abel (1823–1826), I. Liouville (1832–1873), B. Riemann (1847), H. Holmgren (1865–1867), A.K. Grünwald (1867–1872), A.V. Letnikov (1868–1872), H. Laurent (1884), P.A. Nekrassov (1888), A. Krug (1890), I. Hadamard (1892), O. Heaviside (1892–1912), S. Pincherle (1902), G.H. Hardy and I.E. Littlewood (1917–1928), H. Weyl (1917), P. Lévy (1923), A. Marchaud (1927), H.T. Davis (1924–1936), E.L. Post (1930), A. Zygmund (1935–1945), E.R. Love (1938–1996), A. Erdélyi (1939–1965), H. Kober (1940), D.V. Widder (1941), M. Riesz (1949), W. Feller (1952). However, it may be considered a novel topic as well. Only since the 1970s, it has been the object of specialized conferences and treatises. For the first conference the merit is due to B. Ross (1975) who, shortly after his Ph.D. dissertation on fractional calculus, organized the *First Conference on Fractional Calculus and its Applications* at the University of New Haven in June 1974, and edited the proceedings, see [328]. For the first monograph, the merit is ascribed to K.B. Oldham and I. Spanier, see [299] who, after a joint collaboration begun in 1968, published a book devoted to fractional calculus in 1974. Nowadays, the series of texts devoted to fractional calculus and its applications includes over ten titles, including (alphabetically ordered by the first author): Kilbas *et al.* (2006); Kiryakova (1994); Miller and Ross (1993); Magin (2006); Nishimoto (1991); Oldham and Spanier (1974); Podlubny (1999); Rubin (1996); Samko *et al.* (1993); West *et al.* (2003); Zaslavsky (2005); Guo *et al.* (2015); Chakraverty *et al.* (2016); Yang *et al.* (2016);

Anastassiou *et al.* (2018); Georgiev *et al.* (2018); Ray *et al.* (2018); Baleanu *et al.* (2019). This list is expected to grow up in the forthcoming years. We also cite three books in Russian: Nakhushev (2003); Pskhu (2005); Uchaikin (2008).

Furthermore, we call attention to some treatises which contain a detailed analysis of some mathematical aspects and/or physical applications of fractional calculus, although without explicit mention in their titles, see, e.g., Babenko (1986); Caputo (1969); Davis (1936); Dzherbashyan (1966); Dzherbashyan (1993); Erdélyi *et al.* (1953–1954); Gel'fand and Shilov (1964); Gorenflo and Vessella (1991). In recent years considerable interest in fractional calculus has been stimulated by the applications it finds in different areas of applied sciences like physics and engineering, possibly including fractal phenomena. In this respect A. Carpinteri and F. Mainardi have edited a collection of lecture notes entitled Fractals and Fractional Calculus in Continuum Mechanics (Carpinteri and Mainardi, 1997), whereas Hilfer has edited a book devoted to the applications in physics (Hilfer, 2000). In these books, the mathematical theory of fractional calculus was reviewed by Gorenflo and Mainardi (1997) and by Butzer and Westphal (2000). Now there are more books of proceedings and special issues of journals published that refer to the applications of fractional calculus in several scientific areas including special functions, control theory, chemical physics, stochastic processes, anomalous diffusion, rheology.

Reviewing its history of three centuries, we could find that fractional calculus were mainly interesting to mathematicians for a long time, due to its lack of application background. However, in the previous decades more and more researchers have paid their attentions to fractional calculus, since they found that the fractional-order integrals and derivatives were more suitable for the description of the phenomena in the real world, such as viscoelastic systems, dielectric polarization, electromagnetic waves, heat conduction, robotics, biological systems, nance and so on; see, e.g., [238,263,316,341]. Owing to great efforts of researchers, there have been rapid developments on the theory of fractional calculus and its applications, including well-posedness, stability, bifurcation and chaos in fractional differential equations and their control. Several useful tools for solving fractional-order equations have been discovered, of which Laplace transform is frequently applied. Furthermore, it is showed to be most helpful in analysis and applications of fractional-order systems, from which some results could be derived immediately. For instance, in [278,279], the authors investigated stability of fractional-order nonlinear dynamical systems using Laplace transform method and Lyapunov direct method, with the introduction of Mittag-Leffler stability and generalized Mittag-Leffler stability concepts. The Laplace transform was also used in [214,374].

Fractional calculus is relative to the traditional integer-order calculus put forward, which is the order of calculus from integer orders extended to any order of the mathematical promotion. From the theoretical point of view, the fractional differential calculus signal processing order extended to any number from an integer, the ways and means of information processing were extended. Fractional differential equations and inclusions have recently been applied in various areas of engineering, mathematics, physics and bioengineering, and other applied sciences [351]. For some fundamental results in the theory of fractional calculus and fractional differential equations, we refer the reader to the monographs of Abbas *et al.* [19,53,71,72], Ahmad *et al.* [88,90,103], Anastassiou *et al.* [107,109], Atangana [115], Baleanu and Lopes [126,127], Capelas de Oliveira [181], Cao and Chen [182], Chakraverty *et al.* [185], Daftardar-Gejji [195], Dutta *et al.* [201], Francesco [213], Georgiev [217], Geo *et al.* [225], Jin [256], Kilbas *et al.* [263], Kochubei and Luchko [258,259], Kumar [268], Milici *et al.* [287], Ortigueira and Valério [301], Petras [314], Ray *et al.* [327], Saha Ray and Sahoo [339], Shishkina and Sitnik [344], Samko *et al.* [341], Sun and Gao [348], Tarasov [349,350], Tas *et al.* [352], Vyawahare *et al.* [361], Yang *et al.* [372,373] and Zhou [382,383], the papers by Abbas *et al.* [20,21,54,56,62–67,73,83,84], Agarwal *et al.* [85–87], Ahmad *et al.* [89,91–97], Benchohra *et al.* [141,155–161,164,165], Lakshmikantham *et al.* [272–274], Vityuk *et al.* [359], and the references therein.

Recently, considerable attention has been given to the existence of solutions of initial and boundary value problems for fractional differential and partial differential equations and inclusions with Caputo fractional derivative; see [73,141,156,157,164,165], and Hadamard fractional integral equations; see [20,21,62,100,101,142,160,228,260], and the references therein.

In [175], Butzer *et al.* investigated properties of the Hadamard fractional integral and derivative. In [176], they obtained the Mellin transform of the Hadamard fractional integral and differential operators, and in [317], Pooseh *et al.* obtained expansion formulas of the Hadamard operators in terms of integer-order derivatives. Many other interesting properties of those operators and others are summarized in [341], and the references therein.

Impulsive differential equations are well known to model problems from many areas of science and engineering. There has been much research activity concerning the theory of impulsive differential equations; see [23,25,47, 71,72,86,99,158,186,232,288,363,380,381] and the references therein.

Implicit functional differential equations have been considered by many authors [19,22,24,26,71,82,99,155–161]. Our intention is to extend the results to implicit differential equations of fractional order.

The stability of functional equations was originally raised by Ulam [357] and next by Hyers [243]. Thereafter, this type of stability is called the Ulam–Hyers stability. In 1978, Rassias [322] provided a remarkable generalization of the Ulam–Hyers stability of mappings by considering variables. The concept of stability for a functional equation arises when we replace the functional equation by an inequality which acts as a perturbation of the equation. Considerable attention has been given to the study of the Ulam–Hyers and Ulam–Hyers–Rassias stability of all kinds of functional equations; one can see the monographs of [71,247], and the papers of Abbas *et al.* [20,22,64,73,159,160], Benchohra and Lazreg [159,160], Petru *et al.* [315], and Rus [330,331] discussed the Ulam–Hyers stability for operatorial equations and inclusions. More details from historical point of view, and recent developments of such stabilities are reported in [171,248,329].

Chapter 1

Preliminary Background

In this chapter, we introduce notations, definitions, and preliminary facts that will be used in the remainder of this book. Some notations and definitions from the fractional calculus, some definitions and proprieties of measures of noncompactness, measures of weak noncompactness, and some fixed-point theorems are presented.

1.1. Notations and Definitions

Let $I = [0, T]$, $T > 0$, be the compact interval of $\mathbb{R}$. We assume that E is a Banach space and denote by $\mathcal{C} := \mathcal{C}(I)$ the Banach space of all continuous functions v from I into E with the supremum (uniform) norm

$$\|v\|_\infty := \sup_{t \in I} \|v(t)\|_E.$$

As usual, $\mathrm{AC}(I)$ denotes the space of absolutely continuous functions from I into E, and $L^1(I)$ denotes the space of Bochner-integrable functions $v : I \to E$ with the norm

$$\|v\|_1 = \int_0^T \|v(t)\|_E dt.$$

For any $n \in \mathbb{N}^*$, we denote by $\mathrm{AC}^n(I)$ the space defined by

$$\mathrm{AC}^n(I) := \left\{ w : I \to E : \frac{d^n}{dt^n} w(t) \in \mathrm{AC}(I) \right\}.$$

Let

$$\delta = t \frac{d}{dt}, \quad q > 0, \quad n = [q] + 1,$$

where $[q]$ is the integer part of q. Define the space

$$AC_\delta^n := \{u : I \to E : \delta^{n-1}[u(t)] \in AC(I)\}.$$

Let $\gamma \in (0, 1]$. By $C_{\gamma,\ln}([1, T])$, $C_\gamma(I)$ and $C_\gamma^1([1, T])$; $T > 1$, we denote the weighted spaces of continuous functions defined by

$$C_{\gamma,\ln}([1, T]) = \{w(t) : (\ln t)^{1-\gamma}w(t) \in \mathcal{C}([1, T])\},$$

with the norm

$$\|w\|_{C_{\gamma,\ln}} := \sup_{t\in[1,T]} \|(\ln t)^{1-\gamma}w(t)\|_E,$$

$$C_\gamma(I) = \{w : (0, T] \to E : t^{1-\gamma}w(t) \in \mathcal{C}\},$$

with the norm

$$\|w\|_{C_\gamma} := \sup_{t\in I} \|t^{1-\gamma}w(t)\|_E,$$

and

$$C_\gamma^1(I) = \left\{w \in \mathcal{C} : \frac{dw}{dt} \in C_\gamma\right\},$$

with the norm

$$\|w\|_{C_\gamma^1} := \|w\|_\infty + \|w'\|_{C_\gamma}.$$

In what follows we denote $\|w\|_{C_{\gamma,\ln}}$ by $\|w\|_C$.

1.2. Fractional Calculus

Now, we give some results and properties of fractional calculus.

Definition 1.1 (Riemann–Liouville fractional integral [71,263, 341]). The Riemann–Liouville integral of order $r > 0$ of a function $w \in L^1(I)$ is defined by

$$(I_0^r w)(t) = \frac{1}{\Gamma(r)} \int_0^t (t - s)^{r-1}w(s)ds \quad \text{for a.e. } t \in I,$$

where $\Gamma(\cdot)$ is the (Euler's) Gamma function defined by

$$\Gamma(\xi) = \int_0^\infty t^{\xi-1}e^{-t}dt; \quad \xi > 0.$$

Notice that for all $r, r_1, r_2 > 0$ and each $w \in \mathcal{C}$, we have $I_0^r w \in \mathcal{C}$, and

$$(I_0^{r_1} I_0^{r_2} w)(t) = (I_0^{r_1+r_2} w)(t) \quad \text{for a.e.} \quad t \in I.$$

Definition 1.2 (Riemann–Liouville fractional derivative [71,263, 341]). The Riemann–Liouville fractional derivative of order $r > 0$ of a function $w \in AC^n(I)$ is defined by

$$(D_0^r w)(t) = \left(\frac{d^n}{dt^n} I_0^{n-r} w \right)(t)$$

$$= \frac{1}{\Gamma(n-r)} \frac{d^n}{dt^n} \int_0^t (t-s)^{n-r-1} w(s) ds \quad \text{for a.e. } t \in I,$$

where $n = [r] + 1$ and $[r]$ is the integer part of r.

In particular, if $r \in (0, 1]$, then

$$(D_0^r w)(t) = \left(\frac{d}{dt} I_0^{1-r} w \right)(t)$$

$$= \frac{1}{\Gamma(1-r)} \frac{d}{dt} \int_0^t (t-s)^{-r} w(s) ds \quad \text{for a.e. } t \in I.$$

Let $r \in (0, 1]$, $\gamma \in [0, 1)$ and $w \in C_{1-\gamma}(I)$. Then the following expression leads to the left inverse operator as follows:

$$(D_0^r I_0^r w)(t) = w(t) \quad \text{for all } t \in (1, T].$$

Moreover, if $I_0^{1-r} w \in C_{1-\gamma}^1(I)$, then the following composition is proved in [341]

$$(I_0^r D_0^r w)(t) = w(t) - \frac{(I_0^{1-r} w)(0^+)}{\Gamma(r)} t^{r-1} \quad \text{for all } t \in (0, T].$$

Definition 1.3 (Caputo fractional derivative [71,263,341]). The Caputo fractional derivative of order $r > 0$ of a function $w \in AC^n(I)$ is defined by

$$(^c D_0^r w)(t) = \left(I_0^{n-r} \frac{d^n}{dt^n} w \right)(t)$$

$$= \frac{1}{\Gamma(n-r)} \int_0^t (t-s)^{n-r-1} \frac{d^n}{ds^n} w(s) ds \quad \text{for a.e. } t \in I.$$

In particular, if $r \in (0,1]$, then

$$(^cD_0^r w)(t) = \left(I_0^{1-r}\frac{d}{dt}w\right)(t)$$

$$= \frac{1}{\Gamma(1-r)}\int_0^t (t-s)^{-r}\frac{d}{ds}w(s)ds \quad \text{for a.e. } t \in I.$$

Let us recall some definitions and properties of Hadamard fractional integration and differentiation. We refer to [228,263] for a more detailed analysis.

Definition 1.4 (Hadamard fractional integral [228,263]). The Hadamard fractional integral of order $q > 0$ for a function $g \in L^1([1,T],E)$; $T > 1$, is defined as

$$(^HI_1^q g)(x) = \frac{1}{\Gamma(q)}\int_1^x \left(\ln\frac{x}{s}\right)^{q-1}\frac{g(s)}{s}ds,$$

provided the integral exists.

Example 1.1. Let $0 < q < 1$. Then

$$^HI_1^q \ln t = \frac{1}{\Gamma(2+q)}(\ln t)^{1+q} \quad \text{for a.e. } t \in [1,e].$$

Remark 1.1. Let $g \in P_1([1,T],E)$. For every $\varphi \in E^*$, we have

$$\varphi(^HI_1^q g)(t) = (^HI_1^q \varphi g)(t) \quad \text{for a.e. } t \in [1,T].$$

Set

$$\delta = x\frac{d}{dx}, \quad q > 0, \quad n = [q]+1,$$

and

$$AC_\delta^n([1,T]) := \{u : [1,T] \to E : \delta^{n-1}[u(x)] \in AC([1,T])\}.$$

Analogous to the Riemann–Liouville fractional calculus, the Hadamard fractional derivative is defined in terms of the Hadamard fractional integral in the following way.

Definition 1.5 (Hadamard fractional derivative [228,263]). The Hadamard fractional derivative of order $q > 0$ applied to the function $w \in AC_\delta^n([1,T])$ is defined as

$$(^HD_1^q w)(x) = \delta^n(^HI_1^{n-q}w)(x).$$

In particular, if $q \in (0,1]$, then

$$({}^{H}D_1^q w)(x) = \delta({}^{H}I_1^{1-q}w)(x).$$

Example 1.2. Let $0 < q < 1$. Then

$${}^{H}D_1^q \ln t = \frac{1}{\Gamma(2-q)}(\ln t)^{1-q} \quad \text{for a.e. } t \in [1,e].$$

It has been proved (see, e.g., [260, Theorem 4.8]) that in the space $L^1([1,T],E)$, the Hadamard fractional derivative is the left-inverse operator to the Hadamard fractional integral, i.e.,

$$({}^{H}D_1^q)({}^{H}I_1^q w)(x) = w(x).$$

From Theorem 2.3 of [263], we have

$$({}^{H}I_1^q)({}^{H}D_1^q w)(x) = w(x) - \frac{({}^{H}I_1^{1-q}w)(1)}{\Gamma(q)}(\ln x)^{q-1}.$$

Analogous to the Hadamard fractional calculus, the Caputo–Hadamard fractional derivative is defined in the following way.

Definition 1.6 (Caputo–Hadamard fractional derivative). The Caputo–Hadamard fractional derivative of order $q > 0$ applied to the function $w \in AC_\delta^n$ is defined as

$$({}^{HC}D_1^q w)(x) = ({}^{H}I_1^{n-q}\delta^n w)(x).$$

In particular, if $q \in (0,1]$, then

$$({}^{HC}D_1^q w)(x) = ({}^{H}I_1^{1-q}\delta w)(x).$$

In [238], Hilfer studied applications of a generalized fractional operator having the Riemann–Liouville and the Caputo derivatives as specific cases (see also [239,354]).

Definition 1.7 (Hilfer fractional derivative). Let $\alpha \in (0,1)$, $\beta \in [0,1]$, $w \in L^1([1,T])$, $I_1^{(1-\alpha)(1-\beta)}w \in AC([1,T])$. The Hilfer fractional derivative of order α and type β of w is defined as

$$(D_1^{\alpha,\beta}w)(t) = \left(I_1^{\beta(1-\alpha)}\frac{d}{dt}I_1^{(1-\alpha)(1-\beta)}w\right)(t) \quad \text{for a.e. } t \in [1,T]. \quad (1.1)$$

Properties. Let $\alpha \in (0,1)$, $\beta \in [0,1]$, $\gamma = \alpha + \beta - \alpha\beta$, and $w \in L^1([1,T])$.

1. The operator $(D_1^{\alpha,\beta} w)(t)$ can be written as

$$(D_1^{\alpha,\beta} w)(t) = \left(I_1^{\beta(1-\alpha)} \frac{d}{dt} I_1^{1-\gamma} w \right)(t) = (I_1^{\beta(1-\alpha)} D_1^{\gamma} w)(t)$$

for a.e. $t \in [1,T]$.

Moreover, the parameter γ satisfies

$$\gamma \in (0,1], \ \gamma \geq \alpha, \ \gamma > \beta, \ 1 - \gamma < 1 - \beta(1-\alpha).$$

2. The generalization (1.1) for $\beta = 0$ coincides with the Riemann–Liouville derivative and for $\beta = 1$ with the Caputo derivative:

$$D_1^{\alpha,0} = D_1^{\alpha} \text{ and } D_1^{\alpha,1} = {}^c D_1^{\alpha}.$$

3. If $D_1^{\beta(1-\alpha)} w$ exists and in $L^1([1,T])$, then

$$(D_1^{\alpha,\beta} I_1^{\alpha} w)(t) = (I_1^{\beta(1-\alpha)} D_1^{\beta(1-\alpha)} w)(t) \text{ for a.e. } t \in [1,T].$$

Furthermore, if $w \in C_\gamma([1,T])$ and $I_1^{1-\beta(1-\alpha)} w \in C_\gamma^1(I)$, then

$$(D_1^{\alpha,\beta} I_1^{\alpha} w)(t) = w(t) \quad \text{for a.e. } t \in [1,T].$$

4. If $D_1^{\gamma} w$ exists and in $L^1([1,T])$, then

$$(I_1^{\alpha} D_1^{\alpha,\beta} w)(t) = (I_1^{\gamma} D_1^{\gamma} w)(t) = w(t) - \frac{I_1^{1-\gamma}(1^+)}{\Gamma(\gamma)} t^{\gamma-1} \text{ for a.e. } t \in [1,T].$$

Corollary 1.1. *Let* $h \in C_\gamma(I)$. *Then the linear Cauchy problem*

$$\begin{cases} (D_0^{\alpha,\beta} u)(t) = h(t); & t \in I, \\ (I_0^{1-\gamma} u)(t)|_{t=0} = \phi, \end{cases}$$

has a unique solution $u \in L^1(I)$ *given by*

$$u(t) = \frac{\phi}{\Gamma(\gamma)} t^{\gamma-1} + (I_0^{\alpha} h)(t).$$

From the Hadamard fractional integral, the Hilfer–Hadamard fractional derivative (introduced for the first time in [320]) is defined in the following way.

Definition 1.8 (Hilfer–Hadamard fractional derivative). Let $\alpha \in (0,1)$, $\beta \in [0,1]$, $\gamma = \alpha + \beta - \alpha\beta$, $w \in L^1([1,T])$, and $^H I_1^{(1-\alpha)(1-\beta)} w \in AC([1,T])$. The Hilfer–Hadamard fractional derivative of order α and type β applied to the function w is defined as

$$\begin{aligned}
(^H D_1^{\alpha,\beta} w)(t) &= (^H I_1^{\beta(1-\alpha)}(^H D_1^{\gamma} w))(t) \\
&= (^H I_1^{\beta(1-\alpha)} \delta(^H I_1^{1-\gamma} w))(t) \text{ for a.e. } t \in [1,T].
\end{aligned} \tag{1.2}$$

This new fractional derivative (1.8) may be viewed as interpolating the Hadamard fractional derivative and the Caputo–Hadamard fractional derivative. Indeed for $\beta = 0$ this derivative reduces to the Hadamard fractional derivative and when $\beta = 1$, we recover the Caputo–Hadamard fractional derivative:

$$^H D_1^{\alpha,0} = {}^H D_1^{\alpha}, \quad \text{and} \quad {}^H D_1^{\alpha,1} = {}^{HC} D_1^{\alpha}.$$

1.3. Multivalued Analysis

Let $(X, \|\cdot\|)$ be a Banach space and K be a subset of X. Define

$$\begin{aligned}
\mathcal{P}(X) &= \{K \subset X : K \neq \emptyset\}, \\
\mathcal{P}_{cl}(X) &= \{K \subset \mathcal{P}(X) : K \text{ is closed}\}, \\
\mathcal{P}_b(X) &= \{K \subset \mathcal{P}(X) : K \text{ is bounded}\}, \\
\mathcal{P}_{cv}(X) &= \{K \subset \mathcal{P}(X) : K \text{ is convex}\}, \\
\mathcal{P}_{cp}(X) &= \{K \subset \mathcal{P}(X) : K \text{ is compact}\}, \\
\mathcal{P}_{cp,cv}(X) &= \mathcal{P}_{cp}(X) \cap \mathcal{P}_{cv}(X), \\
\mathcal{P}_{cp,cl,cv}(X) &= \mathcal{P}_{cp}(X) \cap \mathcal{P}_{cl}(X) \cap \mathcal{P}_{cv}(X).
\end{aligned}$$

Let $A, B \in \mathcal{P}(X)$. Consider $H_d : \mathcal{P}(X) \times \mathcal{P}(X) \to \mathbb{R}_+ \cup \{\infty\}$ the Hausdorff distance between A and B given by

$$H_d(A,B) = \max\{\sup_{a \in A} d(a,B), \sup_{b \in B} d(A,b)\},$$

where $d(A,b) = \inf_{a \in A} d(a,b)$ and $d(A,B) = \inf_{b \in B} d(a,b)$. As usual, $d(x,\emptyset) = +\infty$.

Then $(\mathcal{P}_{b,cl}(X), H_d)$ is a metric space and $(\mathcal{P}_{cl}(X), H_d)$ is a generalized (complete) metric space; which consists of a set K together with a distance function $d(\cdot,\cdot) : K \times K \to \mathbb{R}_+$ satisfying $d(x,x) = 0$ and $d(x,z) \leq d(x,y) + d(y,z)$ for all x, y and z in K (see [266]).

Definition 1.9. A multivalued operator $N : X \to \mathcal{P}_{cl}(X)$ is called:

(a) γ-Lipschitz if there exists $\gamma > 0$ such that
$$H_d(N(x), N(y)) \leq \gamma d(x, y) \quad \text{for all} \quad x, \, y \in X;$$

(b) a contraction if it is γ-Lipschitz with $\gamma < 1$.

Definition 1.10. A multivalued map $F : I \to \mathcal{P}_{cl}(X)$ is said to be measurable if, for each $y \in X$, the function
$$t \longmapsto d(y, F(t)) = inf\{d(x, z) : z \in F(t)\}$$
is measurable.

Definition 1.11. The selection set of a multivalued map $G : I \to \mathcal{P}(X)$ is defined by
$$S_G = \{u \in L^1(I) : u(t) \in G(t), \text{ a.e. } t \in I\}.$$

For each $u \in \mathcal{C}$, the set S_{Fou} known as the set of selectors from F is defined by
$$S_{Fou} = \{v \in L^1(I) : v(t) \in F(t, u(t)), \text{ a.e. } t \in I\}.$$

Definition 1.12. Let X and Y be two sets. The graph of a set-valued map $N : X \to \mathcal{P}(Y)$ is defined by
$$\text{graph}(N) = \{(x, y) : x \in X, \, y \in N(X)\}.$$

There are more details about multivalued functions in [114,200,205,242].

Definition 1.13. Let $(X, \| \cdot \|)$ be a Banach space. A multivalued map $F : X \to \mathcal{P}(X)$ is convex (closed) if $F(X)$ is convex (closed) for all $x \in X$.

The map F is bounded on bounded sets if $F(\mathcal{B}) = \cup_{x \in \mathcal{B}} F(x)$ is bounded in X for all $\mathcal{B} \in \mathcal{P}_b(X)$, i.e., $\sup_{x \in \mathcal{B}}\{\sup\{|y| : y \in F(x)\}\} < \infty$.

Definition 1.14. A multivalued map F is called upper semicontinuous (u.s.c. for short) on X if for each $x_0 \in X$ the set $F(x_0)$ is a nonempty, closed subset of X, and for each open set U of X containing $F(x_0)$, there exists an open neighborhood V of x_0 such that $F(V) \subset U$. A set-valued map F is said to be upper semicontinuous if it is so at every point $x_0 \in X$. F is said to be completely continuous if $F(\mathcal{B})$ is relatively compact for every $\mathcal{B} \in \mathcal{P}_b(X)$.

If the multivalued map F is completely continuous with nonempty compact values, then F is upper semicontinuous if and only if F has closed graph (i.e., $x_n \to x_*$, $y_n \to y_*$, $y_n \in G(x_n)$ imply $y_* \in F(x_*)$).

The map F has a fixed point if there exists $x \in X$ such that $x \in Gx$. The set of fixed points of the multivalued operator G will be denoted by $\operatorname{Fix} G$.

Definition 1.15. A measurable multivalued function $F : I \to \mathcal{P}_{b,cl}(X)$ is said to be integrably bounded if there exists a function $g \in L^1(I, \mathbb{R}_+)$ such that $|f| \le g(t)$ for a.e. $t \in I$ for all $f \in F(t)$.

Lemma 1.1 ([242]). *Let G be a completely continuous multivalued map with nonempty compact values. Then G is u.s.c. if and only if G has a closed graph (i.e., $u_n \to u$, $w_n \to w$, $w_n \in G(u_n)$ imply $w \in G(u)$).*

Lemma 1.2 ([277]). *Let X be a Banach space. Let $F : I \times X \to \mathcal{P}_{cp,cv}(X)$ be an L^1-Carathéodory multivalued map and let Λ be a linear continuous mapping from $L^1(I, X)$ to $C(I, X)$. Then the operator*

$$\Lambda \circ S_{Fou} : C(I, X) \to \mathcal{P}_{cp,cv}(C(JI, X)),$$

$$w \longmapsto (\Lambda \circ S_{Fou})(w) := (\Lambda S_{Fou})(w)$$

is a closed graph operator in $C(I, X) \times C(I, X)$.

Proposition 1.1 ([242]). *Let $F : X \to Y$ be a u.s.c. map with closed values. Then $Gr(F)$ is closed.*

Definition 1.16. A multivalued map $F : I \times \mathbb{R} \times \mathbb{R} \to \mathcal{P}(\mathbb{R})$ is said to be L^1-Carathéodory if

(i) $t \to F(t, x, y)$ is measurable for each $x, y \in \mathbb{R}$;
(ii) $(x, y) \to F(t, x, y)$ is upper semicontinuous for a.e. $t \in I$;
(iii) for each $q > 0$, there exists $\varphi_q \in L^1(J, \mathbb{R}_+)$ such that

$$\|F(t, x, y)\|_{\mathcal{P}} = \sup\{|f| : f \in F(t, x, y)\} \le \varphi_q(t)$$

for all $|x| \le q$, $|y| \le q$ and for a.e. $t \in I$.

The multivalued map F is said to be Carathéodory if it satisfies (i) and (ii).

Lemma 1.3 ([219]). *Let X be a separable metric space. Then every measurable multivalued map $F : X \to \mathcal{P}_{cl}(X)$ has a measurable selection.*

For more details on multivalued maps and the proof of the known results cited in this section, we refer interested reader to the books of Aubin and Cellina [114], Deimling [198], Gorniewicz [219], and Hu and Papageorgiou [242].

1.4. Measure of Noncompactness

We will define the Kuratowski (1896–1980) and Hausdorf (1868–1942) measures of noncompactness ($\mathcal{MNC}$ for short) and give their basic properties. Let us recall some fundamental facts of the notion of measure of noncompactness in a Banach space.

Let (X, d) be a complete metric space and $\mathcal{P}_{bd}(X)$ be the family of all bounded subsets of X. Analogously denote by $\mathcal{P}_{rcp}(X)$ the family of all relatively compact and nonempty subsets of X. Recall that $B \subset X$ is said to be bounded if B is contained in some ball. If $B \subset \mathcal{P}_{bd}(X)$ is not relatively compact, (precompact) then there exists an $\epsilon > 0$ such that B cannot be covered by a finite number of ϵ-balls, and it is then also impossible to cover B by finitely many sets of diameter $< \epsilon$. Recall that the diameter of B is given by

$$\text{diam}(B) := \begin{cases} \sup_{(x,y) \in B^2} d(x,y) & \text{if } B \neq \phi, \\ 0 & \text{if } B = \phi. \end{cases}$$

Definition 1.17 ([267]). Let (X, d) be a complete metric space and $\mathcal{P}_{bd}(X)$ be the family of bounded subsets of X. For every $B \in \mathcal{P}_{bd}(X)$, we define the Kuratowski measure of noncompactness $\alpha(B)$ of the set B as the infimum of the numbers d such that B admits a finite covering by sets of diameter smaller than d.

Remark 1.2. It is clear that $0 \leq \alpha(B) \leq \text{diam}(B) < +\infty$ for each nonempty bounded subset B of X and that $\text{diam}(B) = 0$ if and only if B is an empty set or consists of exactly one point.

Definition 1.18 ([129]). Let X be a Banach space and $\mathcal{P}_{bd}(X)$ be the family of bounded subsets of X. For every $B \in \mathcal{P}_{bd}(X)$, the Kuratowski measure of noncompactness is the map $\alpha : \mathcal{P}_{bd}(X) \to [0, +\infty)$ defined by

$$\alpha(B) = \inf \left\{ r > 0 : B \subseteq \bigcup_{i=1}^{n} B_i \text{ and } \text{diam}(B_i) < r \right\}.$$

The Kuratowski measure of noncompactness satisfies the following properties.

Proposition 1.2 ([129,131,267]). *Let X be a Banach space. Then for all bounded subsets A, B of X the following assertions hold:*

(1) $\alpha(B) = 0$ *implies* $\overline{B}$ *is compact (B is relatively compact), where $\overline{B}$ denotes the closure of B.*

(2) $\alpha(\phi) = 0$.

(3) $\alpha(B) = \alpha(\overline{B}) = \alpha(\text{conv}\, B)$, *where* conv B *is the convex hull of B.*

(4) *Monotonicity:* $A \subset B$ *implies* $\alpha(A) \leq \alpha(B)$.

(5) *Algebraic semi-additivity:* $\alpha(A + B) \leq \alpha(A) + \alpha(B)$, *where* $A + B = \{x + y : x \in A; y \in B\}$.

(6) *Semihomogencity:* $\alpha(\lambda B) = |\lambda| \alpha(B)$, $\lambda \in \mathbb{R}$, *where* $\lambda(B) = \{\lambda x : x \in B\}$.

(7) *Semiadditivity:* $\alpha(A \cup B) = \max\{\alpha(A), \alpha(B)\}$.

(8) *Semiadditivity:* $\alpha(A \cap B) = \min\{\alpha(A), \alpha(B)\}$.

(9) *Invariance under translations:* $\alpha(B + x_0) = \alpha(B)$ *for any* $x_0 \in X$.

The following definition of measure of noncompactness appeared in [129].

Definition 1.19. A function $\mu : \mathcal{P}_{bd}(X) \to [0, \infty)$ will be called a measure of noncompactness if it satisfies the following conditions:

(1) $\text{Ker}\mu(A) = \{A \in \mathcal{P}_{bd}(X) : \mu(A) = 0\}$ is nonempty and $\ker \mu(A) \subset \mathcal{P}_{rcp}(X)$.

(2) $A \subset B$ implies $\mu(A) \leq \mu(B)$.

(3) $\mu(\overline{A}) = \mu(A)$.

(4) $\mu(\text{conv}\, A) = \mu(A)$.

(5) $\mu(\lambda A + (1 - \lambda)B) \leq \lambda \mu(A) + (1 - \lambda)\mu(B)$ for $\lambda \in [0, 1]$.

(6) If $(A_n)_{n \geq 1}$ is a sequence of closed sets in $\mathcal{P}_{bd}(X)$ such that

$$X_{n+1} \subset A_n \ (n = 1, 2, \ldots.)$$

and

$$\lim_{n \to +\infty} \mu(A_n) = 0,$$

then the intersection set $A_\infty = \bigcap_{n=1}^{\infty} A_n$ is nonempty.

Remark 1.3. The family Ker μ described in 1 is said to be the kernel of the measure of noncompactness μ. Observe that the intersection set A_∞ in condition (6) is a member of the family Ker μ. Since $\mu(A_\infty) \leq \mu(A_n)$

for any n, we infer that $\mu(A_\infty) = 0$. This yields that $\mu(A_\infty) \in \text{Ker}\,\mu$. This simple observation will be essential in our further investigations.

Moreover, by introducing the notion of a measure of noncompactness in $L^1(J)$, we let $\mathcal{P}_{bd}(J)$ be the family of all bounded subsets of $L^1(J)$. Analogously denote by $\mathcal{P}_{rcp}(J)$ the family of all relatively compact and nonempty subsets of $L^1(J)$. In particular, the measure of noncompactness in $L^1(I)$ is defined as follows. Let X be a fixed nonempty and bounded subset of $L^1(I)$. For $x \in X$, set

$$\mu(X) = \lim_{\delta \to 0} \left\{ \sup \left\{ \sup \left(\int_0^T |x(t+h) - x(t)|dt \right), \ |h| \le \delta \right\}, \ x \in X \right\}.$$
(1.3)

It can be easily shown that μ is measure of noncompactness in $L^1(I)$ (see [129]).

For more details on measure of noncompactness and the proof of the known results cited in this section we refer the reader to [98,129,131].

Lemma 1.4 ([169]). *If Y is a bounded subset of Banach space X, then for each $\epsilon > 0$, there is a sequence $\{y_k\}_{k=1}^\infty \subset Y$ such that*

$$\mu(Y) \le 2\mu(\{y_k\}_{k=1}^\infty) + \epsilon.$$

Lemma 1.5 ([291]). *If $\{u_k\}_{k=1}^\infty \subset L^1(I)$ is uniformly integrable, then $\mu(\{u_k\}_{k=1}^\infty)$ is measurable and for each $t \in I$,*

$$\mu\left(\left\{\int_0^t u_k(s)ds\right\}_{k=1}^\infty\right) \le 2 \int_0^t \mu(\{u_k(s)\}_{k=1}^\infty)ds.$$

Lemma 1.6 ([226]). *If $V \subset C(I, E)$ is a bounded and equicontinuous set, then*

(i) *The function $t \to \alpha(V(t))$ is continuous on I, and*

$$\alpha_c(V) = \sup_{0 \le t \le T} \alpha(V(t)).$$

(ii)

$$\alpha\left(\int_0^T x(s)ds : x \in V\right) \le \int_0^T \alpha(V(s))ds,$$

where

$$V(s) = \{x(s) : x \in V\}, \quad s \in I.$$

1.5. Measure of Weak Noncompactness

The measure of weak noncompactness was introduced by DeBlasi [200]. The strong measure of noncompactness was developed first by Banaś and Goebel [129] and subsequently developed and used in many papers; see for example, [105,154,226], and the references therein. In [154,303], the authors considered some existence results by applying the techniques of the measure of noncompactness. Recently, several researchers obtained other results by application of the technique of measure of weak noncompactness; see [72, 145,151], and the references therein.

Let $(E, w) = (E, \sigma(E, E^*))$ be the Banach space E with its weak topology.

Definition 1.20. A Banach space X is called weakly compactly generated (WCG, in short) if it contains a weakly compact set whose linear span is dense in X.

Definition 1.21. A function $h : E \to E$ is said to be weakly sequentially continuous if h takes each weakly convergent sequence in E to a weakly convergent sequence in E (i.e., for any (u_n) in E with $u_n \to u$ in (E, w) then $h(u_n) \to h(u)$ in (E, w)).

Definition 1.22 ([314]). The function $u : I \to E$ is said to be Pettis integrable on I if and only if there is an element $u_J \in E$ corresponding to each $J \subset I$ such that $\phi(u_J) = \int_J \phi(u(s)) ds$ for all $\phi \in E^*$, where the integral on the right-hand side is assumed to exist in the sense of Lebesgue (by definition, $u_J = \int_J u(s) ds$).

Let $P(I, E)$ be the space of all E-valued Pettis integrable functions on I, and $L^1(I, E)$ be the Banach space of Bochner measurable functions $u : I \to E$. Define the class $P_1(I, E)$ by

$$P_1(I, E) = \{u \in P(I, E) : \varphi(u) \in L^1(I, \mathbb{R}); \text{ for every } \varphi \in E^*\}.$$

The space $P_1(I, E)$ is normed by

$$\|u\|_{P_1} = \sup_{\varphi \in E^*, \, \|\varphi\| \leq 1} \int_0^T |\varphi(u(x))| d\lambda x,$$

where λ stands for a Bochner measure on I.

The following result is due to Pettis (see [314, Theorem 3.4 and Corollary 3.41]).

Proposition 1.3 ([314]). *If* $u \in P_1(I, E)$ *and* h *is a measurable and essentially bounded real-valued function, then* $uh \in P_1(J, E)$.

For all what follows, the sign "$\int$" denotes the Pettis integral.

Remark 1.4. Let $g \in P_1([1, T], E)$. For every $\varphi \in E^*$, we have

$$\varphi(^H I_1^q g)(x) = (^H I_1^q \varphi g)(x) \quad \text{for a.e. } x \in [1, T].$$

Definition 1.23 ([200]). Let E be a Banach space, Ω_E be the bounded subsets of E and B_1 be the unit ball of E. The De Blasi measure of weak noncompactness is the map $\beta : \Omega_E \to [0, \infty)$ defined by

$$\beta(X) = \inf\{\epsilon > 0 : \text{ there exists a weakly compact subset } \Omega \text{ of }$$

$$E : X \subset \epsilon B_1 + \Omega\}.$$

The De Blasi measure of weak noncompactness satisfies the following properties:

(a) $A \subset B \Rightarrow \beta(A) \leq \beta(B)$;
(b) $\beta(A) = 0 \Leftrightarrow A$ is relatively weakly compact;
(c) $\beta(A \cup B) = \max\{\beta(A), \beta(B)\}$;
(d) $\beta(\overline{A}^\omega) = \beta(A)$ (where $\overline{A}^\omega$ denotes the weak closure of A);
(e) $\beta(A + B) \leq \beta(A) + \beta(B)$;
(f) $\beta(\lambda A) = |\lambda| \beta(A)$;
(g) $\beta(\text{conv}(A)) = \beta(A)$;
(h) $\beta(\cup_{|\lambda| \leq h} \lambda A) = h \beta(A)$.

The next result follows directly from the Hahn–Banach theorem.

Proposition 1.4. *Let* E *be a normed space, and* $x_0 \in E$ *with* $x_0 \neq 0$. *Then, there exists* $\varphi \in E^*$ *with* $\|\varphi\| = 1$ *and* $|\varphi(x_0)| = \|x_0\|$.

For a given set V of functions $v : I \to E$ let us denote by

$$V(t) = \{v(t) : v \in V\}; \ t \in I,$$

and

$$V(I) = \{v(t) : v \in V, \ t \in I\}.$$

Lemma 1.7 ([226]). *Let $H \subset C$ be a bounded and equicontinuous. Then the function $t \to \beta(H(t))$ is continuous on I, and*

$$\beta_C(H) = \max_{t \in I} \beta(H(t)),$$

and

$$\beta\left(\int_I u(s)ds\right) \leq \int_I \beta(H(s))ds,$$

where β_C is the De Blasi measure of weak noncompactness defined on the bounded sets of C.

1.6. Some Attractivity Concepts

In this section, we present some results concerning the attractivity concepts of fixed-point equations.

Denote by BC $:=$ BC($\mathbb{R}_+$) the Banach space of all bounded and continuous functions from $\mathbb{R}_+$ into $\mathbb{R}$. Let $\emptyset \neq \Omega \subset$ BC, and let $N : \Omega \to \Omega$, and consider the solutions of equation

$$(Nu)(t) = u(t). \tag{1.4}$$

We introduce the following concept of attractivity of solutions for Eq. (1.4).

Definition 1.24. A solution of Eq. (1.4) is locally attractive if there exists a ball $B(u_0, \eta)$ in the space BC such that, for arbitrary solutions $v = v(t)$ and $w = w(t)$ of Eq. (1.4) belonging to $B(u_0, \eta) \cap \Omega$, we have

$$\lim_{t \to \infty} (v(t) - w(t)) = 0. \tag{1.5}$$

When the limit (1.5) is uniform with respect to $B(u_0, \eta) \cap \Omega$, the solutions of Eq. (1.4) are said to be uniformly locally attractive (or equivalently that solutions of Eq. (1.4) are locally asymptotically stable).

Definition 1.25. A solution $v = v(t)$ of Eq. (1.4) is said to be globally attractive if (1.5) holds for each solution $w = w(t)$ of (1.4). If condition (1.5) is satisfied uniformly with respect to the set Ω, the solutions of Eq. (1.4) are said to be globally asymptotically stable (or uniformly globally attractive).

Lemma 1.8 ([191, p. 62]). *Let $D \subset$ BC. Then D is relatively compact in BC if the following conditions hold:*

(a) *D is uniformly bounded in* BC.
(b) *The functions belonging to D are almost equicontinuous on* $\mathbb{R}_+$, *i.e.,*
    equicontinuous on every compact of $\mathbb{R}_+$.
(c) *The functions from D are equiconvergent, i.e., given* $\epsilon > 0$, *there exists*
    $T(\epsilon) > 0$ *such that* $|u(t) - \lim_{t\to\infty} u(t)| < \epsilon$ *for any* $t \geq T(\epsilon)$ *and* $u \in D$.

1.7. Some Ulam Stability Concepts

Now, we consider the Ulam stability for Eq. (1.4). Let $\epsilon > 0$ and $\Phi : I \to [0, \infty)$ be a continuous function. We consider the following inequalities:

$$|u(t) - (Nu)(t)| \leq \epsilon; \quad t \in I, \tag{1.6}$$

$$|u(t) - (Nu)(t)| \leq \Phi(t); \quad t \in I, \tag{1.7}$$

$$|u(t) - (Nu)(t)| \leq \epsilon\Phi(t); \quad t \in I. \tag{1.8}$$

Definition 1.26 ([71,329]). The problem (1.4) is Ulam–Hyers stable if there exists a real number $c_f > 0$ such that for each $\epsilon > 0$ and for each solution $u \in C_\gamma$ of the inequality (1.6) there exists a solution $v \in C_\gamma$ of (1.4) with

$$|u(t) - v(t)| \leq \epsilon c_f; \quad t \in I.$$

Definition 1.27 ([71,329]). The problem (1.4) is generalized Ulam–Hyers stable if there exists $c_f : C([0,\infty), [0,\infty))$ with $c_f(0) = 0$ such that for each $\epsilon > 0$ and for each solution $u \in C_\gamma$ of the inequality (1.6) there exists a solution $v \in C_\gamma$ of (1.4) with

$$|u(t) - v(t)| \leq c_f(\epsilon); \quad t \in I.$$

Definition 1.28 ([71,329]). The problem (1.4) is Ulam–Hyers–Rassias stable with respect to Φ if there exists a real number $c_{f,\Phi} > 0$ such that for each $\epsilon > 0$ and for each solution $u \in C_\gamma$ of the inequality (1.8) there exists a solution $v \in C_\gamma$ of (1.4) with

$$|u(t) - v(t)| \leq \epsilon c_{f,\Phi}\Phi(t); \quad t \in I.$$

Definition 1.29 ([71,329]). The problem (1.4) is generalized Ulam–Hyers–Rassias stable with respect to Φ if there exists a real number $c_{f,\Phi} > 0$ such that for each solution $u \in C_\gamma$ of the inequality (1.7) there exists a solution $v \in C_\gamma$ of (1.4) with

$$|u(t) - v(t)| \leq c_{f,\Phi}\Phi(t); \quad t \in I.$$

Remark 1.5. It is clear that

(i) Definition 1.26 implies Definition 1.27;
(ii) Definition 1.28 implies Definition 1.29;
(iii) Definition 1.28 for $\Phi(\cdot) = 1$ implies Definition 1.26.

One can have similar remarks for the inequalities (1.6) and (1.8).

1.8. Some Fixed-Point Theorems

In this section, we give the main fixed-point theorems that will be used in the following chapters.

Definition 1.30 ([104]). Let (M, d) be a metric space. The map $T : M \to M$ is said to be Lipschitzian if there exists a constant $k \geq 0$ (called Lipschitz constant) such that

$$d(T(x), T(y)) \leq kd(x, y) \text{ for all } x, \ y \in M.$$

A Lipschitzian mapping with a Lipschitz constant $k < 1$ is called a contraction.

Theorem 1.1 (Banach's fixed-point theorem [223]). *Let C be a nonempty closed subset of a Banach space X. Then any contraction mapping T of C into itself has a unique fixed point.*

Theorem 1.2 (Schauder fixed-point theorem [223]). *Let X be a Banach space, Q be a convex and closed subset of X and $T : Q \to Q$ be a compact and continuous map. Then T has at least one fixed point in Q.*

Theorem 1.3 (Burton and Kirk fixed-point theorem [173]). *Let X be a Banach space, and $A, B : X \to X$ be two operators satisfying*

(i) *A is a contraction;*
(ii) *B is completely continuous.*

Then either the operator equation $y = A(y) + B(y)$ admits a solution; or the set $\Omega = \left\{ u \in X : u = \lambda A\left(\frac{u}{\lambda}\right) + \lambda B(u), \ for \ some \ \lambda \in [0, 1] \right\}$ is unbounded.

Theorem 1.4 ([14]). *Let (Ω, d) be a generalized complete metric space and $\Theta : \Omega \to \Omega$ be a strictly contractive operator with a Lipscitz constant $L < 1$. If there exists a nonnegative integer k such that $d(\Theta^{k+1}x, \Theta^k x) < \infty$ for some $x \in \Omega$, then the following propositions hold true:*

(A) *The sequence* $(\Theta^k x)_{n \in N}$ *converges to a fixed point* x^* *of* Θ.
(B) x^* *is the unique fixed point of* Θ *in* $\Omega^* = \{y \in \Omega \mid d(\Theta^k x, y) < \infty\}$.
(C) *If* $y \in \Omega^*$, *then* $d(y, x^*) \leq \frac{1}{1-L} d(y, \Theta x)$.

In the next definition, we will consider a special class of continuous and bounded operators.

Definition 1.31. Let $T : M \subset X \to X$ be a bounded operator from a Banach space X into itself. The operator T is called a k-set contraction if there is a number $k \geq 0$ such that

$$\mu(T(A)) \leq k\mu(A)$$

for all bounded sets A in M. The bounded operator T is called condensing if $\mu(T(A)) < \mu(A)$ for all bounded sets A in M with $\mu(A) > 0$.

Obviously, every k-set contraction for $0 \leq k < 1$ is condensing. Every compact map T is a k-set contraction with $k = 0$.

Theorem 1.5 (Darbo's fixed-point theorem [129]). *Let M be nonempty, bounded, convex and closed subset of a Banach space X and $T : M \to M$ be a continuous operator satisfying $\mu(TA) \leq k\mu(A)$ for any nonempty subset A of M and for some constant $k \in [0, 1)$. Then T has at least one fixed point in M.*

Theorem 1.6 (Mönch's fixed-point theorem [87,291]). *Let D be a bounded, closed and convex subset of a Banach space X such that $0 \in D$, α be the Kuratowski measure of noncompactness and N be a continuous mapping of D into itself. If $[V = \overline{\mathrm{conv}}\, N(V)$ or $V = N(V) \cup \{0\}]$ implies $\alpha(V) = 0$ for every subset V of D, then N has a fixed point.*

Theorem 1.7 ([302]). *Let Q be a nonempty, closed, convex and equicontinuous subset of a metrizable locally convex vector space X such that $0 \in Q$. Suppose $T : Q \to Q$ is weakly-sequentially continuous. If the implication*

$$\overline{V} = \overline{\mathrm{conv}}(\{0\} \cup T(V)) \Rightarrow V \text{ is relatively weakly compact,} \qquad (1.9)$$

holds for every subset $V \subset Q$, then the operator T has a fixed point.

Theorem 1.8 (Nonlinear alternative of Leray Schauder type [223]). *Let X be a Banach space and C a nonempty convex subset of X. Let U a nonempty open subset of C with $0 \in U$ and $T : \overline{U} \to C$ be a continuous and compact operator.*

Then, either

(a) *T has fixed points, or*
(b) *there exist $u \in \partial U$ and $\lambda \in (0,1)$ with $u = \lambda T(u)$.*

Theorem 1.9 (Martelli's fixed-point theorem [286]). *Let X be a Banach space and $N : X \to \mathcal{P}_{cl,cv}(X)$ be a u.s.c. and condensing map. If the set $\Omega := \{u \in X : \lambda u \in N(u) \text{ for some } \lambda > 1\}$ is bounded, then N has a fixed point.*

Theorem 1.10 ([117]). *Let $(X, \| \cdot \|_n)$ be a Fréchet space and let $A, B : X \to X$ be two operators such that*

(a) *A is a compact operator;*
(b) *B is a contraction operator with respect to a family of seminorms $\{\| \cdot \|_n\}$;*
(c) *the set $\left\{ x \in X : x = \lambda A(x) + \lambda B\left(\frac{x}{\lambda}\right), \ \lambda \in (0,1) \right\}$ is bounded.*

Then, the operator equation $A(u) + B(u) = u$ has a solution in X.

Theorem 1.11 (Random fixed-point theorem [246]). *Let K be a non-empty, closed convex bounded subset of the separable Banach space X and let $N : \Omega \times K \to K$ be a compact and continuous random operator. Then, the random equation $N(w)u = u$ has a random solution.*

For more details, see [87,110,219,223,264,375].

Next, we state two multivalued fixed-point theorems.

Theorem 1.12 (Set-valued version of the Mönch fixed-point theorem, [304]). *Let X be Banach space and $K \subset X$ be a closed and convex set. Also, let U be a relatively open subset of K and $N : \overline{U} \to \mathcal{P}_c(K)$. Suppose that N maps compact sets into relatively compact sets, graph(N) is closed and for some $x_0 \in U$, we have*

$$\text{conv}(x_0 \cup N(M)) \supset M \subset \overline{U} \tag{1.10}$$

and $\overline{M} = \overline{U}$ ($C \subset M$ countable) imply $\overline{M}$ is compact and

$$x \notin (1 - \lambda)x_0 + \lambda N(x) \quad \forall x \in \overline{U} \backslash U, \ \lambda \in (0,1). \tag{1.11}$$

Then there exists $x \in \overline{U}$ with $x \in N(x)$.

Lemma 1.9 (Bohnenblust–Karlin, 1950, [168]). *Let X be a Banach space and $K \in \mathcal{P}_{cl,cv}(X)$ and suppose that the operator $G : K \to \mathcal{P}_{cl,cv}(K)$ is upper semicontinuous and the set $G(K)$ is relatively compact in X. Then G has a fixed point in K.*

Lemma 1.10 (Covitz–Nadler, [192]). *Let (X, d) be a complete metric space. If $N : X \to \mathcal{P}_{cl}(X)$ is a contraction, then* Fix $N \neq \phi$.

Theorem 1.13 ([203]). *Let $(\Omega, \mathcal{A})$ be a complete σ-finite measure space, X be a separable Banach space, $\mathcal{M}(\Omega, X)$ be the space of all measurable X-valued functions defined on Ω, and let $N : \Omega \times X \to \mathcal{P}_{cp,cv}(X)$ be a continuous and condensing multivalued random operator. If the set $\{u \in \mathcal{M}(\Omega, X) : \lambda u \in N(w)u\}$ is bounded for each $w \in \Omega$ and all $\lambda > 1$, then $N(w)$ has a random fixed point.*

Theorem 1.14 ([297]). *Let $(\Omega, \mathcal{A})$ be a complete σ-finite measure space, E be a separable Banach space, and let $N : \Omega \times E \to \mathcal{P}_{cl}(E)$ be a random multivalued contraction. Then $N(w)$ has a random fixed point.*

1.9. Auxiliary Lemmas

We state the following generalization of Gronwall's lemma for singular kernel.

Lemma 1.11 ([374]). *Let $v : I \to [0, +\infty)$ be a real function and $w(\cdot)$ be a nonnegative, locally integrable function on $[0, T]$. Assume that there exist constants $a > 0$ and $0 < \alpha < 1$ such that*

$$v(t) \leq w(t) + a \int_0^t (t - s)^{-\alpha} v(s) ds.$$

Then, there exists a constant $K = K(\alpha)$ such that

$$v(t) \leq w(t) + Ka \int_0^t (t - s)^{-\alpha} w(s) ds \text{ for every } t \in I.$$

Bainov and Hristova [121] introduced the following integral inequality of Gronwall type for piecewise continuous functions that can be used in the sequel.

Lemma 1.12. *Let, for $t \geq t_0 \geq 0$, the following inequality hold*

$$x(t) \leq a(t) + \int_{t_0}^t g(t, s) x(s) ds + \sum_{t_0 < t_k < t} \beta_k(t) x(t_k),$$

where $\beta_k(t)$, $(k \in \mathbb{N})$ *are nondecreasing functions for* $t \geq t_0$, $a \in$ PC$([t_0, \infty), \mathbb{R}_+)$, a *is nondecreasing and* $g(t, s)$ *is a continuous nonnegative function for* $t, s \geq t_0$ *and nondecreasing with respect to t for any fixed* $s \geq t_0$. *Then, for* $t \geq t_0$,

$$x(t) \leq a(t) \prod_{t_0 < t_k < t} (1 + \beta_k(t)) \exp\left(\int_{t_0}^t g(t, s) ds\right).$$

We recall an integral inequality which based on an iteration argument.

Lemma 1.13 ([374]). *Suppose* $\beta > 0$, $a(t)$ *is a nonnegative function locally integrable on* $0 < t < T$ *(for some* $T \leq +\infty$) *and* $g(t)$ *is a nonnegative, nondecreasing continuous function defined on* $0 \leq t < T$, $g(t) \leq M$ *(constant), and suppose* $u(t)$ *is nonnegative and locally integrable on* $0 \leq t < T$ *with*

$$u(t) \leq a(t) + g(t) \int_0^t (t - s)^{\beta - 1} u(s) ds$$

on this interval. Then

$$u(t) \leq a(t) + \int_0^t \left[\sum_{n=1}^{\infty} \frac{(g(t)\Gamma(\beta))^n}{\Gamma(n\beta)} (t - s)^{n\beta - 1} a(s)\right] ds, \quad 0 \leq t < T.$$

From the above lemma, we concluded with the following lemma.

Lemma 1.14. *Suppose* $\beta > 0$, $a(t, w)$ *is a nonnegative function locally integrable on* $[0, T) \times \Omega$ *(for some* $T \leq +\infty$) *and* $g(t, w)$ *is a nonnegative, nondecreasing continuous function with respect to t defined on* $[0, T) \times \Omega$, $g(t, w) \leq M$ *(constant), and suppose* $u(t, w)$ *is nonnegative and locally integrable with respect to t on* $[0, T) \times \Omega$ *with*

$$u(t, w) \leq a(t, w) + g(t, w) \int_0^t (t - s)^{\beta - 1} u(s, w) ds$$

on $[0, T) \times \Omega$. *Then for* $(t, w) \in [0, T) \times \Omega$

$$u(t, w) \leq a(t, w) + \int_0^t \left[\sum_{n=1}^{\infty} \frac{(g(t, w)\Gamma(\beta))^n}{\Gamma(n\beta)} (t - s)^{n\beta - 1} a(s, w)\right] ds.$$

Lemma 1.15 (Arzelà–Ascoli, [230]). *Let* $A \subset C(J, \mathbb{R})$. A *is relatively compact (i.e.,* $\overline{A}$ *is compact) if:*

(1) *A is uniformly bounded, i.e., there exists $M > 0$ such that*
$$\|f(x)\| < M \text{ for every } f \in A \quad \text{and } x \in I.$$

(2) *A is equicontinuous, i.e., for every $\epsilon > 0$, there exists $\delta > 0$ such that for each $x, \overline{x} \in I$, $\|x - \overline{x}\| \leq \delta$ implies $\|f(x) - f(\overline{x})\| \leq \epsilon$ for every $f \in A$.*

Set
$$J := [0, a] \times [0, b], \quad J_0 := \{(x, y, s) : 0 \leq s \leq x \leq a, \ y \in [0, b]\},$$

$$J_1 := \{(x, y, s, t) : 0 \leq s \leq x \leq a, \ 0 \leq t \leq y \leq b\},$$

$$D_1 := \frac{\partial}{\partial x}, \quad D_2 := \frac{\partial}{\partial y},$$

and

$$D_1 D_2 := \frac{\partial^2}{\partial x \partial y}.$$

In the sequel, we will make use of the following variant of the inequality for two independent variables due to Pachpatte.

Lemma 1.16 ([307]). *Let $w \in C(J, \mathbb{R}_+)$, p, $D_1 p \in C(J_0, \mathbb{R}_+)$, q, $D_1 q$, $D_2 q$, $D_1 D_2 q \in C(J_1, \mathbb{R}_+)$, and $c > 0$ a constant. If*

$$w(x, y) \leq c + \int_0^x p(x, y, s) w(s, y) ds + \int_0^x \int_0^y q(x, y, s, t) w(s, t) dt ds,$$

for $(x, y) \in [0, a] \times [0, b]$, then

$$w(x, y) \leq c A(x, y) \exp \left(\int_0^x \int_0^y B(s, t) dt ds \right),$$

where

$$A(x, y) = \exp(Q(x, y)),$$

$$Q(x, y) = \int_0^x \left[p(s, y, s) + \int_0^s D_1 p(s, y, \xi) d\xi \right] ds,$$

and

$$B(x, y) = q(x, y, x, y) A(x, y) + \int_0^x D_1 q(x, y, s, y) A(s, y) ds$$

$$+ \int_0^y D_2 q(x, y, x, t) A(x, t) dt$$

$$+ \int_0^x \int_0^y D_1 D_2 q(x, y, s, t) A(s, t) dt ds.$$

From the above lemma and with $p \equiv 0$, we get the following lemma.

Lemma 1.17. *Let* $w \in C(J, \mathbb{R}_+)$, $q, D_1 q, D_2 q, D_1 D_2 q \in C(J_1, \mathbb{R}_+)$ *and* $c > 0$ *be a constant. If*

$$aw(x, y) \leq c + \int_1^x \int_1^y q(x, y, s, t) w(s, t) dt ds,$$

for $(x, y) \in J$, *then*

$$w(x, y) \leq c \exp \left(\int_1^x \int_1^y B(s, t) dt ds \right),$$

where

$$B(x, y) = q(x, y, x, y) + \int_1^x D_1 q(x, y, s, y) ds$$

$$+ \int_1^y D_2 q(x, y, x, t) dt + \int_1^x \int_1^y D_1 D_2 q(x, y, s, t) dt ds.$$

Chapter 2

Hadamard and Hilfer Fractional Differential Equations and Inclusions in Banach Spaces

2.1. Introduction

Recently, in [129,154,165] the authors have applied the measure of noncompactness to some classes of functional Riemann–Liouville or Caputo fractional differential equations in Banach spaces. Motivated by the above papers, in this chapter, we discuss the existence of solutions for various classes of differential equations and inclusions of Hilfer and Hadamard fractional derivatives.

2.2. Caputo–Hadamard Fractional and Partial Fractional Differential Equations in Banach Spaces

2.2.1. Introduction

This section deals with some existence results for some classes of differential equations involving the Caputo–Hadamard fractional derivative. An application is made of the Mönch fixed-point theorem associated with the technique of measure of noncompactness.

Consider the following problem of Caputo–Hadamard fractional differential equation of the form:

$$\begin{cases} (^{HC}D_1^r u)(t) = f(t, u(t)); & t \in I := [1, T], \\ u(t)|_{t=1} = \phi, \end{cases} \tag{2.1}$$

where $r \in (0,1)$, $T > 1$, $\phi \in E$, $f : I \times E \to E$ is a given continuous function, E is a real (or complex) Banach space with a norm $\|\cdot\|$, $^{HC}D_1^r$ is the Caputo–Hadamard fractional derivative of order r.

Next we discuss the existence of solutions for the following problem of Caputo–Hadamard partial fractional differential equation of the form:

$$\begin{cases} (^{HC}D^r_\sigma u)(t,x) = f(t,x,u(t,x)); & (t,x) \in J := [1,T] \times [1,b], \\ u(t,1) = \phi(t); & t \in [1,T], \\ u(1,x) = \psi(x); & x \in [1,b], \end{cases} \qquad (2.2)$$

where $r = (r_1, r_2) \in (0,1] \times (0,1]$, $T, b > 1$, $\sigma = (1,1)$, $f : J \times E \to E$ is a given continuous function, $\phi : [1,T] \to E$ and $\psi : [1,b] \to E$ are given absolutely continuous functions with $\phi(1) = \psi(1)$, and $^{HC}D^r_1$ is the Caputo–Hadamard partial fractional derivative of order r.

2.2.2. *Existence results for Caputo–Hadamard fractional differential equations*

Let us start by defining what we mean by a solution of problem (2.1). We assume that E is a Banach space and denote by $\mathcal{C}$ the Banach space of all continuous functions v from I into E with the supremum (uniform) norm

$$\|v\|_\infty := \sup_{t \in I} \|v(t)\|_E.$$

Definition 2.1. By a solution of the problem (2.1) we mean a function $u \in \mathcal{C}$ that satisfies the condition $u(1^+) = \phi$, and the equation $(^{HC}D^r_1 u)(t) = f(t, u(t))$ on I.

For the existence of solutions for the problem (2.1), we need the following lemma.

Lemma 2.1. *Let $h \in L^1(I)$. Then the Cauchy problem*

$$\begin{cases} (^{HC}D^r_1 u)(t) = h(t); & t \in I, \\ u(t)|_{t=1} = \phi \end{cases} \qquad (2.3)$$

has the following unique solution:

$$u(t) = \phi + (^H I^r_1 h)(t); \quad t \in I. \qquad (2.4)$$

Proof. Let u be a solution of the problem (2.3). Then, taking into account the definition of the Caputo–Hadamard derivative $(^{HC}D_1^r u)(t)$, we have

$$(^H I_1^{1-r})\left(t\frac{d}{dt}u\right)(t) = h(t).$$

Thus

$$(^H I_1^r)\left(^H I_1^{1-r} t\frac{d}{dt}u\right)(t) = (^H I_1^r h)(t).$$

Hence, we obtain

$$\left(^H I_1^1 t\frac{d}{dt}u\right)(t) = (^H I_1^r h)(t).$$

Since

$$\left(^H I_1^1 t\frac{d}{dt}u\right)(t) = \left(I_1^1 \frac{d}{dt}u\right)(t) = \int_1^t \frac{d}{ds}u(s)ds = u(t) - u(1),$$

we get

$$u(t) = \phi + (^H I_1^r h)(t).$$

Now let $u(t)$ satisfy (2.4). It is clear that $u(t)$ satisfies

$$(^{HC}D_1^r u)(t) = h(t); \text{ on } I.$$

$\square$

Now, we shall prove the following theorem concerning the existence of solutions of the problem (2.1).

Theorem 2.1. *Assume that the following hypothese hold:*

(2.1.1) *The function $t \mapsto f(t,u)$ is measurable on I for each $u \in E$, and the function $u \mapsto f(t,u)$ is continuous on E for a.e. $t \in I$.*

(2.1.2) *There exists a continuous function $p : I \to [0,\infty)$ such that*

$$\|f(t,u)\| \le p(t) \quad \text{for a.e. } t \in I, \text{ and each } u \in E.$$

(2.1.3) *For each bounded and measurable set $B \subset E$ and for each $t \in I$, we have*

$$\alpha(f(t,B)) \le p(t)\alpha(B).$$

If

$$L := \frac{p^*(\ln T)^r}{\Gamma(1+r)} < 1, \tag{2.5}$$

where $p^* = \sup_{t\in I} p(t)$, then the problem (2.1) has at least one solution defined on I.

Proof. Consider the operator $N : \mathcal{C} \to \mathcal{C}$ defined by

$$(Nu)(t) = \phi + \int_1^t \left(\ln \frac{t}{s}\right)^{r-1} \frac{f(s, u(s))}{s\Gamma(r)} ds. \qquad (2.6)$$

From Lemma 2.1, the fixed points of the operator N are solution of the problem (2.1).

For any $u \in \mathcal{C}$, and each $t \in I$, we have

$$\|(Nu)(t)\| \leq \|\phi\| + \frac{1}{\Gamma(r)} \int_1^t \left(\ln \frac{t}{s}\right)^{r-1} \|f(s, u(s))\| \frac{ds}{s}$$

$$\leq \|\phi\| + \frac{1}{\Gamma(r)} \int_1^t \left(\ln \frac{t}{s}\right)^{r-1} p(s) \frac{ds}{s}$$

$$\leq \|\phi\| + \frac{p^*}{\Gamma(r)} \int_1^t \left(\ln \frac{t}{s}\right)^{r-1} \frac{ds}{s}$$

$$\leq \|\phi\| + \frac{p^*(\ln T)^r}{\Gamma(1+r)}.$$

Thus

$$\|N(u)\|_\infty \leq |\phi| + \frac{p^*(\ln T)^r}{\Gamma(1+r)} := R. \qquad (2.7)$$

This proves that N transforms the ball $B_R := B(0, R) = \{w \in \mathcal{C} : \|w\|_\infty \leq R\}$ into itself. We shall show that the operator $N : B_R \to B_R$ satisfies all the assumptions of Theorem 1.6. The proof will be given in several steps.

Step 1. $N : B_R \to B_R$ *is continuous.*

Let $\{u_n\}_{n\in\mathbb{N}}$ be a sequence such that $u_n \to u$ in B_R. Then, for each $t \in I$, we have

$$|(Nu_n)(t) - (Nu)(t)| \leq \frac{1}{\Gamma(r)} \int_1^t \left(\ln \frac{t}{s}\right)^{r-1} |f(s, u_n(s)) - f(s, u(s))| \frac{ds}{s}. \qquad (2.8)$$

Since $u_n \to u$ as $n \to \infty$ and f is continuous, then by the Lebesgue dominated convergence theorem, Eq. (2.8) implies

$$\|N(u_n) - N(u)\|_\infty \to 0 \quad \text{as } n \to \infty.$$

Step 2. $N(B_R)$ *is bounded and equicontinuous.*

Since $N(B_R) \subset B_R$ and B_R is bounded, then $N(B_R)$ is bounded.

Let $t_1, t_2 \in I$, $t_1 < t_2$ and let $u \in B_R$. Thus, we have

$$\|(Nu)(t_2) - (Nu)(t_1)\|$$

$$\leq \left\| \int_1^{t_2} \left(\ln \frac{t_2}{s} \right)^{r-1} \frac{f(s, u(s))}{s\Gamma(r)} ds - \int_1^{t_1} \left(\ln \frac{t_1}{s} \right)^{r-1} \frac{f(s, u(s))}{s\Gamma(r)} ds \right\|$$

$$\leq \int_{t_1}^{t_2} \left(\ln \frac{t_2}{s} \right)^{r-1} \frac{|f(s, u(s))|}{s\Gamma(r)} ds$$

$$+ \int_1^{t_1} \left| \left(\ln \frac{t_2}{s} \right)^{r-1} - \left(\ln \frac{t_1}{s} \right)^{r-1} \right| \frac{\|f(s, u(s))\|}{s\Gamma(r)} ds$$

$$\leq \int_{t_1}^{t_2} \left(\ln \frac{t_2}{s} \right)^{r-1} \frac{p(s)}{s\Gamma(r)} ds$$

$$+ \int_1^{t_1} \left| \left(\ln \frac{t_2}{s} \right)^{r-1} - \left(\ln \frac{t_1}{s} \right)^{r-1} \right| \frac{p(s)}{s\Gamma(r)} ds.$$

Hence, we get

$$\|(Nu)(t_2) - (Nu)(t_1)\| \leq \frac{p^*(\ln T)^r}{\Gamma(1+r)} \left(\ln \frac{t_2}{t_1} \right)^r$$

$$+ \frac{p^*}{\Gamma(r)} \int_1^{t_1} \left| \left(\ln \frac{t_2}{s} \right)^{r-1} - \left(\ln \frac{t_1}{s} \right)^{r-1} \right| ds.$$

As $t_1 \to t_2$, the right-hand side of the above inequality tends to zero.

Now let V be a subset of B_r such that $V \subset \overline{N(V)} \cup \{0\}$, V is bounded and equicontinuous and therefore the function $t \to v(t) = \alpha(V(t))$ is continuous on I. By (2.1.3) and the properties of the measure α, for each $t \in I$, we have

$$v(t) \leq \alpha((NV)(t) \cup \{0\})$$

$$\leq \alpha((NV)(t))$$

$$\leq \frac{1}{\Gamma(r)} \int_1^t \left(\ln \frac{t}{s} \right)^{r-1} p(s)\alpha(V(s)) ds$$

$$\leq \frac{1}{\Gamma(r)} \int_1^t \left(\ln \frac{t}{s} \right)^{r-1} p(s)v(s) ds$$

$$\leq \frac{p^*(\ln T)^r}{\Gamma(1+r)} \|v\|_\infty.$$

Thus

$$\|v\|_\infty \leq L\|v\|_\infty.$$

From (2.5), we get $\|v\|_\infty = 0$, that is $v(t) = \alpha(V(t)) = 0$, for each $t \in I$ and then $V(t)$ is relatively compact in E. In view of the Arzelà–Ascoli theorem, V is relatively compact in $\mathcal{C}$. Applying now Theorem 1.6, we conclude that N has a fixed point which is a solution of the problem (2.1).   □

2.2.3. *Caputo–Hadamard partial fractional differential equations*

Now, we are concerned with the existence of solutions of the problem (2.2). Let $C(J)$ be the Banach space of all continuous functions v from J into E with the supremum (uniform) norm

$$\|v\|_\infty := \sup_{(t,x) \in J} \|v(t,x)\|.$$

By $L^1(J)$, we denote the space of Bochner-integrable functions $v : J \to E$ with the norm

$$\|v\|_1 = \int_1^T \int_1^b \|v(t,x)\| dx dt.$$

As usual, $AC(J)$ denotes the space of absolutely continuous functions from J into E.

Now, we give some results and properties of partial fractional calculus.

Definition 2.2 (Riemann–Liouville partial fractional integral [71, 263,341]). The left-sided mixed Riemann–Liouville integral of order $r = (r_1, r_2)$; $r_1, r_2 > 0$ of a function $w \in L^1(J)$ is defined as

$$(I_\sigma^r w)(t,x) = \int_1^t \int_1^x (t-\tau)^{r_1-1}(x-\xi)^{r_2-1}\frac{f(\tau,\xi)}{\Gamma(r_1)\Gamma(r_2)}d\xi d\tau \quad \text{for a.e. } (t,x) \in J.$$

Define by $D_{tx}^2 := \frac{\partial^2}{\partial t \partial x}$ and $1 - r := (1 - r_1, 1 - r_2) \in (0,1] \times (0,1]$.

Definition 2.3 (Caputo partial fractional derivative [71,263]). The Caputo fractional derivative of order r of a function $w \in L^1(J)$ is defined as

$$({}^c D_\sigma^r w)(t) = \left(I_\sigma^{1-r} \frac{d}{dt} w \right)(t)$$

$$= \frac{1}{\Gamma(1-r_1)\Gamma(1-r_2)} \int_1^t \int_1^x (t-\tau)^{-r_1}(x-\xi)^{-r_2}$$

$$D_{\tau\xi}^2 w(\tau,\xi)d\xi d\tau \text{ for a.e. } (t,x) \in J.$$

Let us recall some definitions and properties of Hadamard partial fractional integration and differentiation. We refer to [228,263] for a more detailed analysis.

Definition 2.4 (Hadamard partial fractional integral [263]). The Hadamard fractional integral of order $r = (r_1, r_2)$; $r_1, r_2 > 0$ for a function $w \in L^1(J)$ is defined as

$$(^H I_\sigma^r w)(t, x) = \int_1^t \int_1^x \left(\ln \frac{t}{\tau} \right)^{r_1 - 1} \left(\ln \frac{x}{\xi} \right)^{r_2 - 1} \frac{w(\tau, \xi)}{\tau \xi \Gamma(r_1) \Gamma(r_2)} d\xi d\tau$$

$$\text{for a.e. } (t, x) \in J,$$

provided the integral exists.

Set $\delta_2 = tx D_{tx}^2$; $(t, x) \in J$.

Definition 2.5 (Hadamard partial fractional derivative [263]). The Hadamard fractional derivative of order $r = (r_1, r_2) \in (0, 1] \times (0, 1]$ applied to the function $w \in L^1(J)$ is defined as

$$(^H D_\sigma^r w)(t, x) = \delta_2 (^H I_\sigma^{1-r} w)(t, x).$$

Definition 2.6 (Caputo–Hadamard partial fractional derivative). The Caputo-Hadamard fractional derivative of order $r = (r_1, r_2) \in (0, 1] \times (0, 1]$ applied to the function $w \in L^1(J)$ is defined as

$$(^{HC} D_\sigma^q w)(t, x) = (^H I_\sigma^{1-r} \delta_2 w)(t, x).$$

Definition 2.7. By a solution of the problem (2.2) we mean a function $u \in C(J)$ that satisfies the conditions $u(t, 1) = \phi(t)$, $u(1, x) = \psi(x)$ with $\phi(1) = \psi(1)$, and the equation $(^{HC} D_\sigma^r u)(t, x) = f(t, x, u(t, x))$ on J.

For the existence of solutions for the problem (2.2), we need the following lemma.

Lemma 2.2. *Let $h \in L^1(J)$. Then the Darboux problem*

$$\begin{cases} (^{HC} D_\sigma^r u)(t, x) = h(t, x); & (t, x) \in J, \\ u(t, 1) = \phi(t); & t \in [1, T], \\ u(1, x) = \psi(x); & x \in [1, b], \\ \phi(1) = \psi(1) \end{cases} \qquad (2.9)$$

has the following unique solution:

$$u(t, x) = \mu(t, x) + (^H I_\sigma^r h)(t, x); \quad (t, x) \in J, \qquad (2.10)$$

where $\mu(t, x) = \phi(t) + \psi(x) - \phi(1)$.

Proof. Let $u(t, x)$ be a solution of the problem (2.9). Then, taking into account the definition of the Caputo–Hadamard derivative $({}^{HC}D_\sigma^r u)(t, x)$, we have

$$({}^H I_\sigma^{1-r}) \left(tx D_{tx}^2 u \right)(t, x) = h(t, x).$$

Thus

$$({}^H I_\sigma^r) \left({}^H I_\sigma^{1-r} tx D_{tx}^2 u \right)(t, x) = ({}^H I_\sigma^r h)(t, x).$$

Hence, we obtain

$$\left({}^H I_\sigma^\sigma tx D_{tx}^2 u \right)(t, x) = ({}^H I_\sigma^r h)(t, x).$$

Since

$$\left({}^H I_\sigma^\sigma tx D_{tx}^2 u \right)(t, x) = \left(I_\sigma^\sigma D_{tx}^2 u \right)(t, x)$$

$$= \int_1^t \int_1^x D_{\tau\xi}^2 u(\tau, \xi) d\xi d\tau$$

$$= u(t, x) - u(t, 1) - u(1, x) + u(1, 1),$$

we get

$$u(t, x) = \mu(t, x) + ({}^H I_\sigma^r h)(t, x).$$

Now let $u(t, x)$ satisfy (2.10). It is clear that $u(t, x)$ satisfies

$$({}^{HC}D_\sigma^r u)(t, x) = h(t, x); \text{ on } J. \qquad \square$$

Theorem 2.2. *Assume that the following hypothese hold:*

(2.2.1) *The function $(t, x) \mapsto f(t, x, u)$ is measurable on J for each $u \in E$, and the function $u \mapsto f(t, x, u)$ is continuous on E for a.e. $(t, x) \in J$.*

(2.2.2) *There exist continuous functions $p_1, p_2 : J \to [0, \infty)$ such that*

$$\|f(t, x, u)\| \le p_1(t, x) + p_2(t, x)\|u\|$$

for a.e. $(t, x) \in J$, and each $u \in E$,

(2.2.3) *For each bounded and measurable set $B \subset E$ and for each $(t, x) \in J$, there exists a continuous function $q : J \to [0, \infty)$ such that we have*

$$\alpha(f(t, x, B)) \le q(t, x)\alpha(B).$$

If

$$L' := \frac{q^*(\ln T)^{r_1}(\ln b)^{r_2}}{\Gamma(1 + r_1)\Gamma(1 + r_2)} < 1, \qquad (2.11)$$

where $q^* = \sup_{(t,x) \in J} q(t, x)$, then the problem (2.2) has at least one solution defined on J.

Proof. Consider the operator $G : C \to C$ defined by

$$(Gu)(t,x) = \mu(t,x) + \int_1^t \int_1^x \left(\ln\frac{t}{\tau}\right)^{r_1-1}\left(\ln\frac{x}{\xi}\right)^{r_2-1}\frac{f(\tau,\xi,u(\tau,\xi))}{\tau\xi\Gamma(r_1)\Gamma(r_2)}\,d\xi d\tau.$$
(2.12)

From Lemma 2.2, the fixed points of the operator G are solutions of the problem (2.2).

Set

$$\mu^* = \sup_{(t,x)\in J}|\mu(t,x)|, \quad \text{and} \quad p_i^* = \sup_{(t,x)\in J}p_i(t,x);\ i = 1,2.$$

For any $u \in C(J)$, and each $(t,x) \in J$, we have

$$\|(Gu)(t,x)\|$$

$$\leq \|\mu(t,x)\| + \int_1^t \int_1^x \left(\ln\frac{t}{\tau}\right)^{r_1-1}\left(\ln\frac{x}{\xi}\right)^{r_2-1}\frac{\|f(\tau,\xi,u(\tau,\xi))\|}{\tau\xi\Gamma(r_1)\Gamma(r_2)}\,d\xi d\tau$$

$$\leq \|\mu(t,x)\| + \int_1^t \int_1^x \left(\ln\frac{t}{\tau}\right)^{r_1-1}\left(\ln\frac{x}{\xi}\right)^{r_2-1}\frac{p_1(\tau,\xi)+p_2(\tau,\xi)}{\tau\xi\Gamma(r_1)\Gamma(r_2)}\,d\xi d\tau$$

$$\leq \mu^* + \int_1^t \int_1^x \left(\ln\frac{t}{\tau}\right)^{r_1-1}\left(\ln\frac{x}{\xi}\right)^{r_2-1}\frac{p_1^*+p_2^*}{\tau\xi\Gamma(r_1)\Gamma(r_2)}\,d\xi d\tau$$

$$\leq \mu^* + \frac{(p_1^*+p_2^*)(\ln T)^{r_1}(\ln b)^{r_2}}{\Gamma(1+r_1)\Gamma(1+r_2)}.$$

Thus

$$\|G(u)\|_\infty \leq \mu^* + \frac{(p_1^*+p_2^*)(\ln T)^{r_1}(\ln b)^{r_2}}{\Gamma(1+r_1)\Gamma(1+r_2)} := R.$$
(2.13)

This proves that G transforms the ball $B_R := B(0,R) = \{w \in C(J) : \|w\|_\infty \leq R\}$ into itself. We shall show that the operator $G : B_R \to B_R$ satisfies all the assumptions of Theorem 1.6. The proof will be given in several steps.

Step 1. $G : B_R \to B_R$ *is continuous.*

Let $\{u_n\}_{n\in\mathbb{N}}$ be a sequence such that $u_n \to u$ in B_R. Then, for each $(t,x) \in J$, we have

$$\|(Gu_n)(t,x) - (Gu)(t,x)\|$$

$$\leq \int_1^t \int_1^x \left(\ln \frac{t}{\tau}\right)^{r_1-1} \left(\ln \frac{x}{\xi}\right)^{r_2-1}$$

$$\times \frac{\|f(\tau,\xi,u_n(\tau,\xi)) - f(\tau,\xi,u(\tau,\xi))\|}{\tau\xi\Gamma(r_1)\Gamma(r_2)} d\xi d\tau. \qquad (2.14)$$

Since $u_n \to u$ as $n \to \infty$ and f is continuous, then by the Lebesgue dominated convergence theorem, Eq. (2.14) implies

$$\|G(u_n) - G(u)\|_\infty \to 0 \quad \text{as } n \to \infty.$$

Step 2. $G(B_R)$ *is bounded and equicontinuous.*

Since $G(B_R) \subset B_R$ and B_R is bounded, then $G(B_R)$ is bounded.

Now, let $(t_1, x_1), (t_2, x_2) \in J$, $t_1 < t_2$, $x_1 < x_2$ and let $u \in B_R$. Then, we have

$$\|(Gu)(t_2, x_2) - (Gu)(t_1, x_1)\|$$

$$\leq \|\mu(t_2,x_2) - \mu(t_1,x_1)\|$$

$$+ \left\| \int_1^{t_2} \int_1^{x_2} \left(\ln \frac{t_2}{\tau}\right)^{r_1-1} \left(\ln \frac{x_2}{\xi}\right)^{r_2-1} \frac{f(\tau,\xi,u(\tau,\xi))}{\tau\xi\Gamma(r_1)\Gamma(r_2)} d\xi d\tau \right.$$

$$\left. - \int_1^{t_1} \int_1^{x_1} \left(\ln \frac{t_1}{\tau}\right)^{r_1-1} \left(\ln \frac{x_1}{\xi}\right)^{r_2-1} \frac{f(\tau,\xi,u(\tau,\xi))}{\tau\xi\Gamma(r_1)\Gamma(r_2)} d\xi d\tau \right\|$$

$$\leq \|\mu(t_2,x_2) - \mu(t_1,x_1)\|$$

$$+ \int_1^{t_1} \int_{x_1}^{x_2} \left|\ln \frac{t_2}{\tau}\right|^{r_1-1} \left|\ln \frac{x_2}{\xi}\right|^{r_2-1} \frac{\|f(\tau,\xi,u(\tau,\xi))\|}{\tau\xi\Gamma(r_1)\Gamma(r_2)} d\xi d\tau$$

$$+ \int_{t_1}^{t_2} \int_1^{x_1} \left|\ln \frac{t_2}{\tau}\right|^{r_1-1} \left|\ln \frac{x_2}{\xi}\right|^{r_2-1} \frac{\|f(\tau,\xi,u(\tau,\xi))\|}{\tau\xi\Gamma(r_1)\Gamma(r_2)} d\xi d\tau$$

$$+ \int_{t_1}^{t_2} \int_{x_1}^{x_2} \left|\ln \frac{t_2}{\tau}\right|^{r_1-1} \left|\ln \frac{x_2}{\xi}\right|^{r_2-1} \frac{\|f(\tau,\xi,u(\tau,\xi))\|}{\tau\xi\Gamma(r_1)\Gamma(r_2)} d\xi d\tau$$

$$+ \int_1^{t_1} \int_1^{x_1} \left| \left(\ln \frac{t_2}{\tau}\right)^{r_1-1} \left(\ln \frac{x_2}{\xi}\right)^{r_2-1} \right.$$

$$\left. - \left(\ln \frac{t_1}{\tau}\right)^{r_1-1} \left(\ln \frac{x_1}{\xi}\right)^{r_2-1} \right| \frac{\|f(\tau,\xi,u(\tau,\xi))\|}{\tau\xi\Gamma(r_1)\Gamma(r_2)} d\xi d\tau.$$

Thus, we obtain

$$\|(Gu)(t_2, x_2) - (Gu)(t_1, x_1)\|$$

$$\leq \|\mu(t_2, x_2) - \mu(t_1, x_1)\|$$

$$+ \frac{p_1^* + p_2^*}{\Gamma(r_1)\Gamma(r_2)} \left(\int_1^{t_1} \int_{x_1}^{x_2} \left| \ln \frac{t_2}{\tau} \right|^{r_1 - 1} \left| \ln \frac{x_2}{\xi} \right|^{r_2 - 1} \frac{d\xi d\tau}{\tau \xi} \right.$$

$$+ \int_{t_1}^{t_2} \int_1^{x_1} \left| \ln \frac{t_2}{\tau} \right|^{r_1 - 1} \left| \ln \frac{x_2}{\xi} \right|^{r_2 - 1} \frac{d\xi d\tau}{\tau \xi}$$

$$+ \int_{t_1}^{t_2} \int_{x_1}^{x_2} \left| \ln \frac{t_2}{\tau} \right|^{r_1 - 1} \left| \ln \frac{x_2}{\xi} \right|^{r_2 - 1} \frac{d\xi d\tau}{\tau \xi}$$

$$+ \int_1^{t_1} \int_1^{x_1} \left| \left(\ln \frac{t_2}{\tau} \right)^{r_1 - 1} \left(\ln \frac{x_2}{\xi} \right)^{r_2 - 1} \right.$$

$$\left. - \left(\ln \frac{t_1}{\tau} \right)^{r_1 - 1} \left(\ln \frac{x_1}{\xi} \right)^{r_2 - 1} \right| \frac{d\xi d\tau}{\tau \xi} \right).$$

Hence, we get

$$\|(Gu)(t_2, x_2) - (Gu)(t_1, x_1)\|$$

$$\leq \|\mu(t_2, x_2) - \mu(t_1, x_1)\|$$

$$+ \frac{p_1^* + p_2^*}{\Gamma(1 + r_1)\Gamma(1 + r_2)} \left[\left((\ln t_2)^{r_1} - \left(\ln \frac{t_2}{t_1} \right)^{r_1} \right) \left(\ln \frac{x_2}{x_1} \right)^{r_2} \right.$$

$$+ \left(\ln \frac{t_2}{t_1} \right)^{r_t} \left((\ln x_2)^{r_2} - \left(\ln \frac{x_2}{x_1} \right)^{r_2} \right) + \left(\ln \frac{t_2}{t_1} \right)^{r_1} \left(\ln \frac{x_2}{x_1} \right)^{r_2} \right]$$

$$+ \frac{p_1^* + p_2^*}{\Gamma(r_1)\Gamma(r_2)} \int_1^{t_1} \int_1^{x_1} \left| \left(\ln \frac{t_2}{\tau} \right)^{r_1 - 1} \left(\ln \frac{x_2}{\xi} \right)^{r_2 - 1} \right.$$

$$\left. - \left(\ln \frac{t_1}{\tau} \right)^{r_1 - 1} \left(\ln \frac{x_1}{\xi} \right)^{r_2 - 1} \right| d\xi d\tau.$$

As $t_1 \to t_2$ and $x_1 \to x_2$, the right-hand side of the above inequality tends to zero.

Now let V be a subset of B_r such that $V \subset \overline{G(V)} \cup \{0\}$, V is bounded and equicontinuous and therefore the function $(t, x) \to v(t, x) = \alpha(V(t, x))$ is continuous on J. By (2.2.3) and the properties of the measure α, for each

$(t, x) \in J$, we have

$$v(t, x) \leq \alpha((NV)(t, x) \cup \{0\})$$

$$\leq \alpha((NV)(t, x))$$

$$\leq \frac{1}{\Gamma(r_1)\Gamma(r_2)} \int_1^t \int_1^x \left(\ln \frac{t}{\tau} \right)^{r_1-1} \left(\ln \frac{x}{\xi} \right)^{r_2-1} q(\tau, \xi)\alpha(V(\tau, \xi))d\xi d\tau$$

$$\leq \frac{1}{\Gamma(r_1)\Gamma(r_2)} \int_1^t \int_1^x \left(\ln \frac{t}{\tau} \right)^{r_1-1} \left(\ln \frac{x}{\xi} \right)^{r_2-1} q(\tau, \xi)v(\tau, \xi)d\xi d\tau$$

$$\leq \frac{q^*(\ln T)^{r_1}(\ln b)^{r_2}}{\Gamma(1+r_1)\Gamma(1+r_2)} \|v\|_\infty.$$

Thus

$$\|v\|_\infty \leq L'\|v\|_\infty.$$

From (2.11), we get $\|v\|_\infty = 0$, that is $v(t, x) = \alpha(V(t, x)) = 0$, for each $(t, x) \in J$ and then $V(t, x)$ is relatively compact in E. In view of the Arzelà–Ascoli theorem, V is relatively compact in C. Applying now Theorem 1.6, we conclude that G has a fixed point which is a solution of the problem (2.2).   □

2.2.4. *Examples*

Let

$$E = l^1 = \left\{ u = (u_1, u_2, \ldots, u_n, \ldots), \sum_{n=1}^\infty |u_n| < \infty \right\}$$

be the Banach space with the norm

$$\|u\|_E = \sum_{n=1}^\infty |u_n|.$$

Example 1. Consider the following problem of Caputo–Hadamard fractional differential equation:

$$\begin{cases} (^{HC}D_1^{\frac{1}{2}}u_n)(t) = f_n(t, u(t)); & t \in [1, e], \\ u_n(1) = (1, 0, \ldots, 0, \ldots), \end{cases} \tag{2.15}$$

where

$$\begin{cases} f_n(t, u) = \dfrac{(t-1)^{\frac{-1}{4}}u_n \sin(t-1)}{64(1+\sqrt{t-1})(1+\|u\|)}; & t \in (1, e], \\ f_n(1, u) = 0 \end{cases}.$$

with

$$f = (f_1, f_2, \ldots, f_n, \ldots), \text{ and } u = (u_1, u_2, \ldots, u_n, \ldots).$$

The hypothesis (2.1.2) is satisfied with

$$\begin{cases} p(t) = \dfrac{(t-1)^{\frac{-1}{4}} |\sin(t-1)|}{64(1 + \sqrt{t-1})}; & t \in (1, e], \\ p(1) = 0. \end{cases}$$

A simple computation shows that conditions of Theorem 2.1 are satisfied. Hence, the problem (2.15) has at least one solution defined on $[1, e]$.

Example 2. Consider now the following problem of Caputo–Hadamard partial fractional differential equation

$$\begin{cases} (^{HC}D_\sigma^r u_n)(t, x) = f_n(t, x, u(t, x)); & (t, x) \in [1, e] \times [1, e], \\ u_n(t, 1) = u_n(1, x) = (1, 0, \ldots, 0, \ldots); & t, x \in [1, e], \end{cases} \quad (2.16)$$

where $r = (\frac{1}{2}, \frac{1}{2})$,

$$\begin{cases} f_n(t, x, u) = \dfrac{x(t-1)^{\frac{-1}{4}} u_n \sin(t-1)}{64(x + \sqrt{t-1})(1 + |u|)}; & (t, x) \in (1, e] \times [1, e], \\ f_n(1, x, u) = 0; & x \in [1, e]. \end{cases}$$

The hypothesis (2.2.2) is satisfied with

$$\begin{cases} p_1(t, x) = \dfrac{x(t-1)^{\frac{-1}{4}} |\sin(t-1)|}{64(x + \sqrt{t-1})}; & (t, x) \in (1, e] \times [1, e], \\ p_1(1, x) = 0; & x \in [1, e], \end{cases}$$

and $p_2 \equiv 0$. A simple computation shows that conditions of Theorem 2.2 are satisfied. Hence, the problem (2.16) has at least one solution defined on $[1, e] \times [1, e]$.

2.3. Hilfer and Hilfer–Hadamard Fractional Differential Equations in Banach Spaces

2.3.1. *Introduction*

By using the technique that relies on the concept of measure of noncompactness and the fixed-point theory, we prove some existence and Ulam stability results for some Hilfer and Hilfer–Hadamard differential equations of fractional order. Next, we prove that our problems are generalized Ulam–Hyers–Rassias stable. Consider the following problem of Hilfer fractional differential equations:

$$\begin{cases} (D_0^{\alpha,\beta} u)(t) = f(t, u(t)); & t \in I := [0, T], \\ (I_0^{1-\gamma} u)(0) = \phi, \end{cases} \tag{2.17}$$

where $\alpha \in (0,1)$, $\beta \in [0,1]$, $\gamma = \alpha + \beta - \alpha\beta$, $T > 0$, $\phi \in E$, $f : I \times E \to E$ is a given function, E is a real (or complex) Banach space with a norm $\|\cdot\|$, $I_0^{1-\gamma}$ is the left-sided Riemann–Liouville integral of order $1 - \gamma$, and $D_0^{\alpha,\beta}$ is the generalized Riemann–Liouville derivative operator of order α and type β, introduced by Hilfer in [238].

Next, we consider the following problem of Hilfer–Hadamard fractional differential equations

$$\begin{cases} (^H D_1^{\alpha,\beta} u)(t) = g(t, u(t)); & t \in [1, T], \\ (^H I_1^{1-\gamma} u)(1) = \phi_0, \end{cases} \tag{2.18}$$

where $\alpha \in (0,1)$, $\beta \in [0,1]$, $\gamma = \alpha + \beta - \alpha\beta$, $T > 1$, $\phi_0 \in E$, $g : [1, T] \times E \to E$ is a given function, $^H I_1^{1-\gamma}$ is the left-sided Hadamard integral of order $1 - \gamma$, and $^H D_1^{\alpha,\beta}$ is the Hilfer–Hadamard fractional derivative of order α and type β.

2.3.2. *Hilfer fractional differential equations*

In this section, we are concerned with the existence and the generalized Ulam–Hyers–Rassias stability of our problem (2.17).

Let us start by defining what we mean by a solution of the problem (2.17). Define the space

$$C_\gamma(I) = \{ w : I \to E : t^{1-\gamma} w(t) \in C \},$$

with the norm

$$\|w\|_{C_\gamma} := \sup_{t \in I} \|t^{1-\gamma} w(t)\|_E,$$

where $C = C([0, T])$.

Definition 2.8. By a solution of the problem (2.17) we mean a measurable function $u \in C_\gamma(I)$ that satisfies the condition $(I_0^{1-\gamma} u)(0^+) = \phi$, and the equation $(D_0^{\alpha,\beta} u)(t) = f(t, u(t))$ on I.

The following hypotheses will be used in the sequel.

(2.3.1) The function $t \mapsto f(t, u)$ is measurable on I for each $u \in E$, and the function $u \mapsto f(t, u)$ is continuous on E for a.e. $t \in I$.

(2.3.2) There exists a continuous function $p : I \to [0, \infty)$ such that

$$\|f(t, u) - f(t, v)\| \le p(t)\|u - v\| \text{ for a.e. } t \in I, \text{ and each } u, v \in E.$$

(2.3.3) For each bounded and measurable set $B \subset E$ and for each $t \in I$, we have

$$\mu(f(t, B)) \le p(t)\mu(B).$$

(2.3.4) There exists $\lambda_\Phi > 0$ such that for each $t \in I$, we have

$$(I_0^\alpha \Phi)(t) \le \lambda_\Phi \Phi(t).$$

Set

$$p^* = \sup_{t \in I} p(t), \quad f^* = \operatorname{esssup} |f(t, 0)|.$$

From Theorem 21 in [321], we conclude the following lemma.

Lemma 2.3. *Let $f : I \times E \to E$ be such that $f(\cdot, u(\cdot)) \in C_\gamma$ for any $u \in C_\gamma$. Then problem (2.17) is equivalent to the integral equation*

$$u(t) = \frac{\phi}{\Gamma(\gamma)} t^{\gamma - 1} + (I_0^\alpha f(\cdot, u(\cdot)))(t).$$

Now, we shall prove the following theorem concerning the existence of solutions of the problem (2.17).

Theorem 2.3. *Assume that the hypotheses (2.3.1)–(2.3.3) hold. If*

$$\ell := \frac{4p^* T^\alpha}{\Gamma(1 + \alpha)} < 1, \tag{2.19}$$

then the problem (2.17) has at least one solution defined on I. Furthermore, if the hypothesis (2.3.4) holds, then the problem (2.17) is generalized Ulam–Hyers–Rassias stable.

Proof. Consider the operator $N : C_\gamma \to C_\gamma$ defined by

$$(Nu)(t) = \frac{\phi}{\Gamma(\gamma)} t^{\gamma - 1} + \int_0^t (t - s)^{\alpha - 1} \frac{f(s, u(s))}{\Gamma(\alpha)} ds. \tag{2.20}$$

Clearly, the fixed points of the operator N are solution of the problem (2.17).

For any $u \in C_\gamma$, and each $t \in I$ we have

$$\|t^{1-\gamma}(Nu)(t)\| \leq \frac{\|\phi\|}{\Gamma(\gamma)} + \frac{t^{1-\gamma}}{\Gamma(\alpha)} \int_0^t (t-s)^{\alpha-1} \|f(s, u(s))\| ds$$

$$\leq \frac{\|\phi\|}{\Gamma(\gamma)} + \frac{t^{1-\gamma}}{\Gamma(\alpha)} \int_0^t (t-s)^{\alpha-1} \|f(s, 0)\| ds$$

$$+ \frac{t^{1-\gamma}}{\Gamma(\alpha)} \int_0^t (t-s)^{\alpha-1} \|f(s, u(s)) - f(s, 0)\| ds$$

$$\leq \frac{\|\phi\|}{\Gamma(\gamma)} + \frac{t^{1-\gamma}}{\Gamma(\alpha)} \int_0^t (t-s)^{\alpha-1} (\|f(s, 0)\| + p(s)\|u(s)\|) ds$$

$$\leq \frac{\|\phi\|}{\Gamma(\gamma)} + \frac{(f^* + p^* R)T^{1-\gamma}}{\Gamma(\alpha)} \int_0^t (t-s)^{\alpha-1} ds$$

$$\leq \frac{\|\phi\|}{\Gamma(\gamma)} + \frac{(f^* + p^* R)T^{1-\gamma+\alpha}}{\Gamma(1+\alpha)}.$$

Thus

$$\|N(u)\|_C \leq \frac{\|\phi\|}{\Gamma(\gamma)} + \frac{(f^* + p^* R)T^{1-\gamma+\alpha}}{\Gamma(1+\alpha)} := R. \qquad (2.21)$$

This proves that N transforms the ball $B_R := B(0, R) = \{w \in C_\gamma : \|w\|_C \leq R\}$ into itself for an appropriate R. We shall show that the operator $N : B_R \to B_R$ satisfies all the assumptions of Theorem 1.6. The proof will be given in several steps.

Step 1. $N : B_R \to B_R$ *is continuous.*

Let $\{u_n\}_{n \in \mathbb{N}}$ be a sequence such that $u_n \to u$ in B_R. Then, for each $t \in I$, we have

$$\|t^{1-\gamma}(Nu_n)(t) - t^{1-\gamma}(Nu)(t)\| \leq \frac{t^{1-\gamma}}{\Gamma(\alpha)} \int_0^t (t-s)^{\alpha-1} \|f(s, u_n(s))$$

$$- f(s, u(s))\| ds$$

$$\leq \frac{t^{1-\gamma}}{\Gamma(\alpha)} \int_0^t (t-s)^{\alpha-1} p(s) \|u_n(s) - u(s)\| ds$$

$$\leq \frac{p^* T^{1-\gamma}}{\Gamma(\alpha)} \int_0^t (t-s)^{\alpha-1} \|u_n(s) - u(s)\| ds.$$

Hence

$$\|t^{1-\gamma}(Nu_n)(t) - t^{1-\gamma}(Nu)(t)\| \leq \frac{p^* T^{1-\gamma}}{\Gamma(\alpha)} \int_0^t (t-s)^{\alpha-1} \|u_n(s) - u(s)\| ds.$$

$$(2.22)$$

Since $u_n \to u$ as $n \to \infty$, then Eq. (2.22) implies

$$\|N(u_n) - N(u)\|_C \to 0 \quad \text{as } n \to \infty.$$

Step 2. $N(B_R)$ *is bounded.*

Since $N(B_R) \subset B_R$ and B_R is bounded, then $N(B_R)$ is bounded.

Step 3. *For each bounded subset* D *of* B_R, $\mu_c(N(D)) \leq \ell\mu_c(D)$, *where* μ_c *is a measure of noncompactness of* C_γ *defined by* $\mu(D) = \sup_{t\in I} \mu(D(t))$.

From Lemmas 1.4 and 1.5, for any $D \subset B_R$ and any $\epsilon > 0$, there exists a sequence $\{u_n\}_{n=0}^\infty \subset D$, such that for all $t \in I$, we have

$$\mu((ND)(t)) = \mu\left(\left\{\frac{\phi}{\Gamma(\gamma)}t^{\gamma-1} + \int_0^t (t-s)^{\alpha-1}\frac{f(s,u(s))}{\Gamma(\alpha)}ds; \ u \in D\right\}\right)$$

$$\leq 2\mu\left(\left\{\int_0^t \frac{(t-s)^{\alpha-1}}{\Gamma(\alpha)}f(s,u_n(s))ds\right\}_{n=1}^\infty\right) + \epsilon$$

$$\leq 4\int_0^t \mu\left(\left\{\frac{(t-s)^{\alpha-1}}{\Gamma(\alpha)}f(s,u_n(s))\right\}_{n=1}^\infty\right)ds + \epsilon$$

$$\leq 4\int_0^t \frac{(t-s)^{\alpha-1}}{\Gamma(\alpha)}\mu\left(\{f(s,u_n(s))\}_{n=1}^\infty\right)ds + \epsilon$$

$$\leq 4\int_0^t \frac{(t-s)^{\alpha-1}}{\Gamma(\alpha)}p(s)\mu\left(\{u_n(s)\}_{n=1}^\infty\right)ds + \epsilon$$

$$\leq \left(4\int_0^t \frac{(t-s)^{\alpha-1}}{\Gamma(\alpha)}p(s)ds\right)\mu\left(\{u_n\}_{n=1}^\infty\right) + \epsilon$$

$$\leq \left(4\int_0^t \frac{(t-s)^{\alpha-1}}{\Gamma(\alpha)}p(s)ds\right)\mu_c(D) + \epsilon$$

$$\leq \frac{4p^*T^\alpha}{\Gamma(1+\alpha)}\mu_c(D) + \epsilon$$

$$\leq \ell\mu_c(D) + \epsilon.$$

Since $\epsilon > 0$ is arbitrary, then

$$\mu_c(N(B)) \leq \ell\mu_c(B).$$

As a consequence of steps 1–3 together with Theorem 1.6, we can conclude that N has at least one fixed point in B_R which is a solution of problem (2.17).

Step 4. *The generalized Ulam–Hyers–Rassias stability.*

Let u be a solution of the inequality (1.7), and let us assume that v is a solution of problem (2.17). Thus, we have

$$v(t) = \frac{\phi}{\Gamma(\gamma)} t^{\gamma-1} + \int_0^t (t-s)^{\alpha-1} \frac{f(s, v(s))}{\Gamma(\alpha)} ds.$$

For each $t \in I$, we have

$$\left\| u(t) - \frac{\phi}{\Gamma(\gamma)} t^{\gamma-1} - \int_0^t (t-s)^{\alpha-1} \frac{f(s, u(s))}{\Gamma(\alpha)} ds \right\| \le (I_0^\alpha \Phi)(t).$$

Set

$$q^* = \sup_{t \in I} q(t).$$

From hypotheses (2.3.2) and (2.3.3), for each $t \in I$, we get

$$\|u(t) - v(t)\| \le \left\| u(t) - \frac{\phi}{\Gamma(\gamma)} t^{\gamma-1} - \int_0^t (t-s)^{\alpha-1} \frac{f(s, u(s))}{\Gamma(\alpha)} ds \right\|$$

$$+ \int_0^t (t-s)^{\alpha-1} \frac{\|f(s, u(s)) - f(s, v(s))\|}{\Gamma(\alpha)} ds$$

$$\le (I_0^\alpha \Phi)(t) + \int_0^t (t-s)^{\alpha-1} \frac{p(s)\|u(s) - v(s)\|}{\Gamma(\alpha)} ds$$

$$\le \lambda_\phi \Phi(t) + \frac{p^*}{\Gamma(\alpha)} \int_0^t (t-s)^{\alpha-1} \|u(s) - v(s)\| ds.$$

From Lemma 1.11, there exists a constant $\delta = \delta(\alpha)$ such that

$$\|u(t) - v(t)\| \le \lambda_\phi [\Phi(t) + \frac{\delta p^*}{\Gamma(\alpha)} + \int_0^t (t-s)^{\alpha-1} \Phi(s) ds]$$

$$\le [1 + \delta p^* \lambda_\Phi] \lambda_\phi \Phi(t)$$

$$:= c_{f,\Phi} \Phi(t).$$

Hence, the problem (2.17) is generalized Ulam–Hyers–Rassias stable.   □

2.3.3. *Hilfer–Hadamard fractional differential equations*

Now, we are concerned with the existence and the generalized Ulam–Hyers–Rassias stability of our problem (2.18).

Set $C := C([1,T])$, and denote the weighted space of continuous functions by

$$C_{\gamma,\ln}([1,T]) = \{w(t) : (\ln t)^{1-\gamma} w(t) \in C\},$$

with the norm

$$\|w\|_{C_{\gamma,\ln}} := \sup_{t \in [1,T]} |(\ln t)^{1-\gamma} w(t)|.$$

Set

$$\delta = x \frac{d}{dx}, \quad q > 0, \quad n = [q] + 1,$$

and

$$AC_\delta^n := \{u : [1,T] \to E : \delta^{n-1}[u(x)] \in AC([1,T])\}.$$

From Theorem 21 in [321], we conclude the following lemma.

Lemma 2.4. *Let $g : [1,T] \times E \to E$ be such that $g(\cdot, u(\cdot)) \in C_{\gamma,\ln}([1,T])$ for any $u \in C_{\gamma,\ln}([1,T])$. Then the problem (2.18) is equivalent to the following Volterra integral equation:*

$$u(t) = \frac{\phi_0}{\Gamma(\gamma)} (\ln t)^{\gamma-1} + ({}^H I_1^\alpha g(\cdot, u(\cdot)))(t).$$

Definition 2.9. By a solution of the problem (2.18) we mean a measurable function $u \in C_{\gamma,\ln}$ that satisfies the condition $({}^H I_1^{1-\gamma} u)(1^+) = \phi_0$, and the equation $({}^H D_1^{\alpha,\beta} u)(t) = g(t, u(t))$ on $[1,T]$.

Now we give (witnout proof) similar existence and Ulam stability results for problem (2.18). Let us introduce the following hypotheses:

(2.4.1) The function $t \mapsto g(t,u)$ is measurable on $[1,T]$ for each $u \in E$, and the function $u \mapsto g(t,u)$ is continuous on E for a.e. $t \in [1,T]$.

(2.4.2) There exists a continuous function $q : [1,T] \to [0,\infty)$ such that

$$\|g(t,u) - g(t,v)\| \le q(t)\|u-v\| \quad \text{for a.e. } t \in [1,T], \text{ and each } u,v \in E.$$

(2.4.3) For each bounded and measurable set $B \subset E$ and for each $t \in [1,T]$, we have

$$\mu(g(t,B)) \le q(t)\mu(B).$$

(2.4.4) There exists $\lambda_\Phi > 0$ such that for each $t \in [1,T]$, we have

$$({}^H I_1^\alpha \Phi)(t) \le \lambda_\Phi \Phi(t).$$

Theorem 2.4. *Assume that the hypotheses (2.4.1)–(2.4.3) hold. If*

$$\ell^* := \frac{4q^*(\ln T)^\alpha}{\Gamma(1+\alpha)} < 1, \tag{2.23}$$

where $q^* = \sup_{t \in I} q(t)$, then the problem (2.18) has at least one solution defined on I. Furthermore, if the hypothesis (2.4.4) holds, then the problem (2.18) is generalized Ulam–Hyers–Rassias stable.

2.3.4. An example

Let

$$E = l^1 = \left\{ u = (u_1, u_2, \ldots, u_n, \ldots), \sum_{n=1}^{\infty} |u_n| < \infty \right\}$$

be the Banach space with the norm

$$\|u\|_E = \sum_{n=1}^{\infty} |u_n|.$$

Consider the following problem of Hilfer fractional differential equation

$$\begin{cases} (D_0^{\frac{1}{2}, \frac{1}{2}} u_n)(t) = f_n(t, u(t)); & t \in [0, 1], \\ (I_0^{\frac{1}{4}} u_n)(t)|_{t=0} = (1, 0, \ldots, 0, \ldots), \end{cases} \tag{2.24}$$

where

$$\begin{cases} f_n(t, u) = \dfrac{t^{\frac{-1}{4}} u_n \sin t}{64(1 + \sqrt{t})(1 + \|u\|_E)}; & t \in (0, 1], \\ f_n(0, u) = 0 \end{cases}$$

with

$$f = (f_1, f_2, \ldots, f_n, \ldots), \quad u = (u_1, u_2, \ldots, u_n, \ldots), \quad \text{and } c := \frac{e^3}{8} \Gamma\left(\frac{1}{2}\right).$$

Set $\alpha = \beta = \frac{1}{2}$, then $\gamma = \frac{3}{4}$. The hypothesis ((2.4.2)) is satisfied with

$$\begin{cases} p(t) = \dfrac{t^{\frac{-1}{4}} |\sin t|}{64(1 + \sqrt{t})}; & t \in (0, 1], \\ p(0) = 0. \end{cases}$$

Hence, Theorem 2.4 implies that the problem (2.24) has at least one solution defined on $[0, 1]$. Also, the hypothesis (2.3.4) is satisfied with

$$\Phi(t) = e^3, \quad \text{and} \quad \lambda_\Phi = \frac{1}{\Gamma(1 + \alpha)}.$$

Indeed, for each $t \in [0, 1]$ we get

$$(I_0^\alpha \Phi)(t) \leq \frac{e^3}{\Gamma(1 + \alpha)}$$

$$= \lambda_\Phi \Phi(t).$$

Consequently, the problem (2.24) is generalized Ulam–Hyers–Rassias stable.

2.4. Hilfer and Hilfer–Hadamard Fractional Differential Inclusions in Banach Spaces

2.4.1. *Introduction*

We deal in this section with some existence results in Banach spaces for some Hilfer and Hilfer–Hadamard differential inclusions of fractional order. The Mönch's fixed-point theorem and the concept of measure of noncompactness are the main tools used to carry out our results.

Recently, in [9,14] Abbas *et al.* have considered the existence and stability of some classes on integral equations of involving the Hadamard fractional operator. In this section, we discuss the existence of solutions for the following problem of Hilfer fractional differential inclusions:

$$\begin{cases} (D_0^{\alpha,\beta} u)(t) \in F(t, u(t)); & t \in I := [0, T], \\ (I_0^{1-\gamma} u)(0) = \phi, \end{cases} \qquad (2.25)$$

where $\alpha \in (0,1)$, $\beta \in [0,1]$, $\gamma = \alpha + \beta - \alpha\beta$, $T > 0$, $\phi \in E$, $F : I \times E \to \mathcal{P}(E)$ is a given multivalued map, E is a real (or complex) separable Banach space with a norm $\|\cdot\|$, $\mathcal{P}(E)$ is the family of all nonempty subsets of E, $I_0^{1-\gamma}$ is the left-sided mixed Riemann–Liouville integral of order $1 - \gamma$, and $D_0^{\alpha,\beta}$ is the generalized Riemann–Liouville derivative operator of order α and type β, introduced by Hilfer in [238].

Next, we consider the following problem of Hilfer–Hadamard fractional differential inclusions:

$$\begin{cases} (^H D_1^{\alpha,\beta} u)(t) \in G(t, u(t)); & t \in [1, T], \\ (^H I_1^{1-\gamma} u)(1) = \phi_0, \end{cases} \qquad (2.26)$$

where $\alpha \in (0,1)$, $\beta \in [0,1]$, $\gamma = \alpha + \beta - \alpha\beta$, $T > 1$, $\phi_0 \in E$, $G : [1, T] \times E \to \mathcal{P}(E)$ is a given multivalued map, $^H I_1^{1-\gamma}$ is the left-sided mixed Hadamard integral of order $1 - \gamma$, and $^H D_1^{\alpha,\beta}$ is the Hilfer–Hadamard fractional derivative of order α and type β.

2.4.2. *Hilfer fractional differential inclusions*

First, we state the definition of a solution of the problem (2.25). Define the space

$$C_\gamma(I) = \{w : I \to E : t^{1-\gamma} w(t) \in C\},$$

with the norm

$$\|w\|_{C_\gamma} := \sup_{t \in I} \|t^{1-\gamma} w(t)\|_E,$$

where $C = C([0, T])$.

Definition 2.10. By a solution of the problem (2.25) we mean a measurable function $u \in C_\gamma$ that satisfies the condition $(I_0^{1-\gamma}u)(0^+) = \phi$, and the equation $(D_0^{\alpha,\beta}u)(t) = v(t)$ on I, where $v \in S_{Fou}$.

In the sequel, we need the following hypotheses:

(2.5.1) The multivalued map $F : I \times E \to \mathcal{P}_{cp,c}(E)$ is Carathéodory.
(2.5.2) There exists a function $p \in L^\infty(I, [0, \infty))$ such that

$$\|F(t,u)\|_\mathcal{P} = \sup\{\|v\|_C : v(t) \in F(t,u)\} \leq p(t)$$

  for a.e. $t \in I$, and each $u \in E$.
(2.5.3) For each bounded and measurable set $B \subset C_\gamma$ and for each $t \in I$, we have

$$\mu(F(t, B(t)) \leq p(t)\mu(B(t)),$$

  where $B(t) = \{u(t) : u \in B\}$,
(2.5.4) The function $\phi \equiv 0$ is the unique solution in C_γ of the inequality

$$\Phi(t) \leq 2p^*(I_0^\alpha\Phi)(t),$$

  where

$$p^* = \operatorname*{ess\,sup}_{t \in I} p(t).$$

Now, we shall prove the following theorem concerning the existence of solutions of the problem (2.25).

Remark 2.1. In (2.5.3), μ is the Kuratowski measure of noncompactness on the space E.

From Corollary 1.1, we have the following lemma.

Lemma 2.5. *Let $F : I \times E \to \mathcal{P}(E)$ be such that $S_{Fou} \subset C_\gamma$ for any $u \in C_\gamma$. Then the problem (2.25) is equivalent of the integral equation*

$$u(t) = \frac{\phi}{\Gamma(\gamma)}t^{\gamma-1} + (I_0^\alpha v(t),$$

where $v \in S_{Fou}$.

Theorem 2.5. *Assume that the hypotheses (2.5.1)–(2.5.4) hold. Then the problem (2.25) has at least one solution defined on I.*

Proof. Consider the multivalued operator $N : C_\gamma \to \mathcal{P}(C_\gamma)$ defined by

$$N(u) = \left\{ h \in C_\gamma : h(t) = \frac{\phi}{\Gamma(\gamma)} t^{\gamma-1} + \int_0^t (t-s)^{\alpha-1} \frac{v(s)}{\Gamma(\alpha)} ds; \ v \in S_{Fou} \right\}.$$

$$(2.27)$$

Clearly, the fixed points of N are solution of the problem (2.25). We shall show that the multivalued operator N satisfies all the assumptions of Theorem 1.12. The proof will be given in several steps.

Step 1. $N(u)$ *is convex for each* $u \in C_\gamma$.

Indeed, if h_1, h_2 belong to $N(u)$, then there exist $v_1, v_2 \in S_{Fou}$ such that for each $t \in I$ we have

$$h_i(t) = \frac{\phi}{\Gamma(\gamma)} t^{\gamma-1} + \int_0^t (t-s)^{\alpha-1} \frac{v_i(s)}{\Gamma(\alpha)} ds; \quad i = 1, 2.$$

Let $0 \le \lambda \le 1$. Then, for each $t \in I$, we have

$$(\lambda h_1 + (1-\lambda)h_2)(t) = \frac{\phi}{\Gamma(\gamma)} t^{\gamma-1} + \int_0^t (t-s)^{\alpha-1} \frac{\lambda v_1(s) + (1-\lambda)v_2(s)}{\Gamma(\alpha)} ds.$$

Since S_{Fou} is convex (because F has convex values), we have $\lambda h_1 + (1-\lambda)h_2 \in N(u)$.

Step 2. *For each compact* $M \subset C_\gamma$, $N(M)$ *is relatively compact.*

Let (h_n) be any sequence in $N(M)$, where $M \subset C_\gamma$ is compact. By the Arzelà–Ascoli compactness criterion in C_γ, we show (h_n) has a convergent subsequence. Since $h_n \in N(M)$ there exist $u_n \in M$ and $v_n \in S_{Fou_n}$ such that

$$h_n(t) = \frac{\phi}{\Gamma(\gamma)} t^{\gamma-1} + \int_0^t (t-s)^{\alpha-1} \frac{v_n(s)}{\Gamma(\alpha)} ds.$$

Using Lemma 1.5 and the properties of the Kuratowski measure of non-compactness, we have

$$\mu(\{h_n(t)\}) \le \frac{2}{\Gamma(\alpha)} \int_0^t \mu(\{(t-s)^{\alpha-1} v_n(s)\}) ds. \qquad (2.28)$$

On the other hand, since M is compact, the set $\{v_n(s) : n \ge 1\}$ is compact. Consequently, $\mu(\{v_n(s) : n \ge 1\}) = 0$ for a.e. $s \in I$. Furthermore

$$\mu(\{(t-s)^{\alpha-1} v_n(s)\}) = (t-s)^{\alpha-1} \mu(\{v_n(s) : n \ge 1\}) = 0$$

for a.e. $t, s \in I$. Now (2.28) implies that $\{h_n(t) : n \ge 1\}$ is relatively compact for each $t \in I$. In addition, for each t_1 and t_2 from I, with $t_1 < t_2$, we have

$$\|t_2^{1-\gamma} h_n(t_2) - t_1^{1-\gamma} h_n(t_1)\|$$

$$\leq \left\| t_2^{1-\gamma} \int_0^{t_2} (t_2 - s)^{\alpha-1} \frac{v_n(s)}{\Gamma(\alpha)} ds - t_1^{1-\gamma} \int_0^{t_1} (t_1 - s)^{\alpha-1} \frac{v_n(s)}{\Gamma(\alpha)} ds \right\|$$

$$\leq t_2^{1-\gamma} \int_{t_1}^{t_2} (t_2 - s)^{\alpha-1} \frac{p(s)}{\Gamma(\alpha)} ds$$

$$+ \int_0^{t_1} |t_2^{1-\gamma} (t_2 - s)^{\alpha-1} - t_1^{1-\gamma} (t_1 - s)^{\alpha-1}| \frac{p(s)}{\Gamma(\alpha)} ds$$

$$\leq \frac{p^* T^{1-\gamma+\alpha}}{\Gamma(1+\alpha)} (t_2 - t_1)^{\alpha}$$

$$+ \frac{p^*}{\Gamma(\alpha)} \int_0^{t_1} |t_2^{1-\gamma} (t_2 - s)^{\alpha-1} - t_1^{1-\gamma} (t_1 - s)^{\alpha-1}| ds. \tag{2.29}$$

As $t_1 \to t_2$, the right-hand side of the above inequality tends to zero. This shows that $\{h_n : n \geq 1\}$ is equicontinuous. Consequently, $\{h_n : n \geq 1\}$ is relatively compact in C_γ.

Step 3. *The graph of N is closed.*

Let $(u_n, h_n) \in \text{graph}(N)$, Then $n \geq 1$, with $\|u_n - u\|_C, \|h_n - h\|_C \to 0$, as $n \to \infty$. We must show that $(u, h) \in \text{graph}(N)$. Then $(u_n, h_n) \in \text{graph}(N)$ means that $h_n \in N(u_n)$, which means that there exists $v_n \in S_{F \circ u_n}$, such that for each $t \in I$,

$$h_n(t) = \frac{\phi}{\Gamma(\gamma)} t^{\gamma-1} + \int_0^t (t - s)^{\alpha-1} \frac{v_n(s)}{\Gamma(\alpha)} ds.$$

Consider the continuous linear operator $\Theta : L^1(I) \to C_\gamma$,

$$\Theta(v)(t) \mapsto h_n(t) = \frac{\phi}{\Gamma(\gamma)} t^{\gamma-1} + \int_0^t (t - s)^{\alpha-1} \frac{v_n(s)}{\Gamma(\alpha)} ds.$$

Clearly, $\|h_n - h\|_C \to 0$ as $n \to \infty$. From Lemma 1.2, it follows that $\Theta \circ S_F$ is a closed graph operator. Moreover, $h_n(t) \in \Theta(S_{F \circ u_n})$. Since $u_n \to u$, Lemma 1.2 implies

$$h(t) = \frac{\phi}{\Gamma(\gamma)} t^{\gamma-1} + \int_0^t (t - s)^{\alpha-1} \frac{v(s)}{\Gamma(\alpha)} ds$$

for some $v \in S_{F \circ u}$.

Step 4. *M is relatively compact in C_γ.*

Let $M \subset \overline{U}$, where $M \subset \text{conv}(\{0\} \cup N(M))$ and for some countable set $C \subset M$ let $\overline{M} = \overline{C}$. Taking into account (2.29), it is easily seen that

$N(M)$ is equicontinuous. Therefore, $M \subset \mathrm{conv}(\{0\} \cup N(M))$ gives us that M is equicontinuous. It remains to apply the Arzelà–Ascoli theorem to show that for each $t \in I$ the set $M(t)$ is relatively compact. By taking into account that C is countable and $C \subset M \subset \mathrm{conv}(\{0\} \cup N(M))$, we can find a countable set $H = \{h_n : n \geq 1\} \subset N(M)$ such that $C \subset \mathrm{conv}(\{0\} \cup H)$. Then, there exist $u_n \in M$ and $v_n \in S_{Fou_n}$ with

$$h_n(t) = \frac{\phi}{\Gamma(\gamma)}t^{\gamma-1} + \int_0^t (t-s)^{\alpha-1}\frac{v_n(s)}{\Gamma(\alpha)}ds.$$

By taking into account Lemma 1.5 and the fact that $M \subset \overline{C} \subset \overline{conv}(\{0\} \cup H))$, we obtain

$$\mu(M(t)) \leq \mu(\overline{C}(t)) \leq \mu(H(t)) = \mu(\{h_n(t) : n \geq 1\}).$$

Using (2.28), we obtain

$$\mu(t^{1-\gamma}M(t)) \leq \frac{2}{\Gamma(\alpha)}\int_0^t \mu(\{t^{1-\gamma}(t-s)^{\alpha-1}v_n(s)\})ds.$$

Now, since $v_n \in S_{Fou_n}$ and $u_n(s) \in M(s)$, we have

$$\mu(t^{1-\gamma}M(t)) \leq \frac{2}{\Gamma(\alpha)}\int_0^t \mu(\{t^{1-\gamma}(t-s)^{\alpha-1}v_n(s) : n \geq 1\})ds.$$

Also, since $v_n \in S_{Fou_n}$ and $u_n(s) \in M(s)$, then from (2.5.3) we have

$$\mu(\{t^{1-\gamma}(t-s)^{\alpha-1}v_n(s); n \geq 1\}) = t^{1-\gamma}(t-s)^{\alpha-1}p(s)\mu(M(s)).$$

It follows that

$$\mu(M(t)) \leq \frac{2p^*}{\Gamma(\alpha)}\int_0^t (t-s)^{\alpha-1}\mu(M(s))ds.$$

Consequently by (2.5.4), the function Φ given by $\Phi(t) = \mu(M(t))$ satisfies $\Phi \equiv 0$; that is, $\mu(M(t)) = 0$ for all $t \in I$. Now, by the Arzelà–Ascoli theorem, M is relatively compact in C_γ.

Step 5. *A priori estimate.*

Let $u \in C_\gamma$ be such that $u \in \lambda N(u)$ for some $\lambda \in (0,1)$. Then, for each $t \in I$, we have

$$u(t) = \frac{\lambda\phi}{\Gamma(\gamma)}t^{\gamma-1} + \frac{\lambda}{\Gamma(\alpha)}\int_0^t (t-s)^{\alpha-1}v(s)ds,$$

for some $v \in S_{Fou}$. On the other hand,

$$t^{1-\gamma}\|u(t)\| \leq \frac{\|\phi\|}{\Gamma(\gamma)} + \frac{t^{1-\gamma}}{\Gamma(\alpha)}\int_0^t (t-s)^{\alpha-1}\|v(s)\|ds$$

$$\leq \frac{\|\phi\|}{\Gamma(\gamma)} + \frac{T^{1-\gamma}}{\Gamma(\alpha)} \int_0^t (t-s)^{\alpha-1} p(s) ds$$

$$\leq \frac{\|\phi\|}{\Gamma(\gamma)} + \frac{p^* T^{1-\gamma+\alpha}}{\Gamma(1+\alpha)}.$$

Then

$$\|u\|_C \leq \frac{\|\phi\|}{\Gamma(\gamma)} + \frac{p^* T^{1-\gamma+\alpha}}{\Gamma(1+\alpha)} := d.$$

Set

$$U = \{u \in C_\gamma : \|u\|_C < d + 1\}.$$

By our choice of the open set U, from above established steps and Theorem 1.12, we conclude that N has at least one fixed point $u \in C_\gamma$ being a solution of problem (2.25). □

2.4.3. *Hilfer–Hadamard fractional differential inclusions*

Now in this section, we study the existence of solutions for problem (2.26).
Set $C := C([1, T])$. Denote by

$$C_{\gamma,\ln}([1, T]) = \{w(t) : (\ln t)^{1-\gamma} w(t) \in C\}$$

the weighted space of continuous functions equipped with the norm

$$\|w\|_{C_{\gamma,\ln}} := \sup_{t \in [1,T]} \|(\ln t)^{1-\gamma} w(t)\|.$$

From Theorem 21 in [321], we conclude the following lemma.

Lemma 2.6. *Let $F : I \times \mathbb{R} \to \mathcal{P}(\mathbb{R})$ be such that $S_{F \circ u} \in C_{\gamma,\ln}([1, T])$ for any $u \in C_{\gamma,\ln}([1, T])$. Then the problem (2.26) is equivalent to the following Volterra integral equation:*

$$u(t) = \frac{\phi_0}{\Gamma(\gamma)} (\ln t)^{\gamma-1} + (^H I_1^\alpha v(t),$$

where $v \in S_{F \circ u}$.

Definition 2.11. By a solution of the problem (2.26) we mean a measurable function $u \in C_{\gamma,\ln}$ that satisfies the condition $(^H I_1^{1-\gamma} u)(1^+) = \phi_0$, and the equation $(^H D_1^{\alpha,\beta} u)(t) = v(t)$ on $[1, T]$, where $v \in S_{F \circ u}$.

Now we give (without proof) existence results for problem (2.26). The following hypotheses will be used in the sequel.

Theorem 2.6. *Assume that the following hypotheses hold:*

(2.6.1) *The multivalued map $G : [1,T] \times E \to \mathcal{P}_{cp,c}(E)$ is Carathéodory.*

(2.6.2) *There exists a function $q \in L^\infty([1,T],[0,\infty))$ such that*
$$\|G(t,u)\|_\mathcal{P} = \sup\{\|v\|_{C_{\gamma,\ln}} : v(t) \in G(t,u)\} \le q(t)$$
for a.e. $t \in [1,T]$, and each $u \in E$.

(2.6.3) *For each bounded and measurable set $B \subset C_{\gamma,\ln}$ and for each $t \in [1,T]$, we have*
$$\mu(F(t,B(t)) \le q(t)\mu(B(t)).$$

(2.6.4) *The function $\Lambda \equiv 0$ is the unique solution in $C_{\gamma,\ln}$ of the inequality*
$$\Lambda(t) \le 2q^*(^H I_1^\alpha \Lambda)(t),$$

where
$$q^* = \operatorname{ess}\sup_{t\in[1,T]} q(t).$$

Then, the problem (2.26) has at least one solution defined on $[1,T]$.

2.4.4. An example

Let
$$E = l^1 = \left\{ u = (u_1, u_2, \ldots, u_n, \ldots), \sum_{n=1}^\infty |u_n| < \infty \right\}$$
be the Banach space with the norm
$$\|u\|_E = \sum_{n=1}^\infty |u_n|.$$
Consider now the following problem of Hilfer fractional differential inclusion:
$$\begin{cases} (^H D_0^{\frac{1}{2},\frac{1}{2}} u_n)(t) \in F_n(t,u(t)); & t \in [0,e], \\ (^H I_0^{\frac{1}{4}} u)(t)|_{t=0} = (1,0,\ldots,0,\ldots), \end{cases} \tag{2.30}$$
where
$$F_n(t,u(t)) = \frac{ct^2 e^{-4-t}}{1 + \|u(t)\|_E}[u_n(t) - 1, u_n(t)]; \quad t \in [0,e]$$
with
$$u = (u_1, u_2, \ldots, u_n, \ldots), \quad \text{and} \quad c := \frac{e^3}{8}\Gamma\left(\frac{1}{2}\right).$$
Set $\alpha = \beta = \frac{1}{2}$, then $\gamma = \frac{3}{4}$,
$$F = (F_1, F_2, \ldots, F_n, \ldots).$$
We assume that F is closed and convex valued.
For each $u \in E$ and $t \in [0,e]$, we have
$$\|F(t,u)\|_\mathcal{P} \le ct^2 e^{-t-4}.$$

Hence, the hypothesis (2.5.2) is satisfied with $p^* = ce^{-2}$. A simple computation shows that conditions of Theorem 2.5 are satisfied. Hence, the problem (2.30) has at least one solution defined on $[0, e]$.

2.5. Notes and Remarks

The results of Chapter 2 are taken from Abbas *et al.* [48,51,55]. Other results may be found in [53,129,154].

Chapter 3

Attractivity Results for Hilfer Fractional Differential Equations

3.1. Introduction

In [20–23,58,71], Abbas *et al.* presented some results on the local and global attractivity of solutions for some classes of fractional differential equations involving both the Riemann–Liouville and the Caputo fractional derivatives by employing some fixed-point theorems. Motivated by the above papers, in this chapter we discuss the existence and the attractivity of solutions for problems of Hilfer fractional differential equations.

In this section, we present some results concerning the question of existence and attractivity of solutions for some differential equations of Hilfer type. An application is made of a Schauder fixed-point theorem for the existence of solutions. Next we prove that all solutions are uniformly locally attractive.

Consider the following problem of Hilfer fractional differential equations:

$$\begin{cases} (D_0^{\alpha,\beta} u)(t) = f(t, u(t)); \ t \in \mathbb{R}_+ := [0, +\infty), \\ (I_0^{1-\gamma} u)(t)|_{t=0} = \phi, \end{cases} \tag{3.1}$$

where $\alpha \in (0,1)$, $\beta \in [0,1]$, $\gamma = \alpha + \beta - \alpha\beta$, $\phi \in \mathbb{R}$, $f : \mathbb{R}_+ \times \mathbb{R} \to \mathbb{R}$ is a given function, $I_0^{1-\gamma}$ is the left-sided mixed Riemann–Liouville integral of order $1 - \gamma$, and $D_0^{\alpha,\beta}$ is the generalized Riemann–Liouville derivative operator of order α and type β, introduced by Hilfer in [238]. This section initiates the concept of local attractivity of solutions of problem (3.1).

3.2. Asymptotic Stability for Implicit Hilfer Fractional Differential Equations

3.2.1. *Existence of solutions*

Let BC be the space of functions $w : (0, +\infty) \to \mathbb{R}$ which are continuous and bounded. By $BC_\gamma := BC_\gamma(\mathbb{R}_+)$ we denote the weighted space of all bounded and continuous functions defined by

$$BC_\gamma = \{w : (0, +\infty) \to \mathbb{R} : t^{1-\gamma}w(t) \in BC\},$$

with the norm

$$\|w\|_{BC_\gamma} := \sup_{t \in \mathbb{R}_+} |t^{1-\gamma}w(t)|.$$

In the sequel we denote $\|w\|_{BC_\gamma}$ by $\|w\|_{BC}$.

From Theorem 21 in [321], we concluded the following lemma

Definition 3.1. By a solution of the problem (3.1) we mean a measurable function $u \in BC_\gamma$ that satisfies the condition $(I_0^{1-\gamma}u)(0^+) = \phi$, and the equation $(D_0^{\alpha,\beta}u)(t) = f(t, u(t))$ on $\mathbb{R}_+$.

Lemma 3.1. *Let* $f : \mathbb{R}_+ \times R \to R$ *be such that* $f(\cdot, u(\cdot)) \in BC_\gamma$ *for any* $u \in BC_\gamma$. *Then problem (3.1) is equivalent to the Volterra integral equation*

$$u(t) = \frac{\phi}{\Gamma(\gamma)}t^{\gamma-1} + (I_0^\alpha f(\cdot, u(\cdot)))(t).$$

The following hypotheses will be used in the sequel.

(3.1.1) The function $t \mapsto f(t, u)$ is measurable on $\mathbb{R}_+$ for each $u \in \mathbb{R}$, and the function $u \mapsto f(t, u)$ is continuous on $\mathbb{R}$ for a.e. $t \in \mathbb{R}_+$.

(3.1.2) There exists a continuous function $p : \mathbb{R}_+ \to \mathbb{R}_+$ such that

$$|f(t, u)| \le p(t), \quad \text{for a.e. } t \in \mathbb{R}_+, \text{ and each } u \in \mathbb{R}.$$

Moreover, assume that

$$\lim_{t \to \infty} t^{1-\gamma}(I_0^\alpha p)(t) = 0.$$

Set

$$p^* = \sup_{t \in \mathbb{R}_+} t^{1-\gamma}(I_0^\alpha p)(t),$$

Now, we shall prove the following theorem concerning the existence and the attractivity of solutions of our problem (3.1).

Theorem 3.1. *Assume that the hypotheses (3.1.1) and (3.1.2) hold. Then the problem (3.1) has at least one solution defined on $\mathbb{R}_+$. Moreover, solutions of problem (3.1) are uniformly locally attractive.*

Proof. Consider the operator N such that, for any $u \in BC_\gamma$,

$$(Nu)(t) = \frac{\phi}{\Gamma(\gamma)}t^{\gamma-1} + \int_0^t (t-s)^{\alpha-1}\frac{f(s,u(s))}{\Gamma(\alpha)}ds. \qquad (3.2)$$

The operator N maps BC_γ into BC_γ. Indeed the map $N(u)$ is continuous on $\mathbb{R}_+$ for any $u \in BC_\gamma$, and for each $t \in \mathbb{R}_+$ we have

$$|t^{1-\gamma}(Nu)(t)| \leq \frac{|\phi|}{\Gamma(\gamma)} + \frac{t^{1-\gamma}}{\Gamma(\alpha)}\int_0^t (t-s)^{\alpha-1}|f(s,u(s))|ds$$

$$\leq \frac{|\phi|}{\Gamma(\gamma)} + \frac{t^{1-\gamma}}{\Gamma(\alpha)}\int_0^t (t-s)^{\alpha-1}p(s)ds$$

$$\leq \frac{|\phi|}{\Gamma(\gamma)} + p^*.$$

Thus

$$\|N(u)\|_{BC} \leq \frac{|\phi|}{\Gamma(\gamma)} + p^* := R. \qquad (3.3)$$

Hence, $N(u) \in BC_\gamma$. This proves that the operator N maps BC_γ into itself. By Lemma 3.1, the problem of finding the solutions of the problem (3.1) is reduced to finding the solutions of the operator equation $N(u) = u$. Inequality (3.3) implies that N transforms the ball $B_R := B(0,R) = \{w \in BC_\gamma : \|w\|_{BC} \leq R\}$ into itself.

We shall show that the operator N satisfies all the assumptions of Theorem 1.2. The proof will be given in several steps.

Step 1. *N is continuous.*

Let $\{u_n\}_{n\in\mathbb{N}}$ be a sequence such that $u_n \to u$ in B_R. Then, for each $t \in \mathbb{R}_+$, we have

$$|t^{1-\gamma}(Nu_n)(t) - t^{1-\gamma}(Nu)(t)|$$
$$\leq \frac{t^{1-\gamma}}{\Gamma(\alpha)}\int_0^t (t-s)^{\alpha-1}|f(s,u_n(s)) - f(s,u(s))|ds. \qquad (3.4)$$

Case 1. If $t \in [0,T]$; $T > 0$, then, since $u_n \to u$ as $n \to \infty$ and f is continuous, by the Lebesgue dominated convergence theorem, Eq. (3.4) implies

$$\|N(u_n) - N(u)\|_{\mathrm{BC}} \to 0 \quad \text{as } n \to \infty.$$

Case 2. If $t \in (T, \infty)$; $T > 0$, then from our hypotheses and (3.4), we get

$$|t^{1-\gamma}(Nu_n)(t) - t^{1-\gamma}(Nu)(t)| \le 2\frac{t^{1-\gamma}}{\Gamma(\alpha)} \int_0^t (t-s)^{\alpha-1} p(s) ds. \qquad (3.5)$$

Since $u_n \to u$ as $n \to \infty$ and $t^{1-\gamma}(I_0^\alpha p)(t) \to 0$ as $t \to \infty$, then (3.5) gives

$$\|N(u_n) - N(u)\|_{\mathrm{BC}} \to 0 \quad \text{as } n \to \infty.$$

Step 2. $N(B_R)$ *is uniformly bounded.*

This is clear since $N(B_R) \subset B_R$ and B_R is bounded.

Step 3. $N(B_R)$ *is equicontinuous on every compact subset* $[0, T]$ *of* $\mathbb{R}_+$; $T > 0$.

Let $t_1, t_2 \in [0, T]$, $t_1 < t_2$ and let $u \in B_R$. Thus we have

$$|t_2^{1-\gamma}(Nu)(t_2) - t_1^{1-\gamma}(Nu)(t_1)|$$

$$\le \left| t_2^{1-\gamma} \int_0^{t_2} (t_2 - s)^{\alpha-1} \frac{f(s, u(s))}{\Gamma(\alpha)} ds - t_1^{1-\gamma} \int_0^{t_1} (t_1 - s)^{\alpha-1} \frac{f(s, u(s))}{\Gamma(\alpha)} ds \right|$$

$$\le t_2^{1-\gamma} \int_{t_1}^{t_2} (t_2 - s)^{\alpha-1} \frac{|f(s, u(s))|}{\Gamma(\alpha)} ds$$

$$+ \int_0^{t_1} |t_2^{1-\gamma}(t_2 - s)^{\alpha-1} - t_1^{1-\gamma}(t_1 - s)^{\alpha-1}| \frac{|f(s, u(s))|}{\Gamma(\alpha)} ds$$

$$\le t_2^{1-\gamma} \int_{t_1}^{t_2} (t_2 - s)^{\alpha-1} \frac{p(s)}{\Gamma(\alpha)} ds$$

$$+ \int_0^{t_1} |t_2^{1-\gamma}(t_2 - s)^{\alpha-1} - t_1^{1-\gamma}(t_1 - s)^{\alpha-1}| \frac{p(s)}{\Gamma(\alpha)} ds.$$

Thus, from the continuity of the function p and by letting $p_* = \sup_{t \in [0,T]} p(t)$, we get

$$|t_2^{1-\gamma}(Nu)(t_2) - t_1^{1-\gamma}(Nu)(t_1)| \le \frac{p_* T^{1-\gamma+\alpha}}{\Gamma(1+\alpha)} (t_2 - t_1)^\alpha$$

$$+ \frac{p_*}{\Gamma(\alpha)} \int_0^{t_1} |t_2^{1-\gamma}(t_2 - s)^{\alpha-1} - t_1^{1-\gamma}(t_1 - s)^{\alpha-1}| ds.$$

As $t_1 \to t_2$, the right-hand side of the above inequality tends to zero.

Step 4. $N(B_R)$ *is equiconvergent.*

Let $t \in \mathbb{R}_+$ and $u \in B_R$. Then we have

$$|t^{1-\gamma}(Nu)(t)| \leq \frac{|\phi|}{\Gamma(\gamma)} + \frac{t^{1-\gamma}}{\Gamma(\alpha)} \int_0^t (t-s)^{\alpha-1}|f(s,u(s))|ds$$

$$\leq \frac{|\phi|}{\Gamma(\gamma)} + \frac{t^{1-\gamma}}{\Gamma(\alpha)} \int_0^t (t-s)^{\alpha-1}p(s)ds$$

$$\leq \frac{|\phi|}{\Gamma(\gamma)} + t^{1-\gamma}(I_0^\alpha p)(t).$$

Since $t^{1-\gamma}(I_0^\alpha p)(t) \to 0$, as $t \to +\infty$, then, we get

$$|(Nu)(t)| \leq \frac{|\phi|}{t^{1-\gamma}\Gamma(\gamma)} + \frac{t^{1-\gamma}(I_0^\alpha p)(t)}{t^{1-\gamma}} \to 0, \ as \ t \to +\infty.$$

Hence,

$$|(Nu)(t) - (Nu)(+\infty)| \to 0 \ as \ t \to +\infty.$$

As a consequence of Steps 1–4 together with the Lemma 1.8, we can conclude that $N : B_R \to B_R$ is continuous and compact. From an application of Schauder's theorem (Theorem 1.2), we deduce that N has a fixed point u which is a solution of the problem (3.1) on $\mathbb{R}_+$.

Step 5. *The uniform local attractivity for solutions.*

Let us assume that u_0 is a solution of problem (3.1) with the conditions of this theorem. Taking $u \in B(u_0, 2p^*)$, we have

$$|t^{1-\gamma}(Nu)(t) - t^{1-\gamma}u_0(t)| = |t^{1-\gamma}(Nu)(t) - t^{1-\gamma}(Nu_0)(t)|$$

$$\leq \frac{t^{1-\gamma}}{\Gamma(\alpha)} \int_0^t (t-s)^{\alpha-1}|f(s,u(s)) - f(s,u_0(s))|ds$$

$$\leq \frac{2t^{1-\gamma}}{\Gamma(\alpha)} \int_0^t (t-s)^{\alpha-1}p(s)ds$$

$$\leq 2p^*.$$

Thus, we get

$$\|N(u) - u_0\|_{\text{BC}} \leq 2p^*.$$

Hence, we obtain that N is a continuous function such that

$$N(B(u_0, 2p^*)) \subset B(u_0, 2p^*).$$

Moreover, if u is a solution of problem (3.1), then

$$|u(t) - u_0(t)| = |(Nu)(t) - (Nu_0)(t)|$$

$$\leq \frac{1}{\Gamma(\alpha)} \int_0^t (t-s)^{\alpha-1} |f(s, u(s)) - f(s, u_0(s))| ds$$

$$\leq 2(I_0^\alpha p)(t).$$

Thus

$$|u(t) - u_0(t)| \leq \frac{2t^{1-\gamma}(I_0^\alpha p)(t)}{t^{1-\gamma}}. \tag{3.6}$$

By using (3.6) and the fact that $\lim_{t \to \infty} t^{1-\gamma}(I_0^\alpha p)(t) = 0$, we deduce that

$$\lim_{t \to \infty} |u(t) - u_0(t)| = 0.$$

Consequently, all solutions of problem (3.1) are uniformly locally attractive.
$\square$

3.2.2. An example

Consider the following problem of Hilfer fractional differential equation:

$$\begin{cases} (D_0^{\frac{1}{2},\frac{1}{2}} u)(t) = f(t, u(t)); & t \in \mathbb{R}_+, \\ (I_0^{\frac{1}{4}} u)(t)|_{t=0} = 1, \end{cases} \tag{3.7}$$

where

$$\begin{cases} f(t, u) = \frac{ct^{\frac{-1}{4}} \sin t}{64(1+\sqrt{t})(1+|u|)}; & t \in (0, \infty) \ u \in \mathbb{R}, \\ f(0, u) = 0; & u \in \mathbb{R}, \end{cases}$$

and $c = \frac{9\sqrt{\pi}}{16}$. Clearly, the function f is continuous.

The hypothesis (3.1.2) is satisfied with

$$\begin{cases} p(t) = \frac{ct^{\frac{-1}{4}}|\sin t|}{64(1+\sqrt{t})}; & t \in (0, \infty), \\ p(0) = 0. \end{cases}$$

Also, we have

$$t^{1-\gamma} I_0^{\frac{1}{2}} p(t) = \frac{t^{\frac{1}{4}}}{\Gamma(\frac{1}{2})} \int_0^t (t-\tau)^{\frac{-1}{2}} p(\tau) d\tau$$

$$\leq \frac{1}{4} t^{\frac{-1}{4}} \to 0 \ as \ t \to \infty.$$

All conditions of Theorem 3.1 are satisfied. Hence, the problem (3.7) has at least one solution defined on $\mathbb{R}_+$, and solutions of this problem are uniformly locally attractive.

3.3. Global Stability for Implicit Hilfer–Hadamard Fractional Differential Equations

3.3.1. *Introduction and motivations*

In this section, we present some results concerning the question of existence and global stability of solutions for the following problem of Hilfer–Hadamard fractional differential equations:

$$\begin{cases} ({}^{H}D_1^{\alpha,\beta}u)(t) = f(t, u(t), ({}^{H}D_1^{\alpha,\beta}u)(t)); \ t \in J := [1, +\infty), \\ ({}^{H}I_1^{1-\gamma}u)(t)|_{t=1} = \phi, \end{cases} \tag{3.8}$$

where $\alpha \in (0,1)$, $\beta \in [0,1]$, $\gamma = \alpha + \beta - \alpha\beta$, $\phi \in \mathbb{R}$, $f : J \times \mathbb{R} \times \mathbb{R} \to \mathbb{R}$ is a given function, ${}^{H}I_1^{1-\gamma}$ is the left-sided mixed Hadamard integral of order $1 - \gamma$, and ${}^{H}D_1^{\alpha,\beta}$ is the Hilfer–Hadamadr derivative operator of order α and type β. We apply a Schauder fixed point theorem for the existence of solutions, and we prove that all solutions are Globally asymptotically stable. In this section, we consider the concept of global asymptotic stability of solutions of implicit fractional differential equations of Hilfer–Hadamard type.

3.3.2. *Existence of solutions*

Denote the weighted space of continuous functions defined by

$$\mathrm{BC}_{\gamma,\ln}([1,T]) = \{w(t) : (\ln t)^{1-\gamma}w(t) \in \mathrm{BC}\},$$

with the norm

$$\|w\|_{C_{\gamma,\ln}} := \sup_{t \in [1,T]} \|(\ln t)^{1-\gamma}w(t)\|.$$

Definition 3.2. By a solution of the problem (3.8) we mean a measurable function $u \in \mathrm{BC}_{\gamma,\ln}$ that satisfies the condition $({}^{H}I_1^{1-\gamma}u)(1^+) = \phi$, and the equation $({}^{H}D_1^{\alpha,\beta}u)(t) = f(t, u(t), ({}^{H}D_1^{\alpha,\beta}u)(t))$ on J.

From Theorem 21 in [321] and Lemma 5.1 in [82], we concluded the following lemma.

Lemma 3.2. *Let* $f : J \times \mathbb{R} \times \mathbb{R} \to \mathbb{R}$ *be such that* $f(\cdot, u(\cdot), v(\cdot)) \in BC_{\gamma,\ln}$ *for any* $u, v \in BC_{\gamma,\ln}$. *Then problem (3.8) is equivalent to the problem of obtaining the solution of the equation*

$$g(t) = f\left(t, \frac{\phi}{\Gamma(\gamma)}(\ln t)^{\gamma-1} + ({}^H I_1^\alpha g)(t), g(t)\right),$$

and if $g(\cdot) \in BC_{\gamma,\ln}$ *is the solution of this equation, then*

$$u(t) = \frac{\phi}{\Gamma(\gamma)}(\ln t)^{\gamma-1} + ({}^H I_1^\alpha g)(t).$$

The following hypotheses will be used in the sequel.

(3.2.1) The function $t \mapsto f(t, u, v)$ is measurable on J for each $u, v \in \mathbb{R}$, and the functions $u \mapsto f(t, u, v)$ and $v \mapsto f(t, u, v)$ are continuous on $\mathbb{R}$ for a.e. $t \in J$,

(3.2.2) There exists a continuous function $p : J \to \mathbb{R}_+$ such that

$$|f(t, u, v)| \le p(t); \quad \text{for a.e. } t \in J, \text{ and each } u, v \in \mathbb{R}.$$

Moreover, assume that

$$\lim_{t\to\infty} (\ln t)^{1-\gamma} ({}^H I_1^\alpha p)(t) = 0.$$

Set

$$p^* = \sup_{t\in J} (\ln t)^{1-\gamma} ({}^H I_1^\alpha p)(t).$$

Now, we shall prove the following theorem concerning the existence and the global stability of solutions of problem (3.8).

Theorem 3.2. *Assume that the hypotheses (3.2.1) and (3.2.2) hold. Then the problem (3.8) has at least one solution defined on* J. *Moreover, solutions of problem (3.8) are globally asymptotically stable.*

Proof. Consider the operator N such that, for any $u \in BC_{\gamma,\ln}$,

$$(Nu)(t) = \frac{\phi}{\Gamma(\gamma)}(\ln t)^{\gamma-1} + \int_1^t \left(\ln \frac{t}{s}\right)^{\alpha-1} \frac{g(s)}{s\Gamma(\alpha)} ds, \tag{3.9}$$

where $g(\cdot) \in BC_{\gamma,\ln}$ such that

$$g(t) = f\left(t, \frac{\phi}{\Gamma(\gamma)}(\ln t)^{\gamma-1} + ({}^H I_1^\alpha g)(t), g(t)\right).$$

The operator N maps $\mathrm{BC}_{\gamma,\ln}$ into $\mathrm{BC}_{\gamma,\ln}$. Indeed the map $N(u)$ is continuous on J for any $u \in \mathrm{BC}_{\gamma,\ln}$, and for each $t \in J$ we have

$$|(\ln t)^{1-\gamma}(Nu)(t)| \leq \frac{|\phi|}{\Gamma(\gamma)} + \frac{(\ln t)^{1-\gamma}}{\Gamma(\alpha)} \int_1^t \left(\ln \frac{t}{s}\right)^{\alpha-1} |g(s)| \frac{ds}{s}$$

$$\leq \frac{|\phi|}{\Gamma(\gamma)} + \frac{(\ln t)^{1-\gamma}}{\Gamma(\alpha)} \int_1^t \left(\ln \frac{t}{s}\right)^{\alpha-1} p(s) \frac{ds}{s}$$

$$\leq \frac{|\phi|}{\Gamma(\gamma)} + p^*.$$

Thus

$$\|N(u)\|_{\mathrm{BC}} \leq \frac{|\phi|}{\Gamma(\gamma)} + p^* := R. \tag{3.10}$$

Hence, $N(u) \in \mathrm{BC}_\gamma$. This proves that the operator N maps BC_γ into itself.

By Lemma 3.2, the problem of finding the solutions of the problem (3.8) is reduced to finding the solutions of the operator equation $N(u) = u$. Inequality (3.10) implies that N transforms the ball $B_R := B(0, R) = \{w \in \mathrm{BC}_{\gamma,\ln} : \|w\|_{\mathrm{BC}} \leq R\}$ into itself.

We shall show that the operator N satisfies all the assumptions of Theorem 1.2. The proof will be given in several steps.

Step 1. *N is continuous.*

Let $\{u_n\}_{n\in\mathbb{N}}$ be a sequence such that $u_n \to u$ in B_R. Then, for each $t \in J$, we have

$$|(\ln t)^{1-\gamma}(Nu_n)(t) - (\ln t)^{1-\gamma}(Nu)(t)|$$

$$\leq \frac{(\ln t)^{1-\gamma}}{\Gamma(\alpha)} \int_1^t \left(\ln \frac{t}{s}\right)^{\alpha-1} |g_n(s)) - g(s)| \frac{ds}{s}, \tag{3.11}$$

where $g, g_n \in \mathrm{BC}_{\gamma,\ln}$ such that

$$g(t) = f\left(t, \frac{\phi}{\Gamma(\gamma)}(\ln t)^{\gamma-1} + (^H I_1^\alpha g)(t), g(t)\right),$$

and

$$g_n(t) = f\left(t, \frac{\phi}{\Gamma(\gamma)}(\ln t)^{\gamma-1} + (^H I_1^\alpha g_n)(t), g_n(t)\right).$$

Case 1. If $t \in [1,T]$; $T > 1$, then, since $u_n \to u$ as $n \to \infty$ and f is continuous, by the Lebesgue dominated convergence theorem, we have

$$|g_n(t)) - g(t)| \to 0 \quad \text{as } n \to \infty.$$

Thus Inequality (3.11) implies

$$\|N(u_n) - N(u)\|_{BC} \to 0 \quad \text{as} \quad n \to \infty.$$

Case 2. If $t \in (T, \infty)$; $T > 1$, then from our hypotheses and (3.11), we get

$$|(\ln t)^{1-\gamma}(Nu_n)(t) - (\ln t)^{1-\gamma}(Nu)(t)|$$

$$\leq 2\frac{(\ln t)^{1-\gamma}}{\Gamma(\alpha)} \int_1^t \left(\ln \frac{t}{s}\right)^{\alpha-1} p(s)\frac{ds}{s}. \qquad (3.12)$$

Since $u_n \to u$ as $n \to \infty$ and $(\ln t)^{1-\gamma}({}^H I_1^\alpha p)(t) \to 0$ as $t \to \infty$, then (3.12) gives

$$\|N(u_n) - N(u)\|_{BC} \to 0 \quad \text{as} \quad n \to \infty.$$

Step 2. $N(B_R)$ *is uniformly bounded.*

This is clear since $N(B_R) \subset B_R$ and B_R is bounded.

Step 3. $N(B_R)$ *is equicontinuous on every compact subset* $[1, T]$ *of* J; $T > 1$.

Let $t_1, t_2 \in [0, T]$, $t_1 < t_2$ and let $u \in B_R$. We have $|(\ln t_2)^{1-\gamma}(Nu)(t_2) - (\ln t_1)^{1-\gamma}(Nu)(t_1)|$

$$\leq \left| (\ln t_2)^{1-\gamma} \int_1^{t_2} \left(\ln \frac{t_2}{s}\right)^{\alpha-1} \frac{g(s)}{s\Gamma(\alpha)} ds \right.$$

$$\left. - (\ln t_1)^{1-\gamma} \int_1^{t_1} \left(\ln \frac{t_1}{s}\right)^{\alpha-1} \frac{g(s)}{s\Gamma(\alpha)} ds \right|,$$

where $g(\cdot) \in BC_{\gamma,\ln}$ such that

$$g(t) = f\left(t, \frac{\phi}{\Gamma(\gamma)}(\ln t)^{\gamma-1} + ({}^H I_1^\alpha g)(t), g(t)\right).$$

Thus we get

$$|(\ln t_2)^{1-\gamma}(Nu)(t_2) - (\ln t_1)^{1-\gamma}(Nu)(t_1)|$$

$$\leq (\ln t_2)^{1-\gamma} \int_{t_1}^{t_2} \left(\ln \frac{t_2}{s}\right)^{\alpha-1} \frac{|g(s)|}{s\Gamma(\alpha)} ds$$

$$+ \int_1^{t_1} \left| (\ln t_2)^{1-\gamma} \left(\ln \frac{t_2}{s}\right)^{\alpha-1} - (\ln t_1)^{1-\gamma} \left(\ln \frac{t_1}{s}\right)^{\alpha-1} \right| \frac{|g(s)|}{s\Gamma(\alpha)} ds$$

$$\leq (\ln t_2)^{1-\gamma} \int_{t_1}^{t_2} \left(\ln \frac{t_2}{s}\right)^{\alpha-1} \frac{p(s)}{s\Gamma(\alpha)} ds$$

$$+ \int_1^{t_1} \left| (\ln t_2)^{1-\gamma} \left(\ln \frac{t_2}{s}\right)^{\alpha-1} - (\ln t_1)^{1-\gamma} \left(\ln \frac{t_1}{s}\right)^{\alpha-1} \right| \frac{p(s)}{s\Gamma(\alpha)} ds.$$

Hence, from the continuity of the function p and by letting $p_* = \sup_{t \in [1,T]} p(t)$, we get

$$|(\ln t_2)^{1-\gamma}(Nu)(t_2) - (\ln t_1)^{1-\gamma}(Nu)(t_1)|$$

$$\leq \frac{p_*(\ln T)^{1-\gamma+\alpha}}{\Gamma(1+\alpha)}\left(\ln\frac{t_2}{t_1}\right)^{\alpha}$$

$$+ \frac{p_*}{\Gamma(\alpha)}\int_1^{t_1}\left|(\ln t_2)^{1-\gamma}\left(\ln\frac{t_2}{s}\right)^{\alpha-1} - (\ln t_1)^{1-\gamma}\left(\ln\frac{t_1}{s}\right)^{\alpha-1}\right| ds.$$

As $t_1 \to t_2$, the right-hand side of the above inequality tends to zero.

Step 4. $N(B_R)$ *is equiconvergent.*

Let $t \in J$ and $u \in B_R$. Then we have

$$|(\ln t)^{1-\gamma}(Nu)(t)| \leq \frac{|\phi|}{\Gamma(\gamma)} + \frac{(\ln t)^{1-\gamma}}{\Gamma(\alpha)}\int_1^t\left(\ln\frac{t}{s}\right)^{\alpha-1}|g(s)|\frac{ds}{s},$$

where $g(\cdot) \in BC_{\gamma,\ln}$ such that

$$g(t) = f\left(t, \frac{\phi}{\Gamma(\gamma)}(\ln t)^{\gamma-1} + ({}^H I_1^\alpha g)(t), g(t)\right).$$

Thus, we get

$$|(\ln t)^{1-\gamma}(Nu)(t)| \leq \frac{|\phi|}{\Gamma(\gamma)} + \frac{(\ln t)^{1-\gamma}}{\Gamma(\alpha)}\int_1^t\left(\ln\frac{t}{s}\right)^{\alpha-1}p(s)\frac{ds}{s}$$

$$\leq \frac{|\phi|}{\Gamma(\gamma)} + (\ln t)^{1-\gamma}({}^H I_1^\alpha p)(t).$$

Since $(\ln t)^{1-\gamma}({}^H I_1^\alpha p)(t) \to 0$, as $t \to +\infty$, then, we get

$$|(Nu)(t)| \leq \frac{|\phi|}{(\ln t)^{1-\gamma}\Gamma(\gamma)} + \frac{(\ln t)^{1-\gamma}({}^H I_1^\alpha p)(t)}{(\ln t)^{1-\gamma}} \to 0, \quad \text{as } t \to +\infty.$$

Hence,

$$|(Nu)(t) - (Nu)(+\infty)| \to 0 \quad \text{as } t \to +\infty.$$

As a consequence of Steps 1–4 together with the Lemma 1.8, we can conclude that $N : B_R \to B_R$ is continuous and compact. From an application of Schauder's theorem (Theorem 1.2), we deduce that N has a fixed-point u which is a solution of the problem (3.8) on J.

Step 5. *Global asymptotic stability of solutions.*

Let us assume that u and v are two solutions of problem (3.8). Then for each $t \in J$, we have

$$|u(t) - v(t)(t)| = |(Nu)(t) - (Nv)(t)|$$

$$\leq \frac{1}{\Gamma(\alpha)} \int_1^t \left(\ln \frac{t}{s} \right)^{\alpha-1} |g(s) - g_v(s)| \frac{ds}{s},$$

where $g, g_v \in BC_{\gamma,\ln}$ such that

$$g(t) = f\left(t, \frac{\phi}{\Gamma(\gamma)}(\ln t)^{\gamma-1} + ({}^H I_1^\alpha g)(t), g(t) \right),$$

and

$$g_v(t) = f\left(t, \frac{\phi}{\Gamma(\gamma)}(\ln t)^{\gamma-1} + ({}^H I_1^\alpha g_v)(t), g_v(t) \right).$$

Thus

$$|u(t) - v(t)| \leq 2({}^H I_1^\alpha p)(t).$$

Hence

$$|u(t) - v(t)| \leq \frac{2(\ln t)^{1-\gamma}({}^H I_1^\alpha p)(t)}{(\ln t)^{1-\gamma}}. \tag{3.13}$$

By using (3.13) and the fact that $\lim_{t\to\infty}(\ln t)^{1-\gamma}({}^H I_1^\alpha p)(t) = 0$, we deduce that

$$\lim_{t\to\infty} |u(t) - v(t)| = 0.$$

Consequently, all solutions of problem (3.8) are globally asymptotically stable. $\qquad\square$

3.3.3. *An example*

Consider the following problem of Hilfer–Hadamard fractional differential equation:

$$\begin{cases} ({}^H D_1^{\frac{1}{2},\frac{1}{2}} u)(t) = f(t, u(t), ({}^H D_1^{\frac{1}{2},\frac{1}{2}} u)(t)); & t \in J, \\ ({}^H I_1^{\frac{1}{4}} u)(t)|_{t=1} = 0, \end{cases} \tag{3.14}$$

where

$$\begin{cases} f(t, u, v) = \dfrac{(t-1)^{\frac{-1}{4}} \sin(t-1)}{64(1+\sqrt{t-1})(1+|u|+|v|)}; & t \in (1, \infty)\ u, v \in \mathbb{R}, \\ f(1, u, v) = 0; & u, v \in \mathbb{R}. \end{cases}$$

Clearly, the function f is continuous.

The hypothesis (3.2.2) is satisfied with

$$\begin{cases} p(t) = \frac{(t-1)^{\frac{-1}{4}}|\sin(t-1)|}{64(1+\sqrt{t-1})}; & t \in (1,\infty), \\ p(1) = 0. \end{cases}$$

Also, we have

$$(\ln t)^{1-\gamma}(^H I_1^{\frac{1}{2}}p)(t) = \frac{(\ln t)^{\frac{1}{4}}}{\Gamma(\frac{1}{2})} \int_1^t \left(\ln \frac{t}{\tau}\right)^{\frac{-1}{2}} p(\tau)\frac{d\tau}{\tau}$$

$$\leq (\ln t)^{\frac{1}{4}} t^{\frac{-1}{2}} \to 0 \quad \text{as } t \to \infty.$$

All conditions of Theorem 3.2 are satisfied. Hence, the problem (3.14) has at least one solution defined on J, and solutions of this problem are globally asymptotically stable.

3.4. Notes and Remarks

The results of Chapter 3 are taken from [49,58,71,80].

Chapter 4

Ulam Stability Results for Hilfer Fractional Differential Equations

4.1. Introduction

This chapter deals with some existence and Ulam stabilities results for some classes of differential equations of Hilfer type. An application is made of Schauder's fixed-point theorem for the existence of solutions. Next we prove that our problem is generalized Ulam–Hyers–Rassias stable.

4.2. Dynamics and Ulam Stability for Hilfer Fractional Differential Equations

4.2.1. Introduction

Recently, considerable attention has been given to the existence of solutions of initial and boundary value problems for fractional differential equations with Hilfer fractional derivative; see [238,239,354]. Motivated by the above papers, in this section we discuss the existence and the Ulam stability of solutions for the following problem of Hilfer fractional differential equations

$$
\begin{cases}
(D_0^{\alpha,\beta} u)(t) = f(t, u(t)); & t \in I := [0, T], \\
(I_0^{1-\gamma} u)(t)|_{t=0} = \phi,
\end{cases}
\tag{4.1}
$$

where $\alpha \in (0, 1)$, $\beta \in [0, 1]$, $\gamma = \alpha + \beta - \alpha\beta$, $T > 0$, $\phi \in \mathbb{R}$, $f : I \times \mathbb{R} \to \mathbb{R}$ is a given function, $I_0^{1-\gamma}$ is the left-sided mixed Riemann–Liouville integral of order $1 - \gamma$, and $D_0^{\alpha,\beta}$ is the generalized Riemann–Liouville derivative operator of order α and type β, introduced by Hilfer in [238].

4.2.2. *Existence and Ulam stability results*

Let $C := C(I)$ be the Banach space of all continuous functions v from I into $\mathbb{R}$ with the supremum (uniform) norm

$$\|v\|_\infty := \sup_{t \in I} \|v(t)\|.$$

By $C_\gamma(I)$ and $C_\gamma^1(I)$, we denote the weighted spaces of continuous functions defined by

$$C_\gamma(I) = \{w : (0, T] \to \mathbb{R} : t^{1-\gamma} w(t) \in C\},$$

with the norm

$$\|w\|_{C_\gamma} := \sup_{t \in I} \|t^{1-\gamma} w(t)\|,$$

and

$$C_\gamma^1(I) = \left\{ w \in C : \frac{dw}{dt} \in C_\gamma \right\},$$

with the norm

$$\|w\|_{C_\gamma^1} := \|w\|_\infty + \|w'\|_{C_\gamma}.$$

In the sequel $\|w\|_C$ stands for $\|w\|_{C_\gamma}$.

Let us start by defining what we mean by a solution of the problem (4.1).

Definition 4.1. By a solution of the problem (4.1) we mean a measurable function $u \in C_\gamma$ that satisfies the condition $(I_0^{1-\gamma} u)(0^+) = \phi$, and the equation $(D_0^{\alpha,\beta} u)(t) = f(t, u(t))$ on I.

Lemma 4.1. *Let $f : I \times \mathbb{R} \to \mathbb{R}$ be such that $f(\cdot, u(\cdot)) \in C_\gamma$ for any $u \in C_\gamma$. Then problem (4.1) is equivalent to the integral equation*

$$u(t) = \frac{\phi}{\Gamma(\gamma)} t^{\gamma-1} + (I_0^\alpha f(\cdot, u(\cdot)))(t).$$

The following hypotheses will be used in the sequel.

(4.1.1) The function $t \mapsto f(t, u)$ is measurable on I for each $u \in \mathbb{R}$, and the function $u \mapsto f(t, u)$ is continuous on $\mathbb{R}$ for a.e. $t \in I$,

(4.1.2) There exists a continuous function $p : I \to [0, \infty)$ such that

$$|f(t, u)| \le p(t), \quad \text{for a.e. } t \in I, \text{ and each } u \in \mathbb{R}.$$

Set

$$p^* = \sup_{t \in I} p(t),$$

Now, we shall prove the following theorem concerning the existence of solutions of problem (4.1).

Theorem 4.1. *Assume that the hypotheses (4.1.1) and (4.1.2) hold. Then the problem (4.1) has at least one solution defined on I.*

Proof. Consider the operator $N : C_\gamma \to C_\gamma$ defined by

$$(Nu)(t) = \frac{\phi}{\Gamma(\gamma)} t^{\gamma-1} + \int_0^t (t-s)^{\alpha-1} \frac{f(s, u(s))}{\Gamma(\alpha)} ds. \qquad (4.2)$$

Clearly, the fixed points of the operator N are solution of the problem (4.1).
For any $u \in C_\gamma$, and each $t \in I$ we have

$$|t^{1-\gamma}(Nu)(t)| \leq \frac{|\phi|}{\Gamma(\gamma)} + \frac{t^{1-\gamma}}{\Gamma(\alpha)} \int_0^t (t-s)^{\alpha-1} |f(s, u(s))| ds$$

$$\leq \frac{|\phi|}{\Gamma(\gamma)} + \frac{t^{1-\gamma}}{\Gamma(\alpha)} \int_0^t (t-s)^{\alpha-1} p(s) ds$$

$$\leq \frac{|\phi|}{\Gamma(\gamma)} + \frac{p^* T^{1-\gamma}}{\Gamma(\alpha)} \int_0^t (t-s)^{\alpha-1} ds$$

$$\leq \frac{|\phi|}{\Gamma(\gamma)} + \frac{p^* T^{1-\gamma+\alpha}}{\Gamma(1+\alpha)}.$$

Thus

$$\|N(u)\|_C \leq \frac{|\phi|}{\Gamma(\gamma)} + \frac{p^* T^{1-\gamma+\alpha}}{\Gamma(1+\alpha)} := R. \qquad (4.3)$$

This proves that N transforms the ball $B_R := B(0, R) = \{w \in C_\gamma : \|w\|_C \leq R\}$ into itself. We shall show that the operator $N : B_R \to B_R$ satisfies all the assumptions of Theorem 1.2. The proof will be given in several steps.

Step 1. $N : B_R \to B_R$ *is continuous.*

Let $\{u_n\}_{n \in \mathbb{N}}$ be a sequence such that $u_n \to u$ in B_R. Then, for each $t \in I$, we have

$$|t^{1-\gamma}(Nu_n)(t) - t^{1-\gamma}(Nu)(t)|$$

$$\leq \frac{t^{1-\gamma}}{\Gamma(\alpha)} \int_0^t (t-s)^{\alpha-1} |f(s, u_n(s)) - f(s, u(s))| ds. \qquad (4.4)$$

Since $u_n \to u$ as $n \to \infty$ and f is continuous, then by the Lebesgue dominated convergence theorem, inequality (4.4) implies

$$\|N(u_n) - N(u)\|_C \to 0 \quad \text{as } n \to \infty.$$

Step 2. $N(B_R)$ *is uniformly bounded.*

This is clear since $N(B_R) \subset B_R$ and B_R is bounded.

Step 3. $N(B_R)$ *is equicontinuous.*

Let $t_1, t_2 \in I$, $t_1 < t_2$ and let $u \in B_R$. Thus we have

$$|t_2^{1-\gamma}(Nu)(t_2) - t_1^{1-\gamma}(Nu)(t_1)|$$

$$\leq \left| t_2^{1-\gamma} \int_0^{t_2} (t_2 - s)^{\alpha-1} \frac{f(s, u(s))}{\Gamma(\alpha)} ds - t_1^{1-\gamma} \int_0^{t_1} (t_1 - s)^{\alpha-1} \frac{f(s, u(s))}{\Gamma(\alpha)} ds \right|$$

$$\leq t_2^{1-\gamma} \int_{t_1}^{t_2} (t_2 - s)^{\alpha-1} \frac{|f(s, u(s))|}{\Gamma(\alpha)} ds$$

$$+ \int_0^{t_1} |t_2^{1-\gamma}(t_2 - s)^{\alpha-1} - t_1^{1-\gamma}(t_1 - s)^{\alpha-1}| \frac{|f(s, u(s))|}{\Gamma(\alpha)} ds$$

$$\leq t_2^{1-\gamma} \int_{t_1}^{t_2} (t_2 - s)^{\alpha-1} \frac{p(s)}{\Gamma(\alpha)} ds$$

$$+ \int_0^{t_1} |t_2^{1-\gamma}(t_2 - s)^{\alpha-1} - t_1^{1-\gamma}(t_1 - s)^{\alpha-1}| \frac{p(s)}{\Gamma(\alpha)} ds.$$

Thus, we get

$$|t_2^{1-\gamma}(Nu)(t_2) - t_1^{1-\gamma}(Nu)(t_1)|$$

$$\leq \frac{p_* T^{1-\gamma+\alpha}}{\Gamma(1+\alpha)}(t_2 - t_1)^{\alpha} + \frac{p_*}{\Gamma(\alpha)} \int_0^{t_1} |t_2^{1-\gamma}(t_2 - s)^{\alpha-1}$$

$$- t_1^{1-\gamma}(t_1 - s)^{\alpha-1}| ds.$$

As $t_1 \to t_2$, the right-hand side of the above inequality tends to zero.

As a consequence of Steps 1–3 together with the Arzelà–Ascoli theorem, we can conclude that N is continuous and compact. From an application of Schauder's theorem, we deduce that N has a fixed point u which is a solution of the problem (4.1). □

Now, we are concerned with the generalized Ulam–Hyers–Rassias stability of our problem (4.1).

Theorem 4.2. *Assume that the hypotheses (4.1.1), (4.1.2) and the following hypotheses hold.*

(4.2.1) *There exists $\lambda_\Phi > 0$ such that for each $t \in I$, we have*

$$(I_0^\alpha \Phi)(t) \le \lambda_\Phi \Phi(t).$$

(4.2.2) *There exists $q \in C(I, [0, \infty))$ such that for each $t \in I$, we have*

$$p(t) \le q(t)\Phi(t).$$

Then the problem (4.1) is generalized Ulam–Hyers–Rassias stable.

Proof. Consider the operator $N : C_\gamma \to C_\gamma$ defined in (4.2). Let u be a solution of the inequality (1.7), and let us assume that v is a solution of problem (4.1). Thus, we have

$$v(t) = \frac{\phi}{\Gamma(\gamma)}t^{\gamma-1} + \int_0^t (t-s)^{\alpha-1}\frac{f(s, v(s))}{\Gamma(\alpha)}ds.$$

From the inequality (1.7) for each $t \in I$, we have

$$\left| u(t) - \frac{\phi}{\Gamma(\gamma)}t^{\gamma-1} - \int_0^t (t-s)^{\alpha-1}\frac{f(s, u(s))}{\Gamma(\alpha)}ds \right| \le \Phi(t).$$

Set

$$q^* = \sup_{t \in I} q(t).$$

From hypotheses (4.2.1) and (4.2.2), for each $t \in I$, we get

$$
\begin{aligned}
|u(t) - v(t)| &\le \left| u(t) - \frac{\phi}{\Gamma(\gamma)}t^{\gamma-1} - \int_0^t (t-s)^{\alpha-1}\frac{f(s, u(s))}{\Gamma(\alpha)}ds \right| \\
&\quad + \int_0^t (t-s)^{\alpha-1}\frac{|f(s, u(s)) - f(s, v(s))|}{\Gamma(\alpha)}ds \\
&\le \Phi(t) + \int_0^t (t-s)^{\alpha-1}\frac{2q^*\Phi(s)}{\Gamma(\alpha)}ds \\
&\le \Phi(t) + 2q^*(I_0^\alpha \Phi)(t) \\
&\le [1 + 2q^*\lambda_\phi]\Phi(t) \\
&:= c_{f,\Phi}\Phi(t).
\end{aligned}
$$

Hence, the problem (4.1) is generalized Ulam–Hyers–Rassias stable. $\square$

Let $X = X(I, E)$ be the metric space, with the metric

$$d(u, v) = \sup_{t \in I} \frac{\|u(t) - v(t)\|_C}{\Phi(t)}.$$

Theorem 4.3. *Assume that (4.2.1) and the following hypothesis hold.*

(4.3.1) *There exists* $\varphi \in C(I, [0, \infty))$ *such that for each* $t \in I$, *and all* $u, v \in \mathbb{R}$, *we have*

$$|f(t, u) - f(t, v)| \leq t^{1-\gamma} \varphi(t) \Phi(t) |u - v|.$$

If

$$L := T^{1-\gamma} \varphi^* \lambda_\phi < 1, \tag{4.5}$$

where $\varphi^* = \sup_{t \in I} \varphi(t)$, *then there exists a unique solution* u_0 *of problem (4.1), and the problem (4.1) is generalized Ulam–Hyers–Rassias stable. Furthermore, we have*

$$|u(t) - u_0(t)| \leq \frac{\Phi(t)}{1 - L}.$$

Proof. Let $N : C_\gamma \to C_\gamma$ be the operator defined in (4.2). Apply Theorem 1.4, we have

$$
\begin{aligned}
|(Nu)(t) - (Nv)(t)| &\leq \int_0^t (t-s)^{\alpha-1} \frac{|f(s, u(s)) - f(s, v(s))|}{\Gamma(\alpha)} ds \\
&\leq \int_0^t (t-s)^{\alpha-1} \frac{\varphi(s)\Phi(s)|s^{1-\gamma}u(s) - s^{1-\gamma}v(s)|}{\Gamma(\alpha)} ds \\
&\leq \int_0^t (t-s)^{\alpha-1} \frac{\varphi^* \Phi(s) \|u - v\|_C}{\Gamma(\alpha)} ds \\
&\leq \varphi^* (I_0^\alpha \Phi)(t) \|u - v\|_C \\
&\leq \varphi^* \lambda_\phi \Phi(t) \|u - v\|_C.
\end{aligned}
$$

Thus

$$|t^{1-\gamma}(Nu)(t) - t^{1-\gamma}(Nv)(t)| \leq T^{1-\gamma} \varphi^* \lambda_\phi \Phi(t) \|u - v\|_C.$$

Hence, we get

$$d(N(u), N(v)) = \sup_{t \in I} \frac{\|(Nu)(t) - (Nv)(t)\|_C}{\Phi(t)} \leq L \|u - v\|_C,$$

from which we conclude the proof of the theorem. $\qquad\square$

4.2.3. *Example*

Consider the following problem of Hilfer fractional differential equation:

$$\begin{cases} (D_0^{\frac{1}{2},\frac{1}{2}}u)(t) = f(t,u(t)); & t \in [0,1], \\ (I_0^{\frac{1}{4}}u)(t)|_{t=0} = 1, \end{cases} \tag{4.6}$$

where

$$\begin{cases} f(t,u) = \frac{ct^{\frac{-1}{4}}\sin t}{64(1+\sqrt{t})(1+|u|)}; & t \in (0,1] \; u \in \mathbb{R}, \\ f(0,u) = 0; & u \in \mathbb{R}, \end{cases}$$

and $c = \frac{9\sqrt{\pi}}{16}$. Clearly, the function f satisfies (4.2.1).
    The hypothesis (4.1.2) is satisfied with

$$\begin{cases} p(t) = \frac{ct^{\frac{-1}{4}}|\sin t|}{64(1+\sqrt{t})}; & t \in (0,1], \\ p(0) = 0. \end{cases}$$

Hence, Theorem 4.1 implies that the problem (4.6) has at least one solution defined on $[0,1]$. Also, the hypothesis (4.4.1) is satisfied with

$$\Phi(t) = e^3 \quad \text{and} \quad \lambda_\Phi = \frac{1}{\Gamma(1+\alpha)}.$$

Indeed, for each $t \in [0,1]$ we get

$$(I_0^\alpha \Phi)(t) \le \frac{e^3}{\Gamma(1+\alpha)}$$
$$= \lambda_\Phi \Phi(t).$$

Consequently, Theorem 4.2 implies that the problem (4.6) is generalized Ulam–Hyers–Rassias stable.

4.3. Ulam Stability for Hilfer Fractional Differential Inclusions via Picard Operators

4.3.1. *Introduction*

The theory of Picard operators was introduced by Ioan A. Rus [331–333] to study problems related to fixed-point theory. This abstract approach was used later on by many mathematicians and it seemed to be a very useful and powerful method in the study of integral equations and inequalities, ordinary and partial differential equations (existence, uniqueness, differentiability of the solutions), etc. We recommend the monograph [333] and the references therein. The theory of Picard operators is a very powerful tool in the study of Ulam–Hyers stability of functional equations. We only have to define a fixed-point equation from the functional equation we want to study, then if the defined operator is c-weakly Picard we also have immediately the Ulam–Hyers stability of the desired equation. Of course it is not always possible to transform a functional equation or a differential equation into a fixed-point problem and actually this point shows a weakness of this theory. The uniform approach with Picard operators to discuss the stability problems of Ulam–Hyers type is due to Rus [330]. This section deals with some existence and Ulam stability results for some differential inclusions of Hilfer fractional derivative. An application is made of the weakly Picard operators theory.

Recently, considerable attention has been given to the existence of solutions of initial and boundary value problems for fractional differential equations with Hilfer fractional derivative; see [238,239,354]. Motivated by the above papers, in this section we discuss the existence and the Ulam stability of solutions for the following problem of Hilfer fractional differential inclusions:

$$\begin{cases} (D_0^{\alpha,\beta} u)(t) \in F(t, u(t)); & t \in I := [0, T], \\ (I_0^{1-\gamma} u)(t)|_{t=0} = \phi, \end{cases} \tag{4.7}$$

where $\alpha \in (0, 1)$, $\beta \in [0, 1]$, $\gamma = \alpha + \beta - \alpha\beta$, $T > 0$, $\phi \in \mathbb{R}$, $F : I \times \mathbb{R} \to \mathcal{P}(\mathbb{R})$ is a given multivalued map, $\mathcal{P}(\mathbb{R})$ is the family of all nonempty subsets of $\mathbb{R}$, $I_0^{1-\gamma}$ is the left-sided mixed Riemann–Liouville integral of order $1 - \gamma$, and $D_0^{\alpha,\beta}$ is the generalized Riemann–Liouville derivative operator of order α and type β, introduced by Hilfer in [238].

4.3.2. *Existence and Ulam stability results*

Definition 4.2. By a solution of the problem (4.7) we mean a measurable function $u \in C_\gamma$ that satisfies the condition $(I_0^{1-\gamma} u)(0^+) = \phi$, and the equation $(D_0^{\alpha,\beta} u)(t) = v(t)$ on I, where $v \in S_{F \circ u}$.

In what follows, we will give some basic definitions and results on Picard operators [333,334].

Let (X, d) be a metric space and $A : X \to X$ be an operator. We denote by F_A the set of the fixed points of A. We also denote $A^0 := 1_X$, $A^1 := A, \ldots, A^{n+1} := A^n \circ A$; $n \in \mathbb{N}$ the iterate operators of the operator A.

Definition 4.3. The operator $A : X \to X$ is a Picard operator (PO) if there exists $x^* \in X$ such that:

(i) $F_A = \{x^*\}$;
(ii) The sequence $(A^n(x_0))_{n \in \mathbb{N}}$ converges to x^* for all $x_0 \in X$.

Definition 4.4. The operator $A : X \to X$ is a weakly Picard operator (WPO) if the sequence $(A^n(x))_{n \in \mathbb{N}}$ converges for all $x \in X$, and its limit (which may depend on x) is a fixed point of A.

Definition 4.5. If A is weakly Picard operator then we consider the operator A^∞ defined by

$$A^\infty : X \to X; \quad A^\infty(x) = \lim_{n \to \infty} A^n(x).$$

Remark 4.1. It is clear that $A^\infty(X) = F_A$.

Definition 4.6. Let A be a weakly Picard operator and $c > 0$. The operator A is c-weakly Picard operator if

$$d(x, A^\infty(x)) \le c\, d(x, A(x)); \quad x \in X.$$

In the multivalued case, we have the following concepts (see [314,336]).

Definition 4.7. Let (X, d) be a metric space, and $F : X \to \mathcal{P}_{cl}(X)$ be a multivalued operator. By definition, F is a multivalued weakly Picard (briefly MWP) operator, if for each $u \in X$ and each $v \in F(x)$, there exists a sequence $(u_n)_{n \in \mathbb{N}}$ such that

(i) $u_0 = u$, $u_1 = v$;
(ii) $u_{n+1} \in F(u_n)$, for each $n \in \mathbb{N}$;
(iii) the sequence $(u_n)_{n \in \mathbb{N}}$ is convergent and its limit is a fixed point of F.

Remark 4.2. A sequence $(u_n)_{n \in \mathbb{N}}$ satisfying condition (i) and (ii) in the above definition is called a sequence of successive approximations of F starting from $(u, v) \in \mathrm{Graph}(F)$.

If $F : X \to \mathcal{P}_{cl}(X)$ is a MWP operator, then we define $F_1 : \mathrm{Graph}(F) \to \mathcal{P}(\mathrm{Fix}(F))$ by the formula $F_1(t) := \{z \in \mathrm{Fix}(F) :$ there exists a sequence of successive approximations of F starting from (u, v) that converges to $z\}$.

Definition 4.8. Let (X, d) be a metric space and let $\Psi : [0, \infty) \to [0, \infty)$ be an increasing function which is continuous at 0 and $\Psi(0) = 0$. Then $F : X \to \mathcal{P}_{cl}(X)$ is said to be a Ψ-multivalued weakly Picard operator (briefly $\Psi-$MWP operator) if it is a multivalued weakly Picard operator and there exists a selection $A^\infty : \text{Graph}(F) \to \text{Fix}(F)$ of F^∞ such that

$$d(u, A^\infty(u, v)) \leq \Psi(d(u, v)); \text{ for all } (u, v) \in \text{Graph}(F).$$

If there exists $c > 0$ such that $\Psi(t) = ct$, for each $t \in [0.\infty)$, then F is called a c-multivalued weakly Picard operator c-MWP operator.

Let us recall the notion of comparison.

Definition 4.9. A function $\varphi : [0, \infty) \to [0, \infty)$ is said to be a comparison function (see [333]) if it is increasing and $\varphi^n \to 0$ as $n \to \infty$.

As a consequence, we have $\varphi(t) < t$, for each $t > 0$, $\varphi(0) = 0$ and φ is continuous at 0.

Definition 4.10. A function $\varphi : [0, \infty) \to [0, \infty)$ is said to be a strict comparison function (see [333]) if it is strictly increasing and $\sum_{n=1}^\infty \varphi^n(t) < \infty$, for each $t > 0$.

Example 4.1. The mappings $\varphi_1, \varphi_2 : [0, \infty) \to [0, \infty)$ given by $\varphi_1(t) = ct$; $c \in [0, 1)$, and $\varphi_2(t) = \frac{t}{1+t}$; $t \in [0, \infty)$, are strict comparison functions.

Definition 4.11. A multivalued operator $N : X \to \mathcal{P}_{cl}(X)$ is called

(a) γ-Lipschitz if and only if there exists $\gamma \geq 0$ such that

$$H_d(N(u), N(v)) \leq \gamma d(u, v); \text{ for each } u, v \in X;$$

(b) a multivalued γ-contraction if and only if it is γ-Lipschitz with $\gamma \in [0, 1)$;

(c) a multivalued φ-contraction if and only if there exists a strict comparison function $\varphi : [0, \infty) \to [0, \infty)$ such that

$$H_d(N(u), N(v)) \leq \varphi(d(u, v)) \text{ for each } u, v \in X.$$

Let us give the definition of Ulam–Hyers stability of the fixed-point inclusion $u \in N(u)$, see, for instance, [59]. Let ϵ be a positive real number and $\Phi : I \to \mathbb{R}$ be a continuous function.

Definition 4.12. The fixed-point inclusion $u \in N(u)$ is said to be Ulam–Hyers stable if there exists a real number $c_N > 0$ such that for each $\epsilon > 0$

and for each solution $u \in C_\gamma$ of the inequality $H_d(u, Nu) \leq \epsilon$, there exists a solution $v \in C_\gamma$ of the inclusion $u \in N(u)$ with

$$|u(t) - v(t)| \leq \epsilon c_N; \ t \in I.$$

Definition 4.13. The fixed-point inclusion $u \in N(u)$ is said to be generalized Ulam–Hyers stable if there exists an increasing function $\theta_N \in C([0, \infty), [0, \infty))$, $\theta_N(0) = 0$ such that for each $\epsilon > 0$ and for each solution $u \in C_\gamma$ of the inequality $H_d(u, Nu) \leq \epsilon$, there exists a solution $v \in C_\gamma$ of the inclusion $u \in N(u)$ with

$$|u(t) - v(t)| \leq \theta_N(\epsilon); \quad t \in I.$$

Definition 4.14. The fixed-point inclusion $u \in N(u)$ is said to be Ulam–Hyers–Rassias stable with respect to Φ if there exists a real number $c_{N,\Phi} > 0$ such that for each $\epsilon > 0$ and for each solution $u \in C_\gamma$ of the inequality $H_d(u, Nu) \leq \epsilon \Phi(t); \ t \in I$, there exists a solution $v \in C_\gamma$ of the inclusion $u \in N(u)$ with

$$|u(t) - v(t)| \leq \epsilon c_{N,\Phi} \Phi(t); \ t \in I.$$

Definition 4.15. The fixed-point inclusion $u \in N(u)$ is said to be generalized Ulam–Hyers–Rassias stable with respect to Φ if there exists a real number $c_{N,\Phi} > 0$ such that for each solution $u \in C_\gamma$ of the inequality $H_d(u, Nu) \leq \Phi(t); \ t \in I$, there exists a solution $v \in C_\gamma$ of the inclusion $u \in N(u)$ with

$$|u(t) - v(t)| \leq c_{N,\Phi} \Phi(t); \quad t \in I.$$

Remark 4.3. It is clear that

(i) Definition 4.12 implies Definition 4.13;
(ii) Definition 4.14 implies Definition 4.15;
(iii) Definition 4.14 for $\Phi(t) = 1$ implies Definition 4.12.

The following result, a generalization of Covitz–Nadler fixed-point principle (see [192,293]), is known in the literature as Węgrzyk's fixed-point theorem.

Lemma 4.2 ([369]). *Let (X, d) be a complete metric space. If $A : X \to \mathcal{P}_{cl}(X)$ is a φ-contraction, then $\mathrm{Fix}(A)$ is nonempty and for any $u_0 \in X$, there exists a sequence of successive approximations of A starting from u_0 which converges to a fixed point of A.*

In particular, in terms of multivalued weakly Picard operator theory, we have the following result.

Lemma 4.3 ([369]). *Let (X, d) be a complete metric space. If an operator $A : X \to \mathcal{P}_{cl}(X)$ is a multivalued φ-contraction, then A is a MWP operator.*

Now we present an important characterization lemma from the point of view of Ulam–Hyers stability.

Lemma 4.4 ([315]). *Let (X, d) be a metric space. If $A : X \to \mathcal{P}_{cp}(X)$ is a Ψ-MWP operator, then the fixed-point inclusion $u \in A(u)$ is generalized Ulam–Hyers stable. In particular, if A is c-MWP operator, then the fixed-point inclusion $u \in A(u)$ is Ulam–Hyers stable.*

Another Ulam–Hyers stability result, more efficient for applications, was proved in [276].

Theorem 4.4 ([276]). *Let (X, d) be a complete metric space and $A : X \to \mathcal{P}_{cp}(X)$ be a multivalued φ-contraction. Then:*

(i) *(Existence of the fixed point) A is a MWP operator;*
(ii) *(Ulam–Hyers stability for the fixed-point inclusion) If additionally $\varphi(ct) \le c\varphi(t)$ for every $t \in [0, \infty)$ (where $c > 1$), then A is a Ψ-MWP operator, with $\Psi(t) := t + \sum_{n=1}^{\infty} \varphi^n(t)$, for each $t \in [0, \infty)$;*
(iii) *(Data dependence of the fixed-point set) Let $S : X \to \mathcal{P}_{cl}(X)$ be a multivalued φ-contraction and $\eta > 0$ be such that $H_d(S(x), A(x)) \le \eta$, for each $x \in X$. Suppose that $\varphi(ct) \le c\varphi(t)$ for every $t \in [0, \infty)$ (where $c > 1$). Then,*

$$H_d(\mathrm{Fix}(S), \mathrm{Fix}(F)) \le \Psi(\eta).$$

Lemma 4.5. *Let $F : I \times \mathbb{R} \to \mathbb{R}$ be such that $S_{F \circ u} \subset C_\gamma$ for any $u \in C_\gamma$. Then problem (4.7) is equivalent to the problem of the solutions of the operator inclusion $u \in N(u)$, where $N : C_\gamma \to \mathcal{P}(C_\gamma)$ is the multifunction defined by*

$$(Nu)(t) = \left\{ \frac{\phi}{\Gamma(\gamma)} t^{\gamma-1} + (I_0^\alpha v)(t); \ v \in S_{F \circ u} \right\}.$$

Now, we present conditions for the existence and Ulam stability of the Hilfer problem (4.7). The following hypotheses will be used in the sequel.

(4.4.1) The function $t \longmapsto F(t, u)$ is measurable for each $u \in \mathbb{R}$.
(4.4.2) The function $u \longmapsto F(t, u)$ is lower semicontinuous for a.e. $t \in I$.

(4.4.3) There exist $p \in L^\infty(I, \mathbb{R}_+)$ and a strict comparison function $\varphi :$ $\mathbb{R}_+ \to \mathbb{R}_+$ such that for a.e $t \in I$, and each $u, v \in \mathbb{R}$, we have

$$H_d(F(t, u), F(t, \overline{u})) \leq p(t)\varphi(|u - \overline{u}|), \tag{4.8}$$

and

$$\frac{T^{1-\gamma+\alpha}\|p\|_{L^\infty}}{\Gamma(1+\alpha)} \leq 1. \tag{4.9}$$

(4.4.4) There exists an integrable function $q : I \to \mathbb{R}$ such that for almost all $t \in I$ and each $u \in \mathbb{R}$, we have

$$F(t, u) \subset q(t)B(0, 1),$$

where $B(0, 1) = \{u \in C_\gamma : \|u\|_C < 1\}$.

Theorem 4.5. *Assume that the hypotheses (4.4.1)–(4.4.4) hold, then*

(a) *The problem (4.7) has at least one solution and N is an $(MWPO)$;*
(b) *if additionally $\varphi(ct) \leq c\varphi(t)$ for every $t \in [0, \infty)$ (where $c > 1$), then the problem (4.7) is generalized Ulam–Hyers stable, and N is a $(\Psi\text{-}MWPO)$, with the function Ψ defined by $\Psi(t) := t + \sum_{n=1}^\infty \varphi^n(t)$, for each $t \in [0, \infty)$. Moreover, in this case, the continuous data dependence of the solution set of the problem (4.7) holds.*

Remark 4.4. For each $u \in C_\gamma$, the set $S_{F \circ u}$ is nonempty since by (4.4.1), F has a measurable selection (see [180, Theorem III.6]).

Proof of Theorem 4.5. We shall show that N defined in Lemma 4.5 satisfies the assumptions of Theorem 4.4. The proof will be given in two steps.

Step 1. $N(u) \in \mathcal{P}_{cp}(C_\gamma)$ *for each $u \in C_\gamma$.*

From [337, Theorem 2], we have that for each $u \in C_\gamma$ there exists $f \in S_{F \circ u}$, for all $t \in I$. Then the function $v(t) = \frac{\phi}{\Gamma(\gamma)}t^{\gamma-1} + I_0^\alpha f(t)$ has the property $v \in N(u)$. Moreover, from (4.4.1) and (4.4.4), via Theorem 8.6.3 in [116], we get that $N(u)$ is a compact set, for each $u \in C_\gamma$.

Step 2. $H_d(N(u), N(\overline{u})) \leq \varphi(\|u - \overline{u}\|_C)$ *for each $u, \overline{u} \in C_\gamma$.*

Let $u, \overline{u} \in C_\gamma$ and $h \in N(u)$. Then, there exists $f(t) \in F(t, u(t))$ such that for each $t \in I$, we have

$$h(t) = \frac{\phi}{\Gamma(\gamma)}t^{\gamma-1} + I_0^\alpha f(t).$$

From (4.4.3) it follows that

$$H_d(F(t, u(t)), F(t, \overline{u}(t))) \leq p(t)\varphi(\|u - \overline{u}\|_C).$$

Hence, there exists $w(t) \in F(t, \overline{u}(t))$ such that

$$|t^{1-\gamma}f(t) - t^{1-\gamma}w(t)| \leq p(t)\varphi(\|u - \overline{u}\|_C); \quad t \in I.$$

Consider $U : I \to \mathcal{P}(\mathbb{R})$ given by

$$U(t) = \{w \in \mathbb{R} : |t^{1-\gamma}f(t) - t^{1-\gamma}w(t)| \leq p(t)\varphi(\|u - \overline{u}\|_C)\}.$$

Since the multivalued operator $u(t) = U(t) \cap F(t, \overline{u}(t))$ is measurable (see [180, Proposition III.4] in), then there exists a function $\overline{f}(t)$ which is a measurable selection for u. So, $\overline{f}(t) \in F(t, \overline{u}(t))$, and for each $t \in I$,

$$|t^{1-\gamma}f(t) - t^{1-\gamma}\overline{f}(t)| \leq p(t)\varphi(\|u - \overline{u}\|_C).$$

Let us define for each $t \in I$,

$$\overline{h}(t) = \frac{\phi}{\Gamma(\gamma)}t^{\gamma-1} + I_0^\alpha \overline{f}(t).$$

Then for each $t \in I$, we have

$$\begin{aligned}
|t^{1-\gamma}h(t) - t^{1-\gamma}\overline{h}(t)| &\leq t^{1-\gamma}I_0^\alpha |f(t) - \overline{f}(t)| \\
&\leq T^{1-\gamma}I_0^\alpha(p(t)\varphi(\|u - \overline{u}\|_C)) \\
&\leq T^{1-\gamma}\|p\|_{L^\infty}\varphi(\|u - \overline{u}\|_C) \left(\int_0^t \frac{|t - s|^{\alpha-1}}{\Gamma(\alpha)} ds \right) \\
&\leq \frac{T^{1-\gamma+\alpha}\|p\|_{L^\infty}}{\Gamma(1+\alpha)}\varphi(\|u - \overline{u}\|_C).
\end{aligned}$$

Thus, by (4.9), we get

$$\|h - \overline{h}\|_C \leq \varphi(\|u - \overline{u}\|_C).$$

By an analogous relation, obtained by interchanging the roles of u and $\overline{u}$, it follows that

$$H_d(N(u), N(\overline{u})) \leq \varphi(\|u - \overline{u}\|_C).$$

Hence, N is a φ-contraction.

(a) By Lemma 4.3, N has a fixed point which is a solution of the inclusion (4.7) on I, and by Theorem 4.4(i), N is an (MWPO).

(b) We will prove that the problem (4.7) is generalized Ulam–Hyers stable. Indeed, let $\epsilon > 0$ and $v \in C_\gamma$ for which there exists $u \in C_\gamma$ such that

$$u(t) \in \frac{\phi}{\Gamma(\gamma)}t^{\gamma-1} + (I_0^\alpha F)(t, v(t)); \quad t \in I,$$

and

$$\|u - v\|_C \leq \epsilon,$$

where

$$(I_0^\alpha F)(t, v(t)) = \{(I_0^\alpha w)(t); w \in S_{F \circ v}\}; \quad t \in I.$$

Then $H_d(v, N(v)) \leq \epsilon$. Moreover, by the above proof we have that N is a multivalued φ-contraction and using Theorem 4.4(i) and (ii), we obtain that N is a (Ψ-MWPO). Then, the fixed-point problem $u \in N(u)$ is generalized Ulam–Hyers stable. Thus, the problem (4.7) is generalized Ulam–Hyers stable. The conclusion of the theorem follows from Theorem 4.4(iii). $\quad\square$

4.3.3. *Example*

Consider the following problem of Hilfer fractional differential inclusion:

$$\begin{cases} (D_0^{\frac{1}{2}, \frac{1}{2}} u)(t) \in F(t, u(t)); \ t \in [0, 1], \\ (I_0^{\frac{1}{4}} u)(t)|_{t=0} = 1, \end{cases} \tag{4.10}$$

where

$$F(t, u(t)) = \{v \in C([0, 1], \mathbb{R}) : |f_1(t, u(t))| \leq |v| \leq |f_2(t, u(t))|\}; \quad t \in [0, 1],$$

with $f_1, f_2 : [0, 1] \times \mathbb{R} \to \mathbb{R}$, such that

$$f_1(t, u(t)) = \frac{t^2 u(t)}{(1 + |u(t)|)e^{10+t}},$$

and

$$f_2(t, u(t)) = \frac{t^2 u(t)}{e^{10+t}}.$$

Set $\alpha = \beta = \frac{1}{2}$, then $\gamma = \frac{3}{4}$. We assume that F is closed and convex valued. We can see that the solutions of the problem (4.10) are solutions of the fixed-point inclusion $u \in A(u)$ where $A : C([0,1], \mathbb{R}) \to \mathcal{P}(C([0,1], \mathbb{R}))$ is the multifunction operator defined by

$$(Au)(t) = \left\{ \mu(x,y) + (I_0^{\frac{1}{2}} f)(t); \ f \in S_{Fou} \right\}; \quad t \in [0,1].$$

For each $t \in [0,1]$ and all $z_1, z_2 \in C_{\frac{3}{4}}$, we have

$$\|f_2(t, z_2) - f_1(t, z_1)\|_C \le t^2 e^{-10-t} \|z_2 - z_1\|_C.$$

Thus, the hypotheses (4.4.1)–(4.4.3) are satisfied with $p(t) = t^2 e^{-10-t}$. We shall show that condition (4.9) holds with $T = 1$. Indeed, $\|p\|_{L^\infty} = e^{-9}$, $\Gamma(1 + \frac{1}{2}) > \frac{1}{2}$. A simple computation shows that

$$\zeta := \frac{T^{\frac{3}{4}} \|p\|_{L^\infty}}{\Gamma(1 + \frac{1}{2})} < 2e^{-9} < 1.$$

The condition (4.4.4) is satisfied with $q(t) = \frac{t^2 e^{-10-t}}{\|F\|_{\mathcal{P}}}$; $t \in [0,1]$, where

$$\|F\|_{\mathcal{P}} = \sup\{\|f\|_C : f \in S_{Fou}\}; \quad \text{for all } u \in C_{\frac{3}{4}}.$$

Consequently, by Theorem 4.5 we concluded that:

(a) The problem (4.10) has at least one solution and A is an (MWPO).
(b) The function $\varphi : \mathbb{R}_+ \to \mathbb{R}_+$ defined by $\varphi(t) = \zeta t$ satisfies

$$\varphi(\zeta t) \le \zeta \varphi(t)$$

for every $t \in [0, \infty)$. Then the problem (4.10) is generalized Ulam–Hyers stable, and A is a (Ψ-MWPO), with the function Ψ defined by $\Psi(t) := t + (1 - \zeta t)^{-1}$, for each $t \in [0, \zeta^{-1})$. Moreover, the continuous data dependence of the solution set of the problem (4.10) holds.

4.4. Ulam Stability for Hilfer–Hadamard Fractional Differential Equations

4.4.1. *Introduction*

This section deals with some existence and Ulam–Hyers–Rassias stability results for a class of differential equations involving the Hilfer–Hadamard fractional derivative. An application is made of Schauders fixed-point theorem for the existence of solutions. Next we prove that our problem is generalized Ulam–Hyers–Rassias stable.

Recently, considerable attention has been given to the existence of solutions of initial and boundary value problems for fractional differential equations with Hilfer fractional derivative; see [238,239,354,368]. Motivated by the Hilfer fractional derivative (which interpolates the Riemann–Liouville derivative and the Caputo derivative), Qassim *et al.* [320,321] considered a new type of fractional derivative (which interpolates the Hadamard derivative and its Caputo counterpart). Motivated by the above papers, in this section we discuss the existence and the Ulam stability of solutions for the following problem of Hilfer–Hadamard fractional differential equations:

$$\begin{cases} (^{H}D_1^{\alpha,\beta}u)(t) = f(t,u(t)); & t \in I := [1,T], \\ (^{H}I_1^{1-\gamma}u)(t)|_{t=1} = \phi, \end{cases} \tag{4.11}$$

where $\alpha \in (0,1)$, $\beta \in [0,1]$, $\gamma = \alpha+\beta-\alpha\beta$, $T > 1$, $\phi \in \mathbb{R}$, $f : I \times \mathbb{R} \to \mathbb{R}$ is a given function, $^{H}I_1^{1-\gamma}$ is the left-sided mixed Hadamard integral of order $1 - \gamma$, and $^{H}D_1^{\alpha,\beta}$ is the Hilfer–Hadamard fractional derivative of order α and type β, introduced by Hilfer in [238].

4.4.2. *Existence and Ulam–Hyers–Rassias stability results*

Set $C := C([1,T])$, and denote the weighted space of continuous functions defined by

$$C_{\gamma,\ln}([1,T]) = \{w(t) : (\ln t)^{1-\gamma}w(t) \in C\},$$

with the norm

$$\|w\|_{C_{\gamma,\ln}} := \sup_{t\in[1,T]} |(\ln t)^{1-\gamma}w(t)|.$$

Let us start by defining what we mean by a solution of the problem (4.11).

Definition 4.16. By a solution of the problem (4.11) we mean a measurable function $u \in C_{\gamma,\ln}$ that satisfies the condition $(^{H}I_1^{1-\gamma}u)(1^{+}) = \phi$, and the equation $(^{H}D_1^{\alpha,\beta}u)(t) = f(t,u(t))$ on I.

The following hypotheses will be used in the sequel.

(4.5.1) The function $t \mapsto f(t,u)$ is measurable on I for each $u \in \mathbb{R}$, and the function $u \mapsto f(t,u)$ is continuous on $\mathbb{R}$ for a.e. $t \in I$.

(4.5.2) There exists a continuous function $p : I \to [0,\infty)$ such that

$$|f(t,u)| \leq p(t); \quad \text{for a.e. } t \in I, \text{ and each } u \in \mathbb{R}.$$

Set

$$p^* = \sup_{t \in I} p(t),$$

Now, we shall prove the following theorem concerning the existence of solutions of problem (4.11).

Theorem 4.6. *Assume that the hypotheses (4.5.1) and (4.5.2) hold. Then the problem (4.11) has at least one solution defined on I.*

Proof. Consider the operator $N : C_{\gamma,\ln} \to C_{\gamma,\ln}$ defined by

$$(Nu)(t) = \frac{\phi}{\Gamma(\gamma)}(\ln t)^{\gamma-1} + \int_1^t \left(\ln \frac{t}{s}\right)^{\alpha-1} \frac{f(s, u(s))}{s\Gamma(\alpha)} ds. \tag{4.12}$$

Clearly, the fixed points of the operator N are solution of the problem (4.11).

For any $u \in C_{\gamma,\ln}$, and each $t \in I$ we have

$$|(\ln t)^{1-\gamma}(Nu)(t)| \leq \frac{|\phi|}{\Gamma(\gamma)} + \frac{(\ln t)^{1-\gamma}}{\Gamma(\alpha)} \int_1^t \left(\ln \frac{t}{s}\right)^{\alpha-1} |f(s, u(s))| \frac{ds}{s}$$

$$\leq \frac{|\phi|}{\Gamma(\gamma)} + \frac{(\ln t)^{1-\gamma}}{\Gamma(\alpha)} \int_1^t \left(\ln \frac{t}{s}\right)^{\alpha-1} p(s) \frac{ds}{s}$$

$$\leq \frac{|\phi|}{\Gamma(\gamma)} + \frac{p^*(\ln T)^{1-\gamma}}{\Gamma(\alpha)} \int_1^t \left(\ln \frac{t}{s}\right)^{\alpha-1} \frac{ds}{s}$$

$$\leq \frac{|\phi|}{\Gamma(\gamma)} + \frac{p^*(\ln T)^{1-\gamma+\alpha}}{\Gamma(1+\alpha)}.$$

Thus

$$\|N(u)\|_C \leq \frac{|\phi|}{\Gamma(\gamma)} + \frac{p^*(\ln T)^{1-\gamma+\alpha}}{\Gamma(1+\alpha)} := R. \tag{4.13}$$

This proves that N transforms the ball $B_R := B(0, R) = \{w \in C_{\gamma,\ln} : \|w\|_C \leq R\}$ into itself. We shall show that the operator $N : B_R \to B_R$ satisfies all the assumptions of Theorem 1.2. The proof will be given in several steps.

Step 1. $N : B_R \to B_R$ *is continuous.*

Let $\{u_n\}_{n \in \mathbb{N}}$ be a sequence such that $u_n \to u$ in B_R. Then, for each $t \in I$, we have

$$|(\ln t)^{1-\gamma}(Nu_n)(t) - (\ln t)^{1-\gamma}(Nu)(t)|$$

$$\leq \frac{(\ln t)^{1-\gamma}}{\Gamma(\alpha)} \int_1^t \left(\ln \frac{t}{s}\right)^{\alpha-1} |f(s, u_n(s)) - f(s, u(s))| \frac{ds}{s}. \qquad (4.14)$$

Since $u_n \to u$ as $n \to \infty$ and f is continuous, then by the Lebesgue dominated convergence theorem, Eq. (4.14) implies

$$\|N(u_n) - N(u)\|_C \to 0 \quad \text{as } n \to \infty.$$

Step 2. $N(B_R)$ *is uniformly bounded.*

This is clear since $N(B_R) \subset B_R$ and B_R is bounded.

Step 3. $N(B_R)$ *is equicontinuous.*

Let $t_1, t_2 \in I$, $t_1 < t_2$ and let $u \in B_R$. Thus, we have

$$|(\ln t_2)^{1-\gamma}(Nu)(t_2) - (\ln t_1)^{1-\gamma}(Nu)(t_1)|$$

$$\leq \left| (\ln t_2)^{1-\gamma} \int_1^{t_2} \left(\ln \frac{t_2}{s}\right)^{\alpha-1} \frac{f(s, u(s))}{s\Gamma(\alpha)} ds \right.$$

$$\left. - (\ln t_1)^{1-\gamma} \int_1^{t_1} \left(\ln \frac{t_1}{s}\right)^{\alpha-1} \frac{f(s, u(s))}{s\Gamma(\alpha)} ds \right|$$

$$\leq (\ln t_2)^{1-\gamma} \int_{t_1}^{t_2} \left(\ln \frac{t_2}{s}\right)^{\alpha-1} \frac{|f(s, u(s))|}{s\Gamma(\alpha)} ds$$

$$+ \int_1^{t_1} \left| (\ln t_2)^{1-\gamma} \left(\ln \frac{t_2}{s}\right)^{\alpha-1} \right.$$

$$\left. - (\ln t_1)^{1-\gamma} \left(\ln \frac{t_1}{s}\right)^{\alpha-1} \right| \frac{|f(s, u(s))|}{s\Gamma(\alpha)} ds$$

$$\leq (\ln t_2)^{1-\gamma} \int_{t_1}^{t_2} \left(\ln \frac{t_2}{s}\right)^{\alpha-1} \frac{p(s)}{s\Gamma(\alpha)} ds$$

$$+ \int_1^{t_1} \left| (\ln t_2)^{1-\gamma} \left(\ln \frac{t_2}{s}\right)^{\alpha-1} \right.$$

$$\left. - (\ln t_1)^{1-\gamma} \left(\ln \frac{t_1}{s}\right)^{\alpha-1} \right| \frac{p(s)}{s\Gamma(\alpha)} ds.$$

Hence, we get

$$
|(\ln t_2)^{1-\gamma}(Nu)(t_2) - (\ln t_1)^{1-\gamma}(Nu)(t_1)|
$$

$$
\leq \frac{p_*(\ln T)^{1-\gamma+\alpha}}{\Gamma(1+\alpha)} \left(\ln \frac{t_2}{t_1}\right)^{\alpha}
$$

$$
+ \frac{p_*}{\Gamma(\alpha)} \int_1^{t_1} \left| (\ln t_2)^{1-\gamma} \left(\ln \frac{t_2}{s}\right)^{\alpha-1} - (\ln t_1)^{1-\gamma} \left(\ln \frac{t_1}{s}\right)^{\alpha-1} \right| ds.
$$

As $t_1 \to t_2$, the right-hand side of the above inequality tends to zero.

As a consequence of Steps 1–3 together with the Arzelà–Ascoli theorem, we can conclude that N is continuous and compact. From an application of Schauder's theorem, we deduce that N has at least a fixed point u which is a solution of the problem (4.11). □

Now, we are concerned with the generalized Ulam–Hyers–Rassias stability of our problem (4.11).

Theorem 4.7. *Assume that the hypotheses (4.5.1), (4.5.2) and the following hypotheses hold.*

(4.6.1) *There exists $\lambda_\Phi > 0$ such that for each $t \in I$, we have*

$$
({}^H I_1^\alpha \Phi)(t) \leq \lambda_\Phi \Phi(t).
$$

(4.6.2) *There exists $q \in C(I, [0, \infty))$ such that for each $t \in I$, we have*

$$
p(t) \leq q(t)\Phi(t).
$$

Then the problem (4.11) is generalized Ulam–Hyers–Rassias stable.

Proof. Consider the operator $N : C_{\gamma,\ln} \to C_{\gamma,\ln}$ defined in (4.12). Let u be a solution of the inequality (1.7), and let us assume that v is a solution of problem (4.11). Thus, we have

$$
v(t) = \frac{\phi}{\Gamma(\gamma)}(\ln t)^{\gamma-1} + \int_1^t \left(\ln \frac{t}{s}\right)^{\alpha-1} \frac{f(s, v(s))}{s\Gamma(\alpha)} ds.
$$

From the inequality (1.7) for each $t \in I$, we have

$$
\left| u(t) - \frac{\phi}{\Gamma(\gamma)}(\ln t)^{\gamma-1} - \int_1^t \left(\ln \frac{t}{s}\right)^{\alpha-1} \frac{f(s, u(s))}{s\Gamma(\alpha)} ds \right| \leq ({}^H I_1^\alpha \Phi)(t).
$$

Set

$$q^* = \sup_{t \in I} q(t).$$

From hypotheses (4.6.1) and (4.6.2), for each $t \in I$, we get

$$|u(t) - v(t)| \leq \left| u(t) - \frac{\phi}{\Gamma(\gamma)}(\ln t)^{\gamma-1} - \int_1^t \left(\ln \frac{t}{s}\right)^{\alpha-1} \frac{f(s, u(s))}{s\Gamma(\alpha)} ds \right|$$

$$+ \int_1^t \left(\ln \frac{t}{s}\right)^{\alpha-1} \frac{|f(s, u(s)) - f(s, v(s))|}{s\Gamma(\alpha)} ds$$

$$\leq ({}^H I_1^\alpha \Phi)(t) + \int_1^t \left(\ln \frac{t}{s}\right)^{\alpha-1} \frac{2q^*\Phi(s)}{s\Gamma(\alpha)} ds$$

$$\leq \lambda_\phi \Phi(t) + 2q^*({}^H I_1^\alpha \Phi)(t)$$

$$\leq [1 + 2q^*]\lambda_\phi \Phi(t)$$

$$:= c_{f,\Phi}\Phi(t).$$

Hence, the problem (4.11) is generalized Ulam–Hyers–Rassias stable.   □

In the sequel, we will use of the following theorem.

Let $X = X(I, \mathbb{R})$ be the metric space, with the metric

$$d(u, v) = \sup_{t \in I} \frac{\|u(t) - v(t)\|_C}{\Phi(.)}.$$

Theorem 4.8. *Assume that (4.6.2) and the following hypothesis hold.*

(4.7.1) *There exists $\varphi \in C(I, [0, \infty))$ such that for each $t \in I$, and all $u, v \in \mathbb{R}$, we have*

$$|f(t, u) - f(t, v)| \leq (\ln t)^{1-\gamma}\varphi(t)\Phi(t)|u - v|.$$

If

$$L := (\ln T)^{1-\gamma}\varphi^*\lambda_\phi < 1, \tag{4.15}$$

where $\varphi^* = \sup_{t \in I} \varphi(t)$, then there exists a unique solution u_0 of problem (4.11), and the problem (4.11) is generalized Ulam–Hyers–Rassias stable. Furthermore, we have

$$|u(t) - u_0(t)| \leq \frac{\Phi(t)}{1 - L}.$$

Proof. Let $N : C_{\gamma,\ln} \to C_{\gamma,\ln}$ be the operator defined in (4.12). By applying Theorem 1.4, we have

$$
\begin{aligned}
|(Nu)(t) - (Nv)(t)| &\leq \int_1^t \left(\ln \frac{t}{s} \right)^{\alpha-1} \frac{|f(s,u(s)) - f(s,v(s))|}{s\Gamma(\alpha)} ds \\
&\leq \int_1^t \left(\ln \frac{t}{s} \right)^{\alpha-1} \frac{\varphi(s)\Phi(s)|(\ln s)^{1-\gamma}u(s) - (\ln s)^{1-\gamma}v(s)|}{s\Gamma(\alpha)} ds \\
&\leq \int_1^t \left(\ln \frac{t}{s} \right)^{\alpha-1} \frac{\varphi^* \Phi(s)\|u - v\|_C}{s\Gamma(\alpha)} ds \\
&\leq \varphi^* (^H I_1^\alpha \Phi)(t)\|u - v\|_C \\
&\leq \varphi^* \lambda_\phi \Phi(t)\|u - v\|_C.
\end{aligned}
$$

Thus

$$
|(\ln t)^{1-\gamma}(Nu)(t) - (\ln t)^{1-\gamma}(Nv)(t)| \leq (\ln T)^{1-\gamma}\varphi^* \lambda_\phi \Phi(t)\|u - v\|_C.
$$

Hence, we get

$$
d(N(u), N(v)) = \sup_{t \in I} \frac{\|(Nu)(t) - (Nv)(t)\|_C}{\Phi(t)} \leq L\|u - v\|_C,
$$

from which we conclude the proof. □

4.4.3. *Example*

Consider the following problem of Hilfer–Hadamard fractional differential equation

$$
\begin{cases}
(^H D_1^{\frac{1}{2},\frac{1}{2}} u)(t) = f(t, u(t)); & t \in [1, e], \\
(^H I_1^{\frac{1}{4}} u)(t)|_{t=1} = 0,
\end{cases}
\tag{4.16}
$$

where

$$
\begin{cases}
f(t, u) = \frac{(t-1)^{\frac{-1}{4}} \sin(t-1)}{64(1+\sqrt{t-1})(1+|u|)}; & t \in (1, e],\ u \in \mathbb{R}, \\
f(1, u) = 0; & u \in \mathbb{R}.
\end{cases}
$$

Clearly, the function f satisfies (4.5.1).
The hypothesis (4.5.2) is satisfied with

$$
\begin{cases}
p(t) = \frac{(t-1)^{\frac{-1}{4}} |\sin(t-1)|}{64(1+\sqrt{t-1})}; & t \in (1, e], \\
p(1) = 0.
\end{cases}
$$

Hence, Theorem 4.6 implies that the problem (4.16) has at least one solution defined on $[1, e]$. Also, the hypothesis (4.6.1) is satisfied with

$$\Phi(t) = e^3 \text{ and } \lambda_\Phi = \frac{2}{\sqrt{\pi}}.$$

Indeed, for each $t \in [1, e]$ we get

$$({}^H I_1^\alpha \Phi)(t) \leq \frac{2e^3}{\sqrt{\pi}}$$

$$= \lambda_\Phi \Phi(t).$$

Consequently, Theorem 4.7 implies that the problem (4.16) is generalized Ulam–Hyers–Rassias stable.

4.5. Ulam Stabilities for Hilfer Fractional Differential Equations in Banach Spaces

4.5.1. *Introduction*

This section deals with some existence and Ulam stability results for some Hilfer and Hilfer–Hadamard differential equations of fractional order. The technique relies on the concept of measure of noncompactness and Mönch's fixed-point theorem, and we prove that our problems are generalized Ulam–Hyers–Rassias stable. In the last section, we give an illustrative example.

In this section, we discuss the existence and the Ulam stability of solutions for the following problem of Hilfer fractional differential equations

$$\begin{cases} (D_0^{\alpha,\beta} u)(t) = f(t, u(t)); & t \in I := [0, T], \\ (I_0^{1-\gamma} u)(0) = \phi, \end{cases} \tag{4.17}$$

where $\alpha \in (0, 1)$, $\beta \in [0, 1]$, $\gamma = \alpha + \beta - \alpha\beta$, $T > 0$, $\phi \in E$, $f : I \times E \to E$ is a given function, E is a real (or complex) Banach space with a norm $\| \cdot \|$, $I_0^{1-\gamma}$ is the left-sided mixed Riemann–Liouville integral of order $1 - \gamma$, and $D_0^{\alpha,\beta}$ is the generalized Riemann–Liouville derivative operator of order α and type β, introduced by Hilfer in [238].

Next, we consider the following problem of Hilfer–Hadamard fractional differential equations

$$\begin{cases} ({}^H D_1^{\alpha,\beta} u)(t) = g(t, u(t)); & t \in [1, T], \\ ({}^H I_1^{1-\gamma} u)(1) = \phi_0, \end{cases} \tag{4.18}$$

where $\alpha \in (0,1)$, $\beta \in [0,1]$, $\gamma = \alpha+\beta-\alpha\beta$, $T > 1$, $\phi_0 \in E$, $g : [1,T] \times E \to E$ is a given function, $^{H}I_1^{1-\gamma}$ is the left-sided mixed Hadamard integral of order $1 - \gamma$, and $^{H}D_1^{\alpha,\beta}$ is the Hilfer–Hadamard fractional derivative of order α and type β.

4.5.2. *Existence and Ulam stability results*

Set $C := C([1,T])$. Denote the weighted space of continuous functions defined by

$$C_{\gamma,\ln}([1,T]) = \{w(t) : (\ln t)^{1-\gamma}w(t) \in C\},$$

with the norm

$$\|w\|_{C_{\gamma,\ln}} := \sup_{t \in [1,T]} \|(\ln t)^{1-\gamma}w(t)\|.$$

Definition 4.17. By a solution of the problem (4.17) we mean a measurable function $u \in C_\gamma$ that satisfies the condition $(I_0^{1-\gamma}u)(0^+) = \phi$, and the equation $(D_0^{\alpha,\beta}u)(t) = f(t,u(t))$ on I.

The following hypotheses will be used in the sequel.

(4.8.1) The function $t \mapsto f(t,u)$ is measurable on I for each $u \in E$, and the function $u \mapsto f(t,u)$ is continuous on E for a.e. $t \in I$.

(4.8.2) There exists a continuous function $p : I \to [0,\infty)$ such that

$$\|f(t,u)\| \le p(t), \quad \text{for a.e. } t \in I, \text{ and each } u \in E,$$

(4.8.3) For each bounded and measurable set $B \subset E$ and for each $t \in I$, we have

$$\mu(f(t,B)) \le t^{1-\gamma}p(t)\mu(B).$$

Set

$$p^* = \sup_{t \in I} p(t),$$

Now, we shall prove the following theorem concerning the existence of solutions of problem (4.17).

Theorem 4.9. *Assume that the hypotheses (4.8.1)–(4.8.3) hold. If*

$$L := \frac{p^* T^{1-\gamma+\alpha}}{\Gamma(1+\alpha)} < 1, \tag{4.19}$$

then the problem (4.17) has at least one solution defined on I.

Proof. Consider the operator $N : C_\gamma \to C_\gamma$ defined by

$$(Nu)(t) = \frac{\phi}{\Gamma(\gamma)} t^{\gamma-1} + \int_0^t (t-s)^{\alpha-1} \frac{f(s, u(s))}{\Gamma(\alpha)} ds. \tag{4.20}$$

Clearly, the fixed points of the operator N are solutions of the problem (4.17).

For any $u \in C_\gamma$ and each $t \in I$, we have

$$\|t^{1-\gamma}(Nu)(t)\| \leq \frac{\|\phi\|}{\Gamma(\gamma)} + \frac{t^{1-\gamma}}{\Gamma(\alpha)} \int_0^t (t-s)^{\alpha-1} \|f(s, u(s))\| ds$$

$$\leq \frac{\|\phi\|}{\Gamma(\gamma)} + \frac{t^{1-\gamma}}{\Gamma(\alpha)} \int_0^t (t-s)^{\alpha-1} p(s) ds$$

$$\leq \frac{\|\phi\|}{\Gamma(\gamma)} + \frac{p^* T^{1-\gamma}}{\Gamma(\alpha)} \int_0^t (t-s)^{\alpha-1} ds$$

$$\leq \frac{\|\phi\|}{\Gamma(\gamma)} + \frac{p^* T^{1-\gamma+\alpha}}{\Gamma(1+\alpha)}.$$

Thus

$$\|N(u)\|_C \leq \frac{\|\phi\|}{\Gamma(\gamma)} + \frac{p^* T^{1-\gamma+\alpha}}{\Gamma(1+\alpha)} := R. \tag{4.21}$$

This proves that N transforms the ball $B_R := B(0, R) = \{w \in C_\gamma : \|w\|_C \leq R\}$ into itself. We shall show that the operator $N : B_R \to B_R$ satisfies all the assumptions of Theorem 1.6. The proof will be given in several steps.

Step 1. $N : B_R \to B_R$ *is continuous.*

Let $\{u_n\}_{n \in \mathbb{N}}$ be a sequence such that $u_n \to u$ in B_R. Then, for each $t \in I$, we have

$$\|t^{1-\gamma}(Nu_n)(t) - t^{1-\gamma}(Nu)(t)\|$$

$$\leq \frac{t^{1-\gamma}}{\Gamma(\alpha)} \int_0^t (t-s)^{\alpha-1} \|f(s, u_n(s)) - f(s, u(s))\| ds. \tag{4.22}$$

Since $u_n \to u$ as $n \to \infty$ and f is continuous, then by the Lebesgue dominated convergence theorem, Eq. (4.22) implies

$$\|N(u_n) - N(u)\|_C \to 0 \quad \text{as } n \to \infty.$$

Step 2. $N(B_R)$ *is bounded and equicontinuous.*

Since $N(B_R) \subset B_R$ and B_R is bounded, then $N(B_R)$ is bounded.

Let $t_1, t_2 \in I$, $t_1 < t_2$ and let $u \in B_R$. Thus we have

$$\|t_2^{1-\gamma}(Nu)(t_2) - t_1^{1-\gamma}(Nu)(t_1)\|$$

$$\leq \left\| t_2^{1-\gamma} \int_0^{t_2} (t_2 - s)^{\alpha-1} \frac{f(s, u(s))}{\Gamma(\alpha)} ds - t_1^{1-\gamma} \int_0^{t_1} (t_1 - s)^{\alpha-1} \frac{f(s, u(s))}{\Gamma(\alpha)} ds \right\|$$

$$\leq t_2^{1-\gamma} \int_{t_1}^{t_2} (t_2 - s)^{\alpha-1} \frac{\|f(s, u(s))\|}{\Gamma(\alpha)} ds$$

$$+ \int_0^{t_1} |t_2^{1-\gamma}(t_2 - s)^{\alpha-1} - t_1^{1-\gamma}(t_1 - s)^{\alpha-1}| \frac{\|f(s, u(s))\|}{\Gamma(\alpha)} ds$$

$$\leq t_2^{1-\gamma} \int_{t_1}^{t_2} (t_2 - s)^{\alpha-1} \frac{p(s)}{\Gamma(\alpha)} ds$$

$$+ \int_0^{t_1} |t_2^{1-\gamma}(t_2 - s)^{\alpha-1} - t_1^{1-\gamma}(t_1 - s)^{\alpha-1}| \frac{p(s)}{\Gamma(\alpha)} ds.$$

Thus, we get

$$\|t_2^{1-\gamma}(Nu)(t_2) - t_1^{1-\gamma}(Nu)(t_1)\|$$

$$\leq \frac{p_* T^{1-\gamma+\alpha}}{\Gamma(1+\alpha)} (t_2 - t_1)^\alpha + \frac{p_*}{\Gamma(\alpha)} \int_0^{t_1} |t_2^{1-\gamma}(t_2 - s)^{\alpha-1} - t_1^{1-\gamma}(t_1 - s)^{\alpha-1}| ds.$$

As $t_1 \to t_2$, the right-hand side of the above inequality tends to zero. Hence, $N(B_R)$ is bounded and equicontinuous.

Now let V be a subset of B_r such that $V \subset \overline{N(V)} \cup \{0\}$, V is bounded and equicontinuous and therefore the function $t \to v(t) = \mu(V(t))$ is continuous on I. By (4.8.3) and the properties of the measure α, for each $t \in I$, we have

$$t^{1-\gamma} v(t) \leq \mu(t^{1-\gamma}(NV)(t) \cup \{0\})$$

$$\leq \mu(t^{1-\gamma}(NV)(t))$$

$$\leq \frac{T^{1-\gamma}}{\Gamma(\alpha)} \int_0^t (t - s)^{\alpha-1} p(s) \mu(V(s)) ds$$

$$\leq \frac{T^{1-\gamma}}{\Gamma(\alpha)} \int_0^t (t - s)^{\alpha-1} s^{1-\gamma} p(s) v(s) ds$$

$$\leq \frac{p^* T^{1-\gamma+\alpha}}{\Gamma(1+\alpha)} \|v\|_C.$$

Thus

$$\|v\|_C \leq L\|v\|_C.$$

From (4.19), we get $\|v\|_C = 0$, that is $v(t) = \mu(V(t)) = 0$, for each $t \in I$ and then $V(t)$ is relatively compact in E. In view of the Ascoli–Arzelà theorem, V is relatively compact in B_r. Applying now Theorem 1.6, we conclude that N has a fixed point which is a solution of the problem (4.17). $\square$

Now, we are concerned with the generalized Ulam–Hyers–Rassias stability of our problem (4.17).

Let $X = X(I, \mathbb{R})$ be the metric space, with the metric

$$d(u, v) = \sup_{t \in I} \frac{\|u(t) - v(t)\|_C}{\Phi(t)}.$$

Theorem 4.10. *Assume that the following hypotheses hold:*

(4.9.1) *There exists $\lambda_\Phi > 0$ such that for each $t \in I$, we have*

$$(I_0^\alpha \Phi)(t) \leq \lambda_\Phi \Phi(t).$$

(4.9.2) *There exists $\varphi \in C(I, [0, \infty))$ such that for each $t \in I$, and all $u, v \in E$, we have*

$$\|f(t, u) - f(t, u)\| \leq t^{1-\gamma} \varphi(t) \Phi(t) \|u - v\|.$$

If

$$L^* := T^{1-\gamma} \varphi^* \lambda_\phi < 1, \tag{4.23}$$

where $\varphi^* = \sup_{t \in I} \varphi(t)$, then there exists a unique solution u_0 of problem (4.17), and the problem (4.17) is generalized Ulam–Hyers–Rassias stable. Furthermore, we have

$$\|u(t) - u_0(t)\| \leq \frac{\Phi(t)}{1 - L^*}.$$

Proof. Let $N : C_\gamma \to C_\gamma$ be the operator defined in (4.20). Apply Theorem 1.4, we have

$$\|(Nu)(t) - (Nv)(t)\| \leq \int_0^t (t - s)^{\alpha-1} \frac{\|f(s, u(s)) - f(s, v(s))\|}{\Gamma(\alpha)} ds$$

$$\leq \int_0^t (t - s)^{\alpha-1} \frac{\varphi(s) \Phi(s) |s^{1-\gamma} u(s) - s^{1-\gamma} v(s)|}{\Gamma(\alpha)} ds$$

$$\leq \int_0^t (t - s)^{\alpha-1} \frac{\varphi^* \Phi(s) \|u - v\|_C}{\Gamma(\alpha)} ds$$

$$\leq \varphi^*(I_0^\alpha \Phi)(t)\|u - v\|_C$$

$$\leq \varphi^* \lambda_\phi \Phi(t)\|u - v\|_C.$$

Thus

$$|t^{1-\gamma}(Nu)(t) - t^{1-\gamma}(Nv)(t)| \leq T^{1-\gamma}\varphi^* \lambda_\phi \Phi(t)\|u - v\|_C.$$

Hence, we get

$$d(N(u), N(v)) = \sup_{t \in I} \frac{\|(Nu)(t) - (Nv)(t)\|_C}{\Phi(t)} \leq L^*\|u - v\|_C,$$

from which we conclude the theorem. □

We are concerned with the existence and the generalized Ulam–Hyers–Rassias stability of our problem (4.18).

Now we give (without proof) existence and Ulam stability results for problem (4.18). The following hypotheses will be used in the sequel.

(4.10.1) The function $t \mapsto g(t, u)$ is measurable on $[1, T]$ for each $u \in E$, and the function $u \mapsto g(t, u)$ is continuous on E for a.e. $t \in [1, T]$.

(4.10.2) There exists a continuous function $q : [1, T] \to [0, \infty)$ such that

$$\|g(t, u)\| \leq q(t), \quad \text{for a.e. } t \in [1, T], \text{ and each } u \in E.$$

(4.10.3) For each bounded and measurable set $B \subset E$ and for each $t \in [1, T]$, we have

$$\beta(g(t, B)) \leq (\ln t)^{1-\gamma} q(t)\beta(B).$$

(4.11.1) There exists $\rho_\Phi > 0$ such that for each $t \in [1, T]$, we have

$$(^H I_1^\alpha \Phi)(t) \leq \rho_\Phi \Phi(t).$$

(4.11.2) There exists $\chi \in C(I, [0, \infty))$ such that for each $t \in I$, and all $u, v \in E$, we have

$$\|g(t, u) - g(t, u)\| \leq (\ln t)^{1-\gamma}\chi(t)\Phi(t)\|u - v\|.$$

Theorem 4.11. *Assume that the hypotheses (4.10.1)–(4.10.3) hold. If*

$$L' := \frac{q^*(\ln T)^{1-\gamma+\alpha}}{\Gamma(1 + \alpha)} < 1, \tag{4.24}$$

where $q^* = \sup_{t \in [1,T]} q(t)$, then the problem (4.18) has at least one solution defined on $[1, T]$.

Theorem 4.12. *Assume that the hypotheses (4.11.1) and (4.11.2) hold. If*

$$L'^* := (\ln T)^{1-\gamma} \chi^* \rho_\phi < 1, \tag{4.25}$$

where $\chi^* = \sup_{t \in [1,T]} \chi(t)$, *then there exists a unique solution* u_1 *of problem (4.18), and the problem (4.18) is generalized Ulam–Hyers–Rassias stable. Furthermore, we have*

$$\|u(t) - u_1(t)\| \le \frac{\Phi(t)}{1 - L'^*}.$$

4.5.3. *Example*

Let

$$E = l^1 = \left\{ u = (u_1, u_2, \dots, u_n, \dots), \sum_{n=1}^{\infty} |u_n| < \infty \right\}$$

be the Banach space with the norm

$$\|u\|_E = \sum_{n=1}^{\infty} |u_n|.$$

Consider the following problem of Hilfer fractional differential equation:

$$\begin{cases} (D_0^{\frac{1}{2}, \frac{1}{2}} u_n)(t) = f_n(t, u(t)); & t \in [0, 1], \\ (I_0^{\frac{1}{4}} u_n)(t)|_{t=0} = (1, 0, \dots, 0, \dots), \end{cases} \tag{4.26}$$

where

$$\begin{cases} f_n(t, u) = \dfrac{ct^{\frac{-1}{4}} u_n \sin t}{64(1 + \sqrt{t})(1 + \|u\|_E)}; & t \in (0, 1], \\ f_n(0, u) = 0, \end{cases}$$

and $c = \frac{9\sqrt{\pi}}{16}$, with

$$f = (f_1, f_2, \dots, f_n, \dots), \quad u = (u_1, u_2, \dots, u_n, \dots).$$

Set $\alpha = \beta = \frac{1}{2}$, then $\gamma = \frac{3}{4}$. The hypothesis (4.11.2) is satisfied with

$$\begin{cases} p(t) = \dfrac{ct^{\frac{-1}{4}} |\sin t|}{64(1 + \sqrt{t})}; & t \in (0, 1], \\ p(0) = 0. \end{cases}$$

Hence, Theorem 4.9 implies that the problem (4.26) has at least one solution defined on $[0, 1]$.

Also, the hypothesis (4.10.4) is satisfied with

$$\Phi(t) = e^3, \text{ and } \lambda_\Phi = \frac{1}{\Gamma(1+\alpha)}.$$

Indeed, for each $t \in [0,1]$, we get

$$(I_0^\alpha \Phi)(t) \leq \frac{e^3}{\Gamma(1+\alpha)}$$

$$= \lambda_\Phi \Phi(t).$$

Consequently, Theorem 4.10 implies that the problem (4.26) is generalized Ulam–Hyers–Rassias stable.

4.6. Ulam–Hyers Stability for Fractional Differential Equations with Maxima via Picard Operators

4.6.1. *Introduction*

In the present section, we investigate some uniqueness and Ulam's-type stability concepts of fixed-point equations due to Rus, for the differential equations with maxima involving the Caputo fractional derivative. Our results are obtained by using weakly Picard operators theory.

Differential equations with Maxima arise naturally when solving practical problems. For example, many problems in the control theory correspond to the maximal deviation of the regulated quantity. The existence and uniqueness of solutions of differential equations with Maxima is considered in [218,221,305,306], and the references therein. By using the theory of weakly Picard operators, in this section, we discuss the uniqueness and Ulam–Hyers–Rassias stability for the fractional differential equation with maxima

$$^cD_0^r u(x) = f(x, u(x)), \max_{0 \leq \xi \leq x} u(\xi)); \quad \text{if } x \in I := [0,a], \tag{4.27}$$

with the initial condition

$$u(0) = u_0 \in \mathbb{R}, \tag{4.28}$$

where $a > 0$, $^cD_0^r$ is the fractional Caputo derivative of order $r \in (0,1]$, $f : I \times \mathbb{R} \times \mathbb{R} \to \mathbb{R}$ is a given continuous function.

Next, we need the following lemma.

Lemma 4.6 ([335]). *Let (X, d) be a Banach space. If an operator $A :$ $X \to X$ is a contraction with the positive constant $q < 1$, then A is c-weakly Picard operator with the positive constant $c_A = \frac{1}{1-q}$. Moreover, the fixed-point equation $x = A(x)$ is Ulam–Hyers stable.*

4.6.2. Uniqueness and stability results

In this section, we present the main results for the uniqueness and the stability of our problem (4.27)–(4.28).

Theorem 4.13. *Assume that the following hypothesis holds*:

(4.12.1) *There exists a constant $l_f > 0$ such that*

$$|f(x, u, v) - f(x, \overline{u}, \overline{v})| \le l_f \max(|u - \overline{u}|, |v - \overline{v}|)$$

$$for \ each \ x \in I, \ u, v, \overline{u}, \overline{u} \in \mathbb{R}.$$

If

$$L_f := \frac{l_f a^r}{\Gamma(1 + r)} < 1, \tag{4.29}$$

then there exists a unique solution for the problem (4.27)–(4.28) on I. Moreover, the operator N is k-weakly Picard operator with the positive constant $k_N = \frac{1}{1-L_f}$ and the fixed-point equation $u = N(u)$ is Ulam–Hyers stable.

Consider the operator $N : \mathcal{C} \to \mathcal{C}$ defined by:

$$(Nu)(t) = u_0 + \int_0^t \frac{(t - s)^{r-1}}{\Gamma(r)} f(s, u(s), \max_{0 \le \xi \le s} u(\xi)) ds; \quad t \in I.$$

Clearly, the fixed points of the operator N are solution of the problem (4.27)–(4.28).

Proof of Theorem 4.13. Let $v, w \in \mathcal{C}$. Then for each $x \in I$, we have

$$|(Nv)(x) - (Nw)(x)|$$

$$\leq \int_0^x \frac{(x-s)^{r-1}}{\Gamma(r)} |f(s, v(s), \max_{0 \leq \xi \leq s} v(\xi)) - f(s, w(s), \max_{0 \leq \xi \leq s} w(\xi))| ds$$

$$\leq \frac{l_f}{\Gamma(r)} \int_0^x (x-s)^{r-1} \max(|v(s) - w(s)|, |\max_{0 \leq \xi \leq s} v(\xi) - \max_{0 \leq \xi \leq s} w(\xi)|) ds.$$

However

$$\max_{0 \leq s \leq a} |\max_{0 \leq \xi \leq s} v(\xi) - \max_{0 \leq \xi \leq s} w(\xi)| \leq \max_{0 \leq s \leq a} |v(s) - w(s)|.$$

So,

$$|(Nv)(x) - (Nw)(x)| \leq \frac{l_f a^r}{\Gamma(1+r)} \|v - w\|_\infty.$$

Thus,

$$\|N(v) - N(w)\|_\infty \leq L_f \|v - w\|_\infty.$$

Hence, by (4.29), N is a contraction. Consequently, by Banach's contraction principle, N has a unique fixed point witch is the unique solution of the problem (4.27)–(4.28). Moreover, Lemma 4.6 implies that the operator N is k-weakly Picard operator with the positive constant $k_N = \frac{1}{1-L_f}$, and the fixed point equation $u = N(u)$ is Ulam–Hyers stable. $\qquad \square$

Now, we present conditions for the generalized Ulam–Hyers–Rassias stability of problem (4.27)–(4.28).

Theorem 4.14. *Assume that (4.12.1) and the following hypothesis hold:*

(4.13.1) $\Phi \in L^1(I, [0, \infty))$ *and there exists* $\lambda_\Phi > 0$ *such that, for each* $x \in I$
we have

$$(I_0^r \Phi)(x) \leq \lambda_\Phi \Phi(x).$$

If the condition (4.29) holds, then the fixed-point equation $u = N(u)$ *is generalized Ulam–Hyers–Rassias stable.*

Proof. Let $u \in \mathcal{C}$ be a solution of the inequality $|u(x) - (Nu)(x)| \leq \Phi(x)$; $x \in I$. By Theorem 4.13, there exists a unique solution v of the fixed point equation $u = N(u)$. Then we have

$$v(x) = (Nv)(x) = u_0 + \int_0^x \frac{(x-s)^{r-1}}{\Gamma(r)} f(s, v(s), \max_{0 \leq \xi \leq s} v(\xi)) ds; \quad x \in I.$$

Thus, for each $x \in I$, it follows that

$$
\begin{aligned}
|u(x) - v(x)| &= |u(x) - (Nv)(x)| \\
&\leq |u(x) - (Nu)(x)| + |(Nu)(x) - (Nv)(x)| \\
&\leq \Phi(x) + \int_0^x \frac{(x-s)^{r-1}}{\Gamma(r)} |f(s, u(s), \max_{0 \leq \xi \leq s} u(\xi)) \\
&\quad - f(s, v(s), \max_{0 \leq \xi \leq s} (\xi))| ds \\
&\leq \Phi(x) + \frac{l_f}{\Gamma(r)} \int_0^x (x-s)^{r-1} \max(|u(s) - v(s)|, \\
&\quad |\max_{0 \leq \xi \leq s} u(\xi) - \max_{0 \leq \xi \leq s} v(\xi)|) ds \\
&\leq \Phi(x) + \frac{l_f}{\Gamma(r)} \int_0^x (x-s)^{r-1} |u(s) - v(s)| ds.
\end{aligned}
$$

From Lemma 1.11, there exists a constant $\delta = \delta(r)$ such that

$$
\begin{aligned}
|u(x) - v(x)| &\leq \Phi(x) + \frac{\delta l_f}{\Gamma(r)} \int_0^x (x-s)^{r-1} \Phi(s) ds \\
&= \Phi(x) + \delta l_f (I_0^r \Phi)(x).
\end{aligned}
$$

Hence, by (4.13.1) for each $x \in I$, we get

$$
\begin{aligned}
|u(x) - v(x)| &\leq (1 + \delta l_f \lambda_\Phi) \Phi(x) \\
&:= c_{f,\Phi} \Phi(x).
\end{aligned}
$$

Finally, the fixed-point equation $u = N(u)$ is generalized Ulam–Hyers–Rassias stable. $\qquad\square$

4.6.3. *Example*

As an application of our results, we consider the following differential equation:

$$
(^c D_0^r u)(x) = \frac{e^{-x-3}}{1 + 5|u(x)| + \max_{0 \leq \xi \leq x} |u(\xi)|}; \quad x \in [0, 1], \qquad (4.30)
$$

with the initial condition

$$
u(0) = 1; \quad x \in [0, 1], \qquad (4.31)
$$

where $r \in (0, 1]$. Set

$$f(x, u(x), \max_{0 \leq \xi \leq x} u(\xi)) = \frac{1}{(1 + 5|u(x)| + \max_{0 \leq \xi \leq x} |u(\xi)|)e^{x+3}}; \quad x \in [0, 1].$$

We can see that the solutions of the problem (4.30)–(4.31) are solutions of the fixed-point equation $u = A(u)$ where $A : C([0, 1], \mathbb{R}) \to C([0, 1], \mathbb{R})$ is the operator defined by

$$(Au)(x) = 1 + \int_0^x \frac{(x - s)^{r-1}}{\Gamma(r)(1 + 5|u(x)| + \max_{0 \leq \xi \leq x} |u(\xi)|)e^{x+3}} ds; \quad x \in [0, 1],$$

For each $u, \overline{u}, \in \mathbb{R}$ and $x \in [0, 1]$, we have

$$|f(x, u(x), \max_{0 \leq \xi \leq x} u(\xi)) - f(x, \overline{u}(x), \max_{0 \leq \xi \leq x} \overline{u}(\xi))| \leq \frac{5}{e^3}|u - \overline{u}|.$$

Thus, condition (4.40.1) is satisfied with $l_f = 5e^{-3}$. We shall show that condition (4.29) holds with $a = 1$. Indeed, $\Gamma(1 + r) > \frac{1}{2}$. A simple computation shows that

$$L_f = \frac{l_f a^r}{\Gamma(1 + r)} = \frac{5}{e^3 \Gamma(1 + r)} < \frac{20}{e^3} < 1.$$

Hence, Theorem 4.13 implies that the problem (4.30)–(4.31) has a unique solution on $[0, 1]$. Moreover, the operator A is k-weakly Picard operator with the positive constant $k_N = \frac{1}{1 - L_f}$ and the fixed point equation $u = A(u)$ is Ulam–Hyers stable.

Finally, we can see that the hypothesis (4.13.1) is satisfied with $\Phi(x) = x^2$ and $\lambda_\Phi \leq \frac{2}{\Gamma(2+r)}$. Indeed, for each $x \in [0, 1]$, we get

$$(I_0^r \Phi)(x) = \frac{2}{\Gamma(2 + r)} x^{1+r} \leq \lambda_\Phi \Phi(x).$$

Consequently, Theorem 4.14 implies that the fixed point equation $u = A(u)$ is generalized Ulam–Hyers–Rassias stable.

4.7. Notes and Remarks

The results of Chapter 4 are taken from [36, 64, 74, 77].

Chapter 5

Random Hilfer Fractional Differential Equations and Inclusions

5.1. Introduction

Let E be a Banach space and $T : \Omega \times E \to E$ be a mapping. Then T is called a random operator if $T(w, u)$ is measurable in w for all $u \in E$ and it expressed as $T(w)u = T(w, u)$. In this case, we also say that $T(w)$ is a random operator on E. A random operator $T(w)$ on E is called continuous (respectively, compact, totally bounded and completely continuous) if $T(w, u)$ is continuous (respectively, compact, totally bounded and completely continuous) in u for all $w \in \Omega$. The details of completely continuous random operators in Banach spaces and their properties appear in [246].

Definition 5.1. Let $\mathcal{P}(Y)$ be the family of all nonempty subsets of Y and C be a mapping from Ω into $\mathcal{P}(Y)$. A mapping $T : \{(w, y) : w \in \Omega, \ y \in C(w)\} \to Y$ is called random operator with stochastic domain C if C is measurable (i.e., for all closed $A \subset Y$, $\{w \in \Omega, C(w) \cap A \neq \emptyset\}$ is measurable) and for all open $D \subset Y$ and all $y \in Y$, $\{w \in \Omega : y \in C(w), T(w, y) \in D\}$ is measurable. T will be called continuous if every $T(w)$ is continuous. For a random operator T, a mapping $y : \Omega \to Y$ is called random (stochastic) fixed point of T if for P-almost all $w \in \Omega$, $y(w) \in C(w)$ and $T(w)y(w) = y(w)$ and for all open $D \subset Y$, $\{w \in \Omega : y(w) \in D\}$ is measurable.

Definition 5.2. A function $T : \Omega \times \mathbb{R} \to \mathbb{R}$ is called jointly measurable if $T(\cdot, u)$ is measurable for all $u \in \mathbb{R}$ and $T(w, \cdot)$ is continuous for all $w \in \Omega$.

Definition 5.3. A function $f : I \times \mathbb{R} \times \Omega \to \mathbb{R}$ is called random Carathéodory if the following conditions are satisfied:

(i) The map $(t, w) \to f(t, u, w)$ is jointly measurable for all $u \in \mathbb{R}$.
(ii) The map $u \to f(t, u, w)$ is continuous for all $t \in I$ and $w \in \Omega$.

5.2. Random Hilfer and Hilfer–Hadamard Fractional Differential Equations

5.2.1. *Introduction*

We discuss the existence and the Ulam stability of solutions for the following problem of Random Hilfer fractional differential equations:

$$\begin{cases} (D_0^{\alpha,\beta}u)(t,w) = f(t,u(t,w),w); \ t \in I := [0,T], \\ (I_0^{1-\gamma}u)(t,w)|_{t=0} = \phi(w), \end{cases} \quad w \in \Omega, \quad (5.1)$$

where $\alpha \in (0,1)$, $\beta \in [0,1]$, $\gamma = \alpha + \beta - \alpha\beta$, $T > 0$, $(\Omega, \mathcal{A})$ is a measurable space, $\phi : \Omega \to \mathbb{R}$ is a measurable function, $f : I \times \mathbb{R} \times \Omega \to \mathbb{R}$ is a given function, $I_0^{1-\gamma}$ is the left-sided mixed Riemann–Liouville integral of order $1 - \gamma$, and $D_0^{\alpha,\beta}$ is the Hilfer fractional derivative of order α and type β.

Next, we consider the following problem of random Hilfer–Hadamard fractional differential equations:

$$\begin{cases} (^H D_1^{\alpha,\beta}u)(t,w) = g(t,u(t,w),w); \ t \in [1,T], \\ (^H I_1^{1-\gamma}u)(1,w) = \phi_0(w), \end{cases} \quad w \in \Omega, \quad (5.2)$$

where $\alpha \in (0,1)$, $\beta \in [0,1]$, $\gamma = \alpha + \beta - \alpha\beta$, $T > 1$, $\phi_0 : \Omega \to \mathbb{R}$ is a measurable function, $g : [1,T] \times \mathbb{R} \times \Omega \to \mathbb{R}$ is a given function, $^H I_1^{1-\gamma}$ is the left-sided mixed Hadamard integral of order $1 - \gamma$, and $^H D_1^{\alpha,\beta}$ is the Hilfer–Hadamard fractional derivative of order α and type β.

The present section initiates the Ulam stability for random differential equations involving Hilfer and Hilfer–Hadamard fractional derivatives.

5.2.2. *Hilfer fractional random differential equations*

In this section, we are concerned with the existence and the Ulam–Hyers–Rassias stability for problem (5.1). Let us start by defining what we mean by a random solution of the problem (5.1).

Definition 5.4. By a random solution of the problem (5.1) we mean a measurable function $u : \Omega \to C_\gamma$ that satisfies the condition $(I_0^{1-\gamma}u)(0^+, w) = \phi(w)$, and the equation $(D_0^{\alpha,\beta}u)(t,w) = f(t,u(t,w),w)$ on $I \times \Omega$.

Lemma 5.1. *Let $f : I \times \mathbb{R} \times \Omega \to \mathbb{R}$ be such that $f(\cdot, u(\cdot, w), w) \in C_\gamma$ for all $w \in \Omega$, and any $u(w) \in C_\gamma$. Then problem (5.1) is equivalent to the integral equation*

$$u(t, w) = \frac{\phi(w)}{\Gamma(\gamma)} t^{\gamma-1} + (I_0^\alpha f(\cdot, u(\cdot, w), w))(t); \quad w \in \Omega.$$

The following hypotheses will be used in the sequel.

(5.1.1) The function f is random Carathéodory on $I \times \mathbb{R} \times \Omega$.

(5.1.2) There exists a measurable and bounded function $p : \Omega \to L^\infty(I, [0, \infty))$ such that

$$|f(t, u, w)| \le p(t, w) \quad \text{for a.e. } t \in I, \text{ and each } u \in \mathbb{R}, \ w \in \Omega.$$

Set

$$p^* = \sup_{w \in \Omega} \|p(w)\|_{L^\infty} \quad \text{and} \quad \phi^* = \sup_{w \in \Omega} |\phi(w)|.$$

Now, we shall prove the following theorem concerning the existence of random solutions of problem (5.1).

Theorem 5.1. *Assume that the hypotheses (5.1.1) and (5.1.2) hold. Then the problem (5.1) has at least one random solution defined on $I \times \Omega$.*

Proof. Define a mapping $N : \Omega \times C_\gamma \to C_\gamma$ by:

$$(N(w)u)(t) = \frac{\phi(w)}{\Gamma(\gamma)} t^{\gamma-1} + \int_0^t (t-s)^{\alpha-1} \frac{f(s, u(s, w), w)}{\Gamma(\alpha)} ds. \quad (5.3)$$

The map ϕ is measurable for all $w \in \Omega$. Again, as the indefinite integral is continuous on I, then $N(w)$ defines a mapping $N : \Omega \times C_\gamma \to C_\gamma$. Thus u is a random solution for the problem (5.1) if and only if $u = N(w)u$.

Next, for any $u \in C_\gamma$, and each $t \in I$ and $w \in \omega$, we have

$$|t^{1-\gamma}(N(w)u)(t)| \le \frac{|\phi(w)|}{\Gamma(\gamma)} + \frac{t^{1-\gamma}}{\Gamma(\alpha)} \int_0^t (t-s)^{\alpha-1} |f(s, u(s, w), w)| ds$$

$$\le \frac{|\phi(w)|}{\Gamma(\gamma)} + \frac{t^{1-\gamma}}{\Gamma(\alpha)} \int_0^t (t-s)^{\alpha-1} p(s, w) ds$$

$$\le \frac{\phi^*}{\Gamma(\gamma)} + \frac{p^* T^{1-\gamma}}{\Gamma(\alpha)} \int_0^t (t-s)^{\alpha-1} ds$$

$$\le \frac{\phi^*}{\Gamma(\gamma)} + \frac{p^* T^{1-\gamma+\alpha}}{\Gamma(1+\alpha)}.$$

Thus

$$\|N(w)u\|_C \leq \frac{\phi^*}{\Gamma(\gamma)} + \frac{p^* T^{1-\gamma+\alpha}}{\Gamma(1+\alpha)} := R. \qquad (5.4)$$

This proves that $N(w)$ transforms the ball $B_R := B(0, R) = \{u \in C_\gamma : \|u\|_C \leq R\}$ into itself. We shall show that the operator $N : \Omega \times B_R \to B_R$ satisfies all the assumptions of Theorem 1.11. The proof will be given in several steps.

Step 1. $N(w)$ *is a random operator on* $\Omega \times B_R$ *into* B_R.

Since $f(t, u, w)$ is random Carathéodory, the map $w \to f(t, u, w)$ is measurable in view of Definition 5.2. Similarly, the product $(t-s)^{\alpha-1} f(s, u(s, w), w)$ of a continuous and a measurable function is again measurable. Further, the integral is a limit of a finite sum of measurable functions, therefore, the map

$$w \mapsto \frac{\phi(w)}{\Gamma(\gamma)} t^{\gamma-1} + \int_0^t \frac{(t-s)^{\alpha-1}}{\Gamma(\alpha)} f(s, u(s, w), w) ds,$$

is measurable. As a result, $N(w)$ is a random operator on $\Omega \times B_R$ into B_R.

Step 2. $N(w)$ *is continuous.*

Let $\{u_n\}_{n\in\mathbb{N}}$ be a sequence such that $u_n \to u$ in B_R. Then, for each $t \in I$ and $w \in \Omega$, we have

$$|t^{1-\gamma}(N(w)u_n)(t) - t^{1-\gamma}(N(w)u)(t)|$$

$$\leq \frac{t^{1-\gamma}}{\Gamma(\alpha)} \int_0^t (t-s)^{\alpha-1} |f(s, u_n(s, w), w) - f(s, u(s, w), w)| ds. \quad (5.5)$$

Since $u_n \to u$ as $n \to \infty$ and f is random Carathéeodory, then by the Lebesgue dominated convergence theorem, inequality (5.5) implies

$$\|N(w)u_n - N(w)u\|_C \to 0 \quad \text{as } n \to \infty.$$

Step 3. $N(w)B_R$ *is uniformly bounded.*

This is clear since $N(w)B_R \subset B_R$ and B_R is bounded.

Step 4. $N(w)B_R$ *is equicontinuous.*

Let $t_1, t_2 \in I$, $t_1 < t_2$ and let $u \in B_R$. Then, for each $w \in \Omega$, we have

$$|t_2^{1-\gamma}(N(w)u)(t_2) - t_1^{1-\gamma}(N(w)u)(t_1)|$$

$$\leq \left| t_2^{1-\gamma} \int_0^{t_2} (t_2 - s)^{\alpha-1} \frac{f(s, u(s, w), w)}{\Gamma(\alpha)} ds \right.$$

$$\left. - t_1^{1-\gamma} \int_0^{t_1} (t_1 - s)^{\alpha-1} \frac{f(s, u(s, w), w)}{\Gamma(\alpha)} ds \right|$$

$$\leq t_2^{1-\gamma} \int_{t_1}^{t_2} (t_2 - s)^{\alpha-1} \frac{|f(s, u(s, w), w)|}{\Gamma(\alpha)} ds$$

$$+ \int_0^{t_1} |t_2^{1-\gamma}(t_2 - s)^{\alpha-1} - t_1^{1-\gamma}(t_1 - s)^{\alpha-1}| \frac{|f(s, u(s, w), w)|}{\Gamma(\alpha)} ds$$

$$\leq t_2^{1-\gamma} \int_{t_1}^{t_2} (t_2 - s)^{\alpha-1} \frac{p(s, w)}{\Gamma(\alpha)} ds$$

$$+ \int_0^{t_1} |t_2^{1-\gamma}(t_2 - s)^{\alpha-1} - t_1^{1-\gamma}(t_1 - s)^{\alpha-1}| \frac{p(s.w)}{\Gamma(\alpha)} ds.$$

Thus, we get

$$|t_2^{1-\gamma}(N(w)u)(t_2) - t_1^{1-\gamma}(N(w)u)(t_1)|$$

$$\leq \frac{p^* T^{1-\gamma+\alpha}}{\Gamma(1+\alpha)}(t_2 - t_1)^\alpha$$

$$+ \frac{p^*}{\Gamma(\alpha)} \int_0^{t_1} |t_2^{1-\gamma}(t_2 - s)^{\alpha-1} - t_1^{1-\gamma}(t_1 - s)^{\alpha-1}| ds.$$

As $t_1 \to t_2$, the right-hand side of the above inequality tends to zero.

As a consequence of Steps 1–4 together with the Arzelà–Ascoli theorem, we can conclude that $N : \Omega \times B_R \to B_R$ is continuous and compact. From an application of Theorem 1.11, we deduce that the operator equation $N(w)u = u$ has a random solution. This implies that the random problem (5.1) has a random solution. ☐

Now, we are concerned with the generalized Ulam–Hyers–Rassias stability of our problem (5.1). Let $\epsilon > 0$ and $\Phi : I \times \Omega \to [0, \infty)$ be a continuous function. We consider the following inequalities

$$|(D_0^{\alpha,\beta}u)(t,w) - f(t,u(t,w),w)| \leq \epsilon; \quad t \in I, \ w \in \Omega, \qquad (5.6)$$

$$|(D_0^{\alpha,\beta}u)(t,w) - f(t,u(t,w),w)| \leq \Phi(t,w); \quad t \in I, \ w \in \Omega, \qquad (5.7)$$

$$|(D_0^{\alpha,\beta}u)(t,w) - f(t,u(t,w),w)| \leq \epsilon\Phi(t,w); \quad t \in I, \ w \in \Omega. \qquad (5.8)$$

Definition 5.5 ([72,329]). The problem (5.1) is Ulam–Hyers stable if there exists a real number $c_f > 0$ such that for each $\epsilon > 0$ and for each random solution $u : \Omega \to C_\gamma$ of the inequality (5.6) there exists a random solution $v : \Omega \to C_\gamma$ of (5.1) with

$$|u(t,w) - v(t,w)| \leq \epsilon c_f; \quad t \in I, \ w \in \Omega.$$

Definition 5.6 ([72,329]). The problem (5.1) is generalized Ulam–Hyers stable if there exists $c_f : C([0,\infty), [0,\infty))$ with $c_f(0) = 0$ such that for each $\epsilon > 0$ and for each random solution $u : \Omega \to C_\gamma$ of the inequality (5.6) there exists a random solution $v : \Omega \to C_\gamma$ of (5.1) with

$$|u(t,w) - v(t,w)| \leq c_f(\epsilon); \quad t \in I, \quad w \in \Omega.$$

Definition 5.7 ([72,329]). The problem (5.1) is Ulam–Hyers–Rassias stable with respect to Φ if there exists a real number $c_{f,\Phi} > 0$ such that for each $\epsilon > 0$ and for each random solution $u : \Omega \to C_\gamma$ of the inequality (5.8) there exists a random solution $v : \Omega \to C_\gamma$ of (5.1) with

$$|u(t,w) - v(t,w)| \leq \epsilon c_{f,\Phi}\Phi(t,w); \quad t \in I, \quad w \in \Omega.$$

Definition 5.8 ([72,329]). The problem (5.1) is generalized Ulam–Hyers–Rassias stable with respect to Φ if there exists a real number $c_{f,\Phi} > 0$ such that for each random solution $u : \Omega \to C_\gamma$ of the inequality (5.7) there exists a random solution $v : \Omega \to C_\gamma$ of (5.1) with

$$|u(t,w) - v(t,w)| \leq c_{f,\Phi}\Phi(t,w); \quad t \in I, \quad w \in \Omega.$$

Remark 5.1. It is clear that

(i) Definition 5.5 implies Definition 5.6;
(ii) Definition 5.7 implies Definition 5.8;
(iii) Definition 5.7 for $\Phi(\cdot,\cdot) = 1$ implies Definition 5.5.

One can have similar remarks for the inequalities (5.6) and (5.8).

Theorem 5.2. *Assume that the hypotheses (5.1.1), (5.1.2) and the following hypotheses hold.*

(5.2.1) *There exists $\lambda_\Phi > 0$ such that for each $t \in I$, and $w \in \Omega$, we have*
$$(I_0^\alpha \Phi)(t, w) \leq \lambda_\Phi \Phi(t, w).$$

(5.2.2) *There exists $q \in C(I, [0, \infty))$ such that for each $t \in I$, and $w \in \Omega$, we have*
$$p(t, w) \leq q(t)\Phi(t, w).$$

Then the problem (5.1) is generalized Ulam–Hyers–Rassias stable.

Proof. Consider the operator N defined in (5.3). Let u be a random solution of the inequality (5.7), and let us assume that v is a random solution of problem (5.1). Thus, we have
$$v(t, w) = \frac{\phi(w)}{\Gamma(\gamma)}t^{\gamma-1} + \int_0^t (t-s)^{\alpha-1}\frac{f(s, v(s, w), w)}{\Gamma(\alpha)}ds.$$
From the inequality (5.7) for each $t \in I$, and $w \in \Omega$, we have
$$\left| u(t, w) - \frac{\phi(w)}{\Gamma(\gamma)}t^{\gamma-1} - \int_0^t (t-s)^{\alpha-1}\frac{f(s, u(s, w), w)}{\Gamma(\alpha)}ds \right| \leq (I_0^\alpha \Phi)(t, w).$$
Set
$$q^* = \sup_{t \in I} q(t).$$
From hypotheses (5.2.1) and (5.2.2), for each $t \in I$, and $w \in \Omega$, we get
$$\begin{aligned}
|u(t, w) - v(t, w)| &\leq \left| u(t, w) - \frac{\phi(w)}{\Gamma(\gamma)}t^{\gamma-1} \right. \\
&\quad \left. - \int_0^t (t-s)^{\alpha-1}\frac{f(s, u(s, w), w)}{\Gamma(\alpha)}ds \right| \\
&\quad + \int_0^t (t-s)^{\alpha-1}\frac{|f(s, u(s, w), w) - f(s, v(s, w), w)|}{\Gamma(\alpha)}ds \\
&\leq (I_0^\alpha \Phi)(t, w) + \int_0^t (t-s)^{\alpha-1}\frac{2q^*\Phi(s, w)}{\Gamma(\alpha)}ds \\
&\leq (I_0^\alpha \Phi)(t) + 2q^*(I_0^\alpha \Phi)(t, w) \\
&\leq [1 + 2q^*]\lambda_\phi \Phi(t, w) \\
&:= c_{f,\Phi}\Phi(t, w).
\end{aligned}$$
Hence, the problem (5.1) is generalized Ulam–Hyers–Rassias stable. $\qquad \square$

Let $X = X(I, \mathbb{R})$ be the metric space, with the metric

$$d(u, v) = \sup_{t \in I} \frac{\|u(\cdot, w) - v(\cdot, w)\|_C}{\Phi(t, w)}.$$

Theorem 5.3. *Assume that (5.2.1) and the following hypothesis hold.*

(5.3.1) *There exists $\varphi \in C(I, [0, \infty))$ such that for each $t \in I$, $w \in \Omega$, and for all $u, v \in \mathbb{R}$, we have*

$$|f(t, u, w) - f(t, v, w)| \leq t^{1-\gamma} \varphi(t) \Phi(t, w)|u - v|.$$

If

$$L := T^{1-\gamma} \varphi^* \lambda_\phi < 1, \tag{5.9}$$

where $\varphi^* = \sup_{t \in I} \varphi(t)$, then there exists a unique random solution u_0 of problem (5.1), and the problem (5.1) is generalized Ulam–Hyers–Rassias stable. Furthermore, we have

$$|u(t, w) - u_0(t, w)| \leq \frac{\Phi(t, w)}{1 - L}.$$

Proof. Let N be the operator defined in (5.3). Apply Theorem 1.4, we have

$$|(N(w)u)(t) - (N(w)v)(t)|$$

$$\leq \int_0^t (t - s)^{\alpha-1} \frac{|f(s, u(s, w), w) - f(s, v(s, w), w)|}{\Gamma(\alpha)} ds$$

$$\leq \int_0^t (t - s)^{\alpha-1} \frac{\varphi(s)\Phi(s, w)|s^{1-\gamma}u(s, w) - s^{1-\gamma}v(s, w)|}{\Gamma(\alpha)} ds$$

$$\leq \int_0^t (t - s)^{\alpha-1} \frac{\varphi^* \Phi(s, w)\|u(w) - v(w)\|_C}{\Gamma(\alpha)} ds$$

$$\leq \varphi^* (I_0^\alpha \Phi)(t, w)\|u(w) - v(w)\|_C$$

$$\leq \varphi^* \lambda_\phi \Phi(t)\|u(w) - v(w)\|_C.$$

Thus

$$|t^{1-\gamma}(N(w)u)(t) - t^{1-\gamma}(N(w)v)(t)| \leq T^{1-\gamma} \varphi^* \lambda_\phi \Phi(t, w)\|u(w) - v(w)\|_C.$$

Hence, we get

$$d(N(u), N(v)) = \sup_{t \in I} \frac{\|(N(w)u)(t) - (N(w)v)(t)\|_C}{\Phi(t, w)} \leq L\|u(w) - v(w)\|_C,$$

from which we conclude the theorem. $\square$

5.2.3. *Hilfer–Hadamard fractional random differential equations*

Now, we are concerned with the existence and the Ulam–Hyers–Rassias stability for problem (5.2).

Set $C := C([1, T])$. Denote the weighted space of continuous functions defined by

$$C_{\gamma,\ln}([1, T]) = \{w(t) : (\ln t)^{1-\gamma} w(t) \in C\},$$

with the norm

$$\|w\|_{C_{\gamma,\ln}} := \sup_{t \in [1,T]} |(\ln t)^{1-\gamma} w(t)|.$$

From [321, Theorem 21], we concluded the following lemma.

Lemma 5.2. *Let* $g : I \times \mathbb{R} \times \Omega \to \mathbb{R}$ *be such that* $g(\cdot, u(\cdot, w), w) \in C_{\gamma,\ln}([1, T])$ *for any* $u(\cdot, w) \in C_{\gamma,\ln}([1, T])$. *Then the problem (5.2) is equivalent to the following volterra integral equation*

$$u(t, w) = \frac{\phi_0(w)}{\Gamma(\gamma)} (\ln t)^{\gamma-1} + ({}^H I_1^\alpha g(\cdot, u(\cdot, w), w))(t); \quad w \in \Omega.$$

Definition 5.9. By a random solution of the problem (5.2) we mean a measurable function $u \in C_{\gamma,\ln}$ that satisfies the condition $({}^H I_1^{1-\gamma} u)(1^+, w) = \phi_0(w)$, and the equation $({}^H D_1^{\alpha,\beta} u)(t, w) = g(t, u(t, w), w)$ on $[1, T] \times \Omega$.

Now we give (without proof) existence and Ulam slability results for problem (5.2). The following hypotheses will be used in the sequel.

(5.4.1) The function g is random Carathéodory on $[1, T] \times \mathbb{R} \times \Omega$.

(5.4.2) There exists a measurable and bounded function $p_1 : \Omega \to L^\infty([1, T], [0, \infty))$ such that

$$|g(t, u, w)| \le p_1(t, w) \quad \text{for a.e. } t \in [1, T], \text{ and each } u \in \mathbb{R}, \ w \in \Omega,$$

(5.4.3) There exists $\lambda_\Phi > 0$ such that for each $t \in [1, T]$, and $w \in \Omega$, we have

$$({}^H I_1^\alpha \Phi)(t, w) \le \lambda_\Phi \Phi(t, w),$$

(5.4.4) There exists $q_1 \in C(I, [0, \infty))$ such that for each $t \in I$, and $w \in \Omega$, we have

$$p_1(t, w) \le q_1(t) \Phi(t, w),$$

(5.4.5) There exists $\varphi_1 \in C([1,T], [0,\infty))$ such that for each $t \in [1,T]$, $w \in \Omega$, and for all $u, v \in \mathbb{R}$, we have

$$|g(t, u, w) - g(t, v, w)| \le (\ln t)^{1-\gamma} \varphi_1(t) \Phi(t, w) |u - v|.$$

Theorem 5.4. *Assume that the hypotheses (5.4.1) and (5.4.2) hold. Then the problem (5.2) has at least one random solution defined on $[1, T] \times \Omega$.*

Theorem 5.5. *Assume that the hypotheses (5.4.1)–(5.4.4) hold. Then the problem (5.2) is generalized Ulam–Hyers–Rassias stable.*

Theorem 5.6. *Assume that the hypotheses (5.4.3) and (5.4.5) hold. If*

$$L_1 := (\ln T)^{1-\gamma} \varphi_1^* \lambda_\phi < 1, \tag{5.10}$$

where $\varphi_1^* = \sup_{t \in [1,T]} \varphi(t)$, then there exists a unique random solution u_1 of problem (5.2), and the problem (5.2) is generalized Ulam–Hyers–Rassias stable. Furthermore, we have

$$|u(t, w) - u_1(t, w)| \le \frac{\Phi(t, w)}{1 - L_1}.$$

5.2.4. *Example*

Consider the following problem of Hilfer fractional differential equation:

$$\begin{cases} (D_0^{\frac{1}{2}, \frac{1}{2}} u)(t, w) = f(t, u(t, w)); & t \in [0, 1], \\ (I_0^{\frac{1}{4}} u)(t, w)|_{t=0} = 1, \end{cases} \tag{5.11}$$

where

$$\begin{cases} f(t, u) = \dfrac{ct^{\frac{-1}{4}} \sin(t, w)}{64(1 + \sqrt{t})(1 + |u|)}; & t \in (0, 1]; \quad u \in \mathbb{R}, \\ f(0, u) = 0; & u \in \mathbb{R}, \end{cases}$$

and $c = \frac{9\sqrt{\pi}}{16}$. Clearly, the function f is continuous.
The hypothesis (5.1.2) is satisfied with

$$\begin{cases} p(t, w) = \dfrac{ct^{\frac{-1}{4}} |\sin(t, w)|}{64(1 + \sqrt{t})}; & t \in (0, 1], \\ p(0, w) = 0. \end{cases}$$

Hence, Theorem 5.1 implies that the problem (5.11) has at least one solution defined on $[0, 1]$. Also, the hypothesis (5.12.1) is satisfied with

$$\Phi(t) = e^3, \text{ and } \lambda_\Phi = \frac{1}{\Gamma(1 + \alpha)}.$$

Indeed, for each $t \in [0, 1]$ we get

$$(I_0^\alpha \Phi)(t) \leq \frac{e^3}{\Gamma(1 + \alpha)}$$

$$= \lambda_\Phi \Phi(t).$$

Consequently, Theorem 5.2 implies that the problem (5.11) is generalized Ulam–Hyers–Rassias stable.

5.3. Random Hilfer and Hilfer–Hadamard Fractional Differential Inclusions

5.3.1. *Introduction*

This section deals with some existence and Ulam stability results for two classes of differential inclusions of Hilfer and Hilfer–Hadamard type with convex and nonconvex right-hand side. We employ some multivalued random fixed-point theorems for the existence of random solutions. Next we prove that our problems are generalized Ulam–Hyers–Rassias stable.

We discuss the existence and the Ulam stability of solutions for the following problem of Random Hilfer fractional differential inclusions:

$$\begin{cases} (D_0^{\alpha,\beta} u)(t, w) \in F(t, u(t, w), w); & t \in I := [0, T], \\ (I_0^{1-\gamma} u)(t, w)|_{t=0} = \phi(w), \end{cases} \quad w \in \Omega, \quad (5.12)$$

where $\alpha \in (0, 1)$, $\beta \in [0, 1]$, $\gamma = \alpha + \beta - \alpha\beta$, $T > 0$, $(\Omega, \mathcal{A})$ is a measurable space, $\phi : \Omega \to \mathbb{R}$ is a measurable function, $F : I \times \mathbb{R} \to \mathcal{P}(\mathbb{R})$ is a given multivalued map, $\mathcal{P}(\mathbb{R})$ is the family of all nonempty subsets of $\mathbb{R}$, $I_0^{1-\gamma}$ is the left-sided mixed Riemann–Liouville integral of order $1 - \gamma$, and $D_0^{\alpha,\beta}$ is the Hilfer fractional derivative of order α and type β.

Next, we consider the following problem of random Hilfer–Hadamard fractional differential inclusions:

$$\begin{cases} (^H D_1^{\alpha,\beta} u)(t, w) \in G(t, u(t, w), w); & t \in [1, T], \\ (^H I_1^{1-\gamma} u)(1, w) = \phi_0(w), \end{cases} \quad w \in \Omega, \quad (5.13)$$

where $\alpha \in (0,1)$, $\beta \in [0,1]$, $\gamma = \alpha + \beta - \alpha\beta$, $T > 1$, $\phi_0 : \Omega \to \mathbb{R}$ is a measurable function, $G : [1,T] \times \mathbb{R} \to \mathcal{P}(\mathbb{R})$ is a given multivalued map, $^H I_1^{1-\gamma}$ is the left-sided mixed Hadamard integral of order $1 - \gamma$, and $^H D_1^{\alpha,\beta}$ is the Hilfer–Hadamard fractional derivative of order α and type β.

5.3.2. *Existence of random solutions*

Let us start by defining what we mean by a random solution of the problem (5.12). Let C be the Banach space of all continuous functions v from I into $\mathbb{R}$ with the supremum (uniform) norm

$$\|v\|_\infty := \sup_{t \in I} |v(t)|.$$

As usual, $\mathrm{AC}(I)$ denotes the space of absolutely continuous functions from I into $\mathbb{R}$. We denote by $\mathrm{AC}^1(I)$ the space defined by

$$\mathrm{AC}^1(I) := \left\{ w : I \to \mathbb{R} : \frac{d}{dt}w(t) \in \mathrm{AC}(I) \right\}.$$

By $L^1(I)$, we denote the space of Bochner integrable functions $v : I \to \mathbb{R}$ with the norm

$$\|v\|_1 = \int_0^T |v(t)| dt.$$

Let $L^\infty(I)$ be the Banach space of measurable functions $u : I \to \mathbb{R}$ which are essentially bounded, equipped with the norm

$$\|u\|_{L^\infty} = \inf\{c > 0 : |u(t)| \le c, \text{ a.e. } t \in I\}.$$

By $C_\gamma(I)$ and $C_\gamma^1(I)$, we denote the weighted spaces of continuous functions defined by

$$C_\gamma(I) = \{w : (0,T] \to \mathbb{R} : t^{1-\gamma}w(t) \in C\},$$

with the norm

$$\|w\|_{C_\gamma} := \sup_{t \in I} |t^{1-\gamma}w(t)|,$$

and

$$C_\gamma^1(I) = \left\{ w \in C : \frac{dw}{dt} \in C_\gamma \right\},$$

with the norm

$$\|w\|_{C_\gamma^1} := \|w\|_\infty + \|w'\|_{C_\gamma}.$$

In what follows, we denote $\|w\|_{C_\gamma}$ by $\|w\|_C$.

For each $u \in C_\gamma$ and $w \in \Omega$, define the set of selections of F by

$$S_{Fou}(w) = \{v : \Omega \to L^1(I) : v(t, w) \in F(t, u(t, w), w); \ t \in I\}.$$

Let E be a Banach space, and denote $P_{cl}(E) = \{A \in \mathcal{P}(E) : A \ \text{closed}\}$, $P_{cp,c}(E) = \{A \in \mathcal{P}(E) : A \ \text{compact and convex}\}$.

Consider $H_d : \mathcal{P}(E) \times \mathcal{P}(E) \to [0, \infty) \cup \{\infty\}$ given by

$$H_d(A, B) = \max\left\{\sup_{a \in A} d(a, B), \sup_{b \in B} d(A, b)\right\},$$

where $d(A, b) = \inf_{a \in A} d(a, b)$, $d(a, B) = \inf_{b \in B} d(a, b)$. Then $(\mathcal{P}_{bd,cl}(E), H_d)$ is a Hausdorff metric space.

Definition 5.10. A multifunction $F : \Omega \to E$ is called $\mathcal{A}$- measurable if, for any open subset B of E, the set $F^{-1}(B) = \{w \in \Omega : F(w) \cap B \neq \emptyset\} \in \mathcal{A}$. Note that if $F(w) \in \mathcal{P}_{cl}(E)$ for all $w \in \Omega$, then F is measurable if and only if $F^{-1}(D) \in \mathcal{A}$ for all $D \in \mathcal{P}_{cl}(E)$. A measurable operator $u : \Omega \to E$ is called a measurable selector for a measurable multifunction $F : \Omega \to E$, if $u(w) \in F(w)$. Let $M \in \mathcal{P}_{cl}(E)$, then a mapping $f : \Omega \times M \to E$ is called a random operator if, for each $u \in M$, the mapping $f(\cdot, u) : \Omega \to E$ is measurable. An operator $u : \Omega \to E$ is said to be a random fixed point of F if u is measurable and $u(w) \in F(w, u(w))$ for all $w \in \Omega$.

Definition 5.11. A multifunction $F : \Omega \times E \to \mathcal{P}(E)$ is called Carathéodory if $F(\cdot, u)$ is measurable for all $u \in E$ and $F(w, \cdot)$ is continuous for all $w \in \Omega$.

Definition 5.12. A multivalued map $F : I \times E \times \Omega \to \mathcal{P}_{cp}(E)$ is said to be random Carathéodory if

(i) $(t, w) \longmapsto F(t, u, w)$ is jointly measurable for each $u \in E$; and
(ii) $u \longmapsto F(t, u, w)$ is Hausdorff continuous for a.e. $t \in I$, $w \in \Omega$.

Definition 5.13 ([223]). Let E be a separable Banach space. If $F : I \times E \to \mathcal{P}_{cp}(E)$ is Carathéodory, then the multivalued mapping $(t, u(t)) \to F(t, u(t))$ is jointly measurable for any measurable E-valued function u on I.

Definition 5.14. A multivalued random operator $N : \Omega \times E \to \mathcal{P}_{cl}(E)$ is called multivalued random contraction if there is a measurable function $k : \Omega \to [0, \infty)$ such that

$$H_d(N(w)u, N(w)v) \leq k(w)\|u - v\|_E,$$

for all $u, v \in E$ and $w \in \Omega$, where $k(w) \in [0, 1)$ on Ω.

Definition 5.15. By a random solution of the problem (5.12) we mean a measurable function $u : \Omega \to C_\gamma$ that satisfies the condition $(I_0^{1-\gamma}u)(0^+, w) = \phi(w)$, and the equation $(D_0^{\alpha,\beta}u)(t, w) = v(t, w)$ on $I \times \Omega$, where $v \in S_{Fou}(w)$.

5.3.2.1. *The convex case*

We present now some existence and Ulam stability results for the problem (5.12) with convex-valued right-hand side.

The following hypotheses will be used in the sequel.

(5.5.1) The multifunction $F : I \times \mathbb{R} \times \Omega \to \mathcal{P}_{cp,cv}(\mathbb{R})$ is strong random Carathéodory on $I \times \mathbb{R} \times \Omega$.

(5.5.2) There exists a measurable and bounded function $l : \Omega \to L^\infty(I, [0, \infty))$ satisfying for each $w \in \Omega$,

$$H_d(F(t, u, w), F(t, \overline{u}, w)) \leq t^{1-\gamma}l(t, w)|u - \overline{u}|$$

for every $t \in I$ and $u, \overline{u} \in \mathbb{R}$,

and

$$d(0, F(t, 0, w)) \leq t^{1-\gamma}l(t, w) \quad \text{for } t \in I.$$

(5.5.3) There exists $\lambda_\Phi > 0$ such that for each $t \in I$, and $w \in \Omega$, we have

$$\int_0^t \left[\sum_{n=1}^\infty \frac{(l^*)^n}{\Gamma(n\alpha)}(t - s)^{n\alpha-1}\Phi(s, w) \right] ds \leq \lambda_\Phi \Phi(t, w).$$

Remark 5.2. For each $u : \Omega \to \mathcal{C}$, the set $S_{F,u}(w)$ is nonempty since by (5.5.1), F has a measurable selection (see [180, Theorem III.6]).

Remark 5.3. The hypothesis (5.5.2) implies that, for every $t \in I$, $u \in \mathbb{R}$ and $w \in \Omega$, we get

$$H_d(F(t, u, w), F(t, 0, w)) \leq l(t, w)|u|,$$

and

$$H_d(0, F(t, u, w)) \leq H_d(0, F(t, 0, w)) + H_d(F(t, u, w), F(t, 0, w))$$
$$\leq l(t, w)(1 + |u|).$$

Theorem 5.7. *Assume that the hypotheses (5.5.1) and (5.5.2) hold. Then, the problem (5.12) has a random solution defined on $I \times \Omega$.*

Proof. Set

$$l^* = \sup_{w \in \Omega} \|l(w)\|_{L^\infty} \text{ and } \phi^* = \sup_{w \in \Omega} |\phi(w)|.$$

Define a multivalued operator $N : \Omega \times C_\gamma \to \mathcal{P}(C_\gamma)$ by

$$(N(w)u)(t) = \left\{ h : \Omega \to C_\gamma : h(t, w) = \frac{\phi(w)}{\Gamma(\gamma)} t^{\gamma-1} \right.$$
$$\left. + (I_0^\alpha v)(t, w); \ t \in I, \ v \in S_{Fou}(w) \right\}. \tag{5.14}$$

The map ϕ is measurable for all $w \in \Omega$. Again, as the indefinite integral is continuous on I, for each $v \in S_{Fou}(w)$, then $N(w)$ defines a multivalued mapping $N : \Omega \times C_\gamma \to \mathcal{P}(C_\gamma)$. Thus u is a random solution for the problem (5.12) if and only if $u \in N(w)u$.

We shall show that the multivalued operator N satisfies all conditions of Theorem 1.13. The proof will be given in several steps.

Step 1. $N(w)$ *is a multivalued random operator on C.*

Since $F(t, u, w)$ is strong random Carathéodory, the map $w \to F(t, u, w)$ is measurable in view of Definition 5.13. Similarly, the product $(t - s)^{\alpha-1} v(s, w)$ of a continuous function and a measurable multifunction is again measurable for each $v \in S_{Fou}(w)$. Further, the integral is a limit of a finite sum of measurable functions, therefore, the map

$$w \mapsto \frac{\phi(w)}{\Gamma(\gamma)} t^{\gamma-1} + \int_0^t \frac{(t - s)^{\alpha-1}}{\Gamma(\alpha)} v(t, w) ds,$$

is measurable. As a result, $N(w)$ is a multivalued random operator on C_γ.

Step 2. $N(w)u \in \mathcal{P}_{cv}(C_\gamma)$ *for each* $u \in C_\gamma$.

Indeed, if h_1, h_2 belong to $N(w)u$, then there exist $v_1, v_2 \in S_{Fou}(w)$ such that for each $t \in I$ and $w \in \Omega$, we have

$$h_i(t, w) = \frac{\phi(w)}{\Gamma(\gamma)} t^{\gamma-1} + (I_0^\alpha v_i)(t, w); \quad i = 1, 2.$$

Let $0 \le d \le 1$. Then, for each $t \in I$ and $w \in \Omega$, we get

$$(dh_1 + (1-d)h_2)(t, w) = \frac{\phi(w)}{\Gamma(\gamma)} t^{\gamma-1} + (I_0^\alpha [dv_1 + (1-d)v_2])(t, w).$$

Since $S_{Fou}(w)$ is convex (because F has convex values), we get

$$dh_1 + (1-d)h_2 \in N(u).$$

Step 3. $N(w)$ *is continuous and* $N(w)u \in \mathcal{P}_{cp}(C_\gamma)$ *for each* $u \in C_\gamma$.

The proof of this step will be given in several claims.

Claim 1: $N(w)$ *is continuous.*
Let $\{u_n\}$ be a sequence such that $u_n \to u$ in C_γ. Then from (5.5.2), for each $t \in I$ and $w \in \Omega$, we have

$$H_d(F(t, u_n(t, w), w), F(t, u(t, w), w)) \le t^{1-\gamma} l(t, w) |u_n(t, w) - u(t, w)|$$

$$\le l^* \|u_n - u\|_C \to 0 \text{ as } n \to \infty.$$

Thus, we obtain

$$H_d(F(t, u_n(t, w), w), F(t, u(t, w), w)) \to 0 \text{ as } n \to \infty.$$

Claim 2: $N(w)$ *maps bounded sets into bounded sets in* C_γ.
Let $B_{\eta^*} = \{u \in C_\gamma : \|u\|_C \le \eta^*\}$ be bounded set in C_γ, and $u \in B_{\eta^*}$. Then for each $h \in N(w)u$, there exists $v \in S_{Fou}(w)$ such that

$$h(t, w) = \frac{\phi(w)}{\Gamma(\gamma)} t^{\gamma-1} + (I_0^\alpha v)(t, w).$$

By (5.5.2), for each $t \in I$ and $w \in \Omega$, we obtain

$$|t^{1-\gamma} h(t, w)| \le \frac{|\phi(w)|}{\Gamma(\gamma)} + T^{1-\gamma} \int_0^t \frac{(t-s)^{\alpha-1}}{\Gamma(\alpha)} |v(s, w)| ds$$

$$\le \frac{|\phi(w)|}{\Gamma(\gamma)} + T^{1-\gamma} \int_0^t \frac{(t-s)^{\alpha-1}}{\Gamma(\alpha)} |s^{1-\gamma} l(s, w)(1 + v(s, w))| ds$$

$$\leq \frac{\phi^*}{\Gamma(\gamma)} + l^* T^{1-\gamma} \int_0^t \frac{(t-s)^{\alpha-1}}{\Gamma(\alpha)} (1 + \|v\|_C) ds$$

$$\leq \frac{\phi^*}{\Gamma(\gamma)} + \frac{l^* T^{1+\alpha-\gamma}}{\Gamma(1+\alpha)} (1 + \eta^*) := \ell.$$

Claim 3: $N(w)$ *maps bounded sets into equicontinuous sets in* C_γ. Let $t_1, t_2 \in I$, $t_1 < t_2$, and let B_{η^*} be a bounded set of C_γ as in claim 2, and let $u \in B_{\eta^*}$ and $h \in N(w)u$. Then, there exists $v \in S_{Fou}(w)$ such that for each $w \in \Omega$, we get

$$|t_2^{1-\gamma} h(t_2, w) - t_1^{1-\gamma} h(t_1, w)|$$

$$\leq \frac{l^* T^{1-\gamma+\alpha}}{\Gamma(1+\alpha)} (t_2 - t_1)^\alpha$$

$$+ \frac{l^*(T^{1-\gamma} + \eta^*)}{\Gamma(\alpha)} \int_0^{t_1} |t_2^{1-\gamma}(t_2 - s)^{\alpha-1} - t_1^{1-\gamma}(t_1 - s)^{\alpha-1} ds.$$

As $t_1 \to t_2$, the right-hand side of the above inequality tends to zero. As a consequence of claims 1–3, together with the Arzelà–Ascoli theorem, we can conclude that $N(w)$ is continuous and completely continuous multi-valued random operator.

Step 4: *The set* $\mathcal{E} := \{u \in C_\gamma : \lambda u \in N(w)u\}$ *is bounded for some* $\lambda > 1$.

Let $u \in \mathcal{M}$ be arbitrary and let $w \in \Omega$ be fixed such that $\lambda u \in N(w)u$ for all $\lambda > 1$. Then, there exists $v \in S_{Fou}(w)$ such that for each $t \in I$, we have

$$u(t, w) = \frac{\phi(w)}{\lambda \Gamma(\gamma)} t^{\gamma-1} + \lambda^{-1}(I_0^\alpha v)(t, w).$$

This implies by (5.5.2) that, for each $t \in I$, we get

$$|t^{1-\gamma} u(t, w)| \leq \frac{\phi^*}{\Gamma(\gamma)} + \int_0^t \frac{(t-s)^{\alpha-1}}{\Gamma(\alpha)} l(s, w)(T^{1-\gamma} + |s^{1-\gamma} v(s, w)|) ds$$

$$\leq \frac{\phi^*}{\Gamma(\gamma)} + \frac{l^* T^{1-\gamma+\alpha}}{\Gamma(1+\alpha)} + l^* \int_0^t \frac{(t-s)^{\alpha-1}}{\Gamma(\alpha)} |s^{1-\gamma} v(s, w)| ds.$$

From Lemma 1.11, for each $(t, w) \in [0, T] \times \Omega$, we have

$$|t^{1-\gamma} u(t, w)| \leq \left[\frac{\phi^*}{\Gamma(\gamma)} + \frac{l^* T^{1-\gamma+\alpha}}{\Gamma(1+\alpha)} \right] \left[1 + \int_0^t \left[\sum_{n=1}^\infty \frac{(l^*)^n}{\Gamma(n\alpha)} (t-s)^{n\alpha-1} \right] ds \right]$$

$$\leq \left[\frac{\phi^*}{\Gamma(\gamma)} + \frac{l^* T^{1-\gamma+\alpha}}{\Gamma(1+\alpha)} \right] \left[1 + \sum_{n=1}^\infty \frac{T^{n\alpha}}{\Gamma(1+n\alpha)} \right] := M.$$

Thus, for all $t \in I$ and $w \in \Omega$, we obtain $\|u\|_\infty \leq M$.

As a consequence of Steps 1–4, together with Theorem 1.13, N has a random fixed point u which is a random solution to problem (5.12). □

Now, we consider the Ulam stability for the problem (5.12). Let $\epsilon > 0$ and $\Phi : I \times \Omega \to [0, \infty)$ be a continuous function. We consider the following inequalities:

$$H_d((D_0^{\alpha,\beta} u)(t, w), F(t, u(t, w), w)) \leq \epsilon; \quad t \in I, \ w \in \Omega, \quad (5.15)$$

$$H_d((D_0^{\alpha,\beta} u)(t, w), F(t, u(t, w), w)) \leq \Phi(t, w); \quad t \in I, \ w \in \Omega, \quad (5.16)$$

$$H_d((D_0^{\alpha,\beta} u)(t, w), F(t, u(t, w), w)) \leq \epsilon\Phi(t, w); \quad t \in I, \ w \in \Omega. \quad (5.17)$$

Definition 5.16 ([71,329]). The problem (5.12) is Ulam–Hyers stable if there exists a real number $c_F > 0$ such that for each $\epsilon > 0$ and for each random solution $u : \Omega \to C_\gamma$ of the inequality (5.15) there exists a random solution $v : \Omega \to C_\gamma$ of (5.12) with

$$|u(t, w) - v(t, w)| \leq \epsilon c_F; \quad t \in I, \ w \in \Omega.$$

Definition 5.17 ([71,329]). The problem (5.12) is generalized Ulam–Hyers stable if there exists $c_F : C([0, \infty), [0, \infty))$ with $c_F(0) = 0$ such that for each $\epsilon > 0$ and for each random solution $u : \Omega \to C_\gamma$ of the inequality (5.15) there exists a random solution $v : \Omega \to C_\gamma$ of (5.12) with

$$|u(t, w) - v(t, w)| \leq c_F(\epsilon); \ t \in I, \ w \in \Omega.$$

Definition 5.18 ([71,329]). The problem (5.12) is Ulam–Hyers–Rassias stable with respect to Φ if there exists a real number $c_{F,\Phi} > 0$ such that for each $\epsilon > 0$ and for each random solution $u : \Omega \to C_\gamma$ of the inequality (5.17) there exists a random solution $v : \Omega \to C_\gamma$ of (5.12) with

$$|u(t, w) - v(t, w)| \leq \epsilon c_{F,\Phi} \Phi(t, w); \quad t \in I, \ w \in \Omega.$$

Definition 5.19 ([71,329]). The problem (5.12) is generalized Ulam–Hyers–Rassias stable with respect to Φ if there exists a real number $c_{F,\Phi} > 0$ such that for each random solution $u : \Omega \to C_\gamma$ of the inequality (5.16), there exists a random solution $v : \Omega \to C_\gamma$ of (5.12) with

$$|u(t, w) - v(t, w)| \le c_{F,\Phi} \Phi(t, w); \ t \in I, \ w \in \Omega.$$

Remark 5.4. It is clear that

(i) Definition 5.16 implies Definition 5.17;
(ii) Definition 5.18 implies Definition 5.19;
(iii) Definition 5.18 for $\Phi(\cdot, \cdot) = 1$ implies Definition 5.16.

One can have similar remarks for the inequalities (5.15) and (5.17).

Theorem 5.8. *Assume that the hypotheses (5.5.1)–(5.5.3) hold. Then the problem (5.12) is generalized Ulam–Hyers–Rassias stable.*

Proof. Let u be a random solution of the inequality (5.16), and let us assume that v is a random solution of problem (5.12). Thus, we have

$$v(t, w) = \frac{\phi(w)}{\Gamma(\gamma)} t^{\gamma-1} + \int_0^t (t - s)^{\alpha-1} \frac{f_v(s, w)}{\Gamma(\alpha)} ds,$$

where $f_v \in S_{Fov}(w)$. From the inequality (5.16) for each $t \in I$, and $w \in \Omega$, we have

$$\left| u(t, w) - \frac{\phi(w)}{\Gamma(\gamma)} t^{\gamma-1} - \int_0^t (t - s)^{\alpha-1} \frac{f(s, w)}{\Gamma(\alpha)} ds \right| \le (I_0^\alpha \Phi)(t, w),$$

where $f \in S_{Fou}(w)$. From hypotheses (5.5.2) and (5.4.3), for each $t \in I$, and $w \in \Omega$, we get

$$|u(t, w) - v(t, w)| \le \left| u(t, w) - \frac{\phi(w)}{\Gamma(\gamma)} t^{\gamma-1} - \int_0^t (t - s)^{\alpha-1} \frac{f(s, w)}{\Gamma(\alpha)} ds \right|$$

$$+ \int_0^t (t - s)^{\alpha-1} \frac{|f(s, w) - f_v(s, w)|}{\Gamma(\alpha)} ds$$

$$\le (I_0^\alpha \Phi)(t, w) + \frac{l^* T^{1-\gamma}}{\Gamma(\alpha)} \int_0^t (t - s)^{\alpha-1} |u(s, w)$$

$$- v(s, w)| ds.$$

From Lemma 1.14, we obtain

$$\|u(t,w) - v(t,w)\| \leq \frac{\lambda_\phi}{l^*} \left[\Phi(t,w) + \int_0^t \left[\sum_{n=1}^\infty \frac{(l^*)^n}{\Gamma(n\alpha)} (t-s)^{n\alpha-1} \Phi(s,w) \right] ds \right]$$

$$\leq \frac{\lambda_\phi}{l^*} [1 + \lambda_\Phi] \Phi(t,w)$$

$$:= c_{F,\Phi} \Phi(t,w).$$

Finally, the problem (5.12) is generalized Ulam–Hyers–Rassias stable. □

5.3.2.2. *The nonconvex case*

We present now some existence and Ulam stabilities results for the problem (5.12) with nonconvex-valued right-hand side.

The following hypotheses will be used in the sequel.

(5.6.1) The multifunction $F : I \times \mathbb{R} \times \Omega \to \mathcal{P}_{cp}(\mathbb{R})$ is strong random Carathéodory on $I \times \mathbb{R} \times \Omega$.

(5.6.2) There exists a measurable and bounded function $l : \Omega \to L^\infty(I, [0, \infty))$ satisfying for each $w \in \Omega$.

$$H_d(F(t,u,w), F(t,\overline{u},w)) \leq t^{1-\gamma} l(t,w)|u - \overline{u}|$$

for every $t \in I$ and $u, \overline{u} \in \mathbb{R}$.

Set

$$l^* = \sup_{w \in \Omega} \|l(w)\|_{L^\infty}.$$

Now, we shall prove the following theorem concerning the existence of random solutions of problem (5.12).

Theorem 5.9. *Assume that the hypotheses (5.6.1) and (5.6.2) hold. If*

$$\frac{l^* T^{1+\alpha-\gamma}}{\Gamma(1+\alpha)} < 1, \tag{5.18}$$

Then the problem (5.12) has at least one random solution defined on $I \times \Omega$.

Proof. Let $N : \Omega \times C_\gamma \to \mathcal{P}(C_\gamma)$ be the multivalued operator defined in (5.14). We know that $N(w)$ is a multivalued random operator on C_γ. We shall show that the multivalued operator N satisfies all conditions of Theorem 1.14. The proof will be given in two steps.

Step 1. $N(w)u \in \mathcal{P}_{cl}(C_\gamma)$ *for each* $u \in C_\gamma$.

Let $\{u_n\}_{n \geq 0} \in N(w)u$ be such that $u_n \to \tilde{u}$ in C_γ. Then, $\tilde{u} \in C_\gamma$ and there exists $f_n(\cdot, \cdot, \cdot) \in S_{Fou}(w)$ such that, for each $t \in I$ and $w \in \Omega$, we have

$$u_n(t, w) = \frac{\phi(w)}{\Gamma(\gamma)} t^{\gamma-1} + (I_0^\alpha f_n)(t, w).$$

Using the fact that F has compact values and from (5.6.1), we may pass to a subsequence if necessary to get that $f_n(\cdot, \cdot, \cdot)$ converges to f in $L^1(I)$, and hence $f \in S_{Fou}(w)$. Then, for each $t \in I$ and $w \in \Omega$, we get

$$u_n(t, w) \to \tilde{u}(t, w) = \frac{\phi(w)}{\Gamma(\gamma)} t^{\gamma-1} + (I_0^\alpha f)(t, w).$$

So, $\tilde{u} \in N(w)u$.

Step 2. *There exists* $0 \leq \lambda < 1$ *such that, for each* $w \in \Omega$,

$$H_d(N(w)u, N(w)\overline{u}) \leq \lambda \|u - \overline{u}\|_C \text{ for each } u, \overline{u} \in C_\gamma.$$

Let $u, \overline{u} \in C_\gamma$ and $h \in N(w)u$. Then, there exists $f(t, w) \in F(t, u(t, w), w)$ such that for each $t \in I$ and $w \in \Omega$, we have

$$h(t, w) = \frac{\phi(w)}{\Gamma(\gamma)} t^{\gamma-1} + (I_0^\alpha f)(t, w).$$

From (5.6.2) it follows that

$$H_d(F(t, u(t, w), w), F(t, \overline{u}(t, w), w)) \leq t^{1-\gamma} l(t, w)|u(t, w) - \overline{u}(t, w)|.$$

Hence, there exists $v \in S_{Fou}$ such that

$$|f(t, w) - v(t, w)| \leq t^{1-\gamma} l(t, w)|u(t, w) - \overline{u}(t, w)|.$$

Consider $U : I \times \Omega \to \mathcal{P}(\mathbb{R})$ given by

$$U(t, w) = \{v(t, w) \in \mathbb{R} : |f(t, w) - v(t, w)| \leq t^{1-\gamma} l(t, w)|u(t, w) - \overline{u}(t, w)|\}.$$

Since the multivalued operator $u(t, w) = U(t, w) \cap F(t, \overline{u}(t, w), w)$ is measurable (see [180, Proposition III.4]), there exists a function $\overline{f}(t, w)$ which is a measurable selection for $\overline{u}$. So, $\overline{f}(t, w) \in F(t, \overline{u}(t, w), w)$, and for each $t \in I$ and $w \in \Omega$, we get

$$|f(t, w) - \overline{f}(t, w)| \leq t^{1-\gamma} l(t, w)|u(t, w) - \overline{u}(t, w)|.$$

Let us define for each $t \in I$ and $w \in \Omega$,

$$\overline{h}(t, w) = \frac{\phi(w)}{\Gamma(\gamma)} t^{\gamma-1} + (I_0^\alpha \overline{f})(t, w).$$

Then for each $t \in I$ and $w \in \Omega$, we obtain

$$|t^{1-\gamma}h(t,w) - t^{1-\gamma}\overline{h}(t,w)| \leq t^{1-\gamma} I_0^\alpha |f(t,w) - \overline{f}(t,w)|$$

$$\leq \frac{T^{1-\gamma}}{\Gamma(\alpha)} \int_0^t (t-s)^{\alpha-1} l(s,w)$$

$$\times |s^{1-\gamma}u(s,w) - s^{1-\gamma}\overline{u}(s,w)| ds$$

$$\leq \frac{l^* T^{1-\gamma} \|u - \overline{u}\|_C}{\Gamma(\alpha)} \int_0^t (t-s)^{\alpha-1} ds.$$

Hence

$$\|h - \overline{h}\|_C \leq \frac{l^* T^{1+\alpha-\gamma}}{\Gamma(1+\alpha)} \|u - \overline{u}\|_C.$$

By an analogous relation, obtained by interchanging the roles of u and $\overline{u}$, it follows that

$$H_d(N(w)u, N(w)\overline{u}) \leq \frac{l^* T^{1+\alpha-\gamma}}{\Gamma(1+\alpha)} \|u - \overline{u}\|_C.$$

So by Eq. (5.18), N is random contraction and thus, by Theorem 1.14, N has a random fixed point u which is a random solution to problem (5.12). $\square$

Now, we can show the following generalized Ulam–Hyers–Rassias stability result.

Theorem 5.10. *Assume that the hypotheses (5.6.1), (5.6.2) and the condition (5.18) hold, then the problem (5.12) is generalized Ulam–Hyers–Rassias stable.*

Set $C := C([1, T])$. Denote the weighted space of continuous functions defined by

$$C_{\gamma, \ln}([1, T]) = \{w(t) : (\ln t)^{1-\gamma} w(t) \in C\},$$

with the norm

$$\|w\|_{C_{\gamma, \ln}} := \sup_{t \in [1, T]} |(\ln t)^{1-\gamma} w(t)|.$$

From [321, Theorem 21], we concluded the following lemma.

Lemma 5.3. *Let* $G : [1, T] \times \mathbb{R} \times \Omega \to \mathcal{P}(\mathbb{R})$ *be such that* $S_{G \circ u}(w) \in C_{\gamma, \ln}([1, T])$ *for any* $u(\cdot, w) \in C_{\gamma, \ln}([1, T])$. *Then problem (5.13) is equivalent to the following volterra integral equation:*

$$u(t, w) = \frac{\phi_0(w)}{\Gamma(\gamma)} (\ln t)^{\gamma - 1} + ({}^H I_1^\alpha g(\cdot, w))(t); \quad w \in \Omega,$$

where $g \in S_{G \circ u}(w)$.

Definition 5.20. By a random solution of the problem (5.13), we mean a measurable function $u \in C_{\gamma, \ln}$ that satisfies the condition $({}^H I_1^{1-\gamma} u)(1^+, w) = \phi_0(w)$, and the equation $({}^H D_1^{\alpha, \beta} u)(t, w) = g(t, w)$ on $[1, T] \times \Omega$, where $g \subset S_{G \circ u}(w)$.

Now we give (without proof) existence and Ulam slability results for problem (5.13). The following hypotheses will be used in the sequel.

(5.7.1) The multifunction $G : [1, T] \times \mathbb{R} \times \Omega \to \mathcal{P}_{cp,cv}(\mathbb{R})$ is strong random Carathéodory.
(5.7.2) There exists a measurable and bounded function $l : \Omega \to L^\infty([1, T], [0, \infty))$ satisfying for each $w \in \Omega$,

$$H_d(G(t, u, w), G(t, \overline{u}, w)) \le t^{1-\gamma} l(t, w) |u - \overline{u}|;$$

$$\text{for every } t \in [1, T] \text{ and } u, \overline{u} \in \mathbb{R},$$

and

$$d(0, G(t, 0, w)) \le (\ln t)^{1-\gamma} l(t, w); \text{ for } t \in [1, T].$$

(5.7.3) There exists $\lambda_\Phi > 0$ such that for each $t \in [1, T]$, and $w \in \Omega$, we have

$$\int_1^t \left[\sum_{n=1}^\infty \frac{(l^*)^n}{\Gamma(n\alpha)} \left(\ln \frac{t}{s} \right)^{n\alpha - 1} \Phi(s, w) \right] \frac{ds}{s} \le \lambda_\Phi \Phi(t, w).$$

Theorem 5.11. *Assume that the hypotheses (5.7.1) and (5.7.2) hold. Then, the problem (5.13) has a random solution defined on* $[1, T] \times \Omega$. *Moreover, if the hypothesis (5.7.3) holds, then the problem (5.13) is generalized Ulam–Hyers–Rassias stable.*

Finally, we give (without proof) existence and Ulam slability results for problem (5.13) with nonconvex-valued right-hand side. The following hypotheses will be used in the sequel.

(5.8.1) The multifunction $G : [1,T] \times \mathbb{R} \times \Omega \to \mathcal{P}_{cp}(\mathbb{R})$ is strong random Carathéodory on $[1,T] \times \mathbb{R} \times \Omega$.

(5.8.2) There exists a measurable and bounded function $p : \Omega \to L^\infty([1,T], [0,\infty))$ satisfying for each $w \in \Omega$,

$$H_d(G(t,u,w), G(t,\overline{u},w)) \le (\ln t)^{1-\gamma} p(t,w)|u - \overline{u}|$$

for every $t \in [1,T]$ and $u, \overline{u} \in \mathbb{R}$.

Set

$$l^* = \sup_{w \in \Omega} \|l(w)\|_{L^\infty}.$$

Theorem 5.12. *Assume that the hypotheses (5.8.1) and (5.8.2) hold. If*

$$\frac{l^* (\ln T)^{1+\alpha-\gamma}}{\Gamma(1+\alpha)} < 1, \tag{5.19}$$

Then, the problem (5.13) has at least one random solution defined on $[1,T] \times \Omega$. Moreover, if the hypothesis (5.7.3) holds, then the problem (5.13) is generalized Ulam–Hyers–Rassias stable.

5.3.3. *Examples*

Let $\Omega = (-\infty, 0)$ be equipped with the usual σ-algebra consisting of Bochner measurable subsets of $(-\infty, 0)$.

Example 1. Consider the following problem of Hilfer fractional differential inclusion:

$$\begin{cases} (D_0^{\frac{1}{2},\frac{1}{2}}u)(t,w) \in F(t, u(t,w), w); \ t \in [0,1], \\ (I_0^{\frac{1}{4}}u)(0,w) = 1, \end{cases} \quad w \in \Omega, \tag{5.20}$$

where

$$F(t, u(t,w), w) = \{v : \Omega \to C([0,1], \mathbb{R}) : |f_1(t, u(t,w), w)|$$

$$\le |v(w)| \le |f_2(t, u(t,w), w)|\}; \quad t \in [0,1], \ w \in \Omega,$$

with $f_1, f_2 : [0,1] \times \mathbb{R} \times \Omega \to \mathbb{R}$, such that

$$f_1(t, u(t,w), w) = \frac{t^2 u}{(1 + w^2 + |u|)e^{10+t}},$$

and

$$f_2(t, u(t,w), w) = \frac{t^2 u}{(1 + w^2)e^{10+t}}.$$

Set $\alpha = \beta = \frac{1}{2}$, then $\gamma = \frac{3}{4}$. We assume that F is closed and convex valued. A simple computation shows that conditions of Theorem 5.7 are satisfied. Hence, the problem (5.20) has at least one random solution defined on $[0, 1]$.

Also, the hypothesis (5.34.3) is satisfied with

$$\Phi(t, w) = \frac{e^3}{1 + w^2} \quad \text{and} \quad \lambda_\Phi = \sum_{n=1}^{\infty} \frac{e^{-10n}}{\Gamma(1 + n\alpha)},$$

$$\Phi(t, w) = \frac{e^3}{1 + w^2} \quad \text{and} \quad \lambda_\Phi = \frac{1}{\Gamma(1 + \alpha)}.$$

Indeed, for each $t \in [0, 1]$, and $w \in \Omega$, we get

$$(I_0^\alpha \Phi)(t, w) \le \frac{e^3}{(1 + w^2)} \sum_{n=1}^{\infty} \frac{e^{-10n}}{\Gamma(1 + n\alpha)}.$$
$$= \lambda_\Phi \Phi(t, w).$$

Consequently, Theorem 5.8 implies that the problem (5.20) is generalized Ulam–Hyers–Rassias stable.

Example 2. Consider now the following problem of Hilfer fractional differential inclusion:

$$\begin{cases} (D_0^{\frac{1}{2}, \frac{1}{2}} u)(t, w) \in F(t, u(t, w), w); \ t \in [0, 1], \\ (I_0^{\frac{1}{4}} u)(0, w) = 1, \end{cases} \quad w \in \Omega, \qquad (5.21)$$

where

$$F(t, u(t, w), w) = \frac{t^2}{(1 + w^2 + |u|)e^{10+t}}[u - 1, u]; \quad t \in [0, 1], \ w \in \Omega.$$

Set $\alpha = \beta = \frac{1}{2}$, then $\gamma = \frac{3}{4}$. We assume that F is closed valued. A simple computation shows that conditions of Theorem 5.9 are satisfied. Hence, the problem (5.21) has at least one random solution defined on $[0, 1]$. Also, Theorem 5.10 implies that the problem (5.21) is generalized Ulam–Hyers–Rassias stable.

5.4. Random Hilfer Fractional Differential Equations in Fréchet Spaces

5.4.1. *Introduction*

This section deals with some existence and Ulam stability results for some fractional random differential equations of Hilfer and Hilfer–Hadamard type in Fréchet spaces. We use a random fixed-point theorem for the existence of random solutions, and we prove that our problems are generalized Ulam–Hyers–Rassias stable.

We discuss the existence and the Ulam stability of solutions for the following problem of Random Hilfer fractional differential equations:

$$\begin{cases} (D_0^{\alpha,\beta}u)(t,w) = f(t,u(t,w),w); \ t \in \mathbb{R}_+ := [0,+\infty), \\ (I_0^{1-\gamma}u)(t,w)|_{t=0} = \phi(w), \end{cases} \quad w \in \Omega, \quad (5.22)$$

where $\alpha \in (0,1)$, $\beta \in [0,1]$, $\gamma = \alpha + \beta - \alpha\beta$, $(\Omega, \mathcal{A})$ is a measurable space, $\phi : \Omega \to \mathbb{R}$ is a measurable function, $f : \mathbb{R}_+ \times \mathbb{R} \times \Omega \to \mathbb{R}$ is a given function, $I_0^{1-\gamma}$ is the left-sided mixed Riemann–Liouville integral of order $1 - \gamma$, and $D_0^{\alpha,\beta}$ is the Hilfer fractional derivative of order α and type β.

Next, we consider the following problem of random Hilfer–Hadamard fractional differential equations:

$$\begin{cases} (^H D_1^{\alpha,\beta}u)(t,w) = g(t,u(t,w),w); \ t \in [1,+\infty), \\ (^H I_1^{1-\gamma}u)(1,w) = \phi_0(w), \end{cases} \quad w \in \Omega, \quad (5.23)$$

where $\alpha \in (0,1)$, $\beta \in [0,1]$, $\gamma = \alpha + \beta - \alpha\beta$, $\phi_0 : \Omega \to \mathbb{R}$ is a measurable function, $g : [1,+\infty) \times \mathbb{R} \times \Omega \to \mathbb{R}$ is a given function, $^H I_1^{1-\gamma}$ is the left-sided mixed Hadamard integral of order $1-\gamma$, and $^H D_1^{\alpha,\beta}$ is the Hilfer–Hadamard fractional derivative of order α and type β.

5.4.2. *Hilfer fractional random differential equations*

In this section, we are concerned with the existence and Ulam–Hyers–Rassias stability for problem (5.22). Let us start by defining what we mean by a random solution of the problem (5.22).

Definition 5.21. By a random solution of the problem (5.22), we mean a measurable function $u : \Omega \to C_\gamma$ that satisfies the condition $(I_0^{1-\gamma}u)(0^+,w) = \phi(w)$, and the equation $(D_0^{\alpha,\beta}u)(t,w) = f(t,u(t,w),w)$ on $\mathbb{R}_+ \times \Omega$.

The following hypotheses will be used in the sequel.

(5.9.1) The function $f : I_p := [0,p] \times \mathbb{R} \times \Omega \mapsto f(t,u,w) \in \mathbb{R}$; $p \in \mathbb{N}\backslash\{0\}$ is random Carathéodory on $I_p \times \mathbb{R} \times \Omega$, and affine with respect to u.

(5.9.2) There exists a measurable and bounded function $l : \Omega \to L^\infty(I_p, [0, \infty))$ such that

$$|f(t, u, w) - f(t, v, w)| \le l(t, w)|u - v|;$$

$$\text{for a.e. } t \in I_p, \text{ and each } u, v \in \mathbb{R}, \; w \in \Omega.$$

(5.9.3) There exists $\lambda_\Phi > 0$ such that for each $t \in I_p$, and $w \in \Omega$, we have

$$\int_0^t \left[\sum_{n=1}^\infty \frac{(l_p^*)^n}{\Gamma(n\alpha)} (t - s)^{n\alpha - 1} \Phi(s, w) \right] ds \le \lambda_\Phi \Phi(t, w),$$

where $l_p^* = \sup\limits_{w \in \Omega} \|l(w)\|_{L^\infty(I_p)}$.

For any $p \in \mathbb{N} \backslash \{0\}$, set

$$f_p^* = \sup_{w \in \Omega} \|f(\cdot, 0, w)\|_{L^\infty(I_p)}, \text{ and } \phi^* = \sup_{w \in \Omega} |\phi(w)|.$$

Now, we shall prove the following theorem concerning the existence of random solutions of problem (5.22).

Theorem 5.13. *Assume that the hypotheses (5.9.1) and (5.9.2) hold. If*

$$\frac{l_p^* p^{1-\gamma+\alpha}}{\Gamma(1 + \alpha)} < 1, \tag{5.24}$$

then problem (5.22) has at least one random solution in the space C_γ. Furthermore, if the hypothesis (5.9.3) holds, then problem (5.22) is generalized Ulam–Hyers–Rassias stable.

Proof. Define a mapping $N : \Omega \times C_\gamma \to C_\gamma$ by

$$(N(w)u)(t) = \frac{\phi(w)}{\Gamma(\gamma)} t^{\gamma-1} + \int_0^t (t - s)^{\alpha-1} \frac{f(s, u(s, w), w)}{\Gamma(\alpha)} ds. \tag{5.25}$$

The map ϕ is measurable for all $w \in \Omega$. Again, as the indefinite integral is continuous on I, then $N(w)$ defines a mapping $N : \Omega \times C_\gamma \to C_\gamma$. Thus u is a random solution for the problem (5.22) if and only if $u = N(w)u$.

For each $p \in \mathbb{N} \backslash \{0\}$ and any $w \in \Omega$, we can show that $N(w)$ transforms the ball $B_\eta := \{u \in C_\gamma : \|u\|_p \le \eta_p\}$ into itself, where

$$\eta_p \ge \frac{\phi^* \Gamma(1 + \alpha) + \Gamma(\gamma) f_p^* p^{1-\gamma+\alpha}}{\Gamma(\gamma)(\Gamma(1 + \alpha) - l_p^* p^{1-\gamma+\alpha})}.$$

Indeed, for any $w \in \Omega$, and each $u \in B_\eta$ and $t \in I_p$, we have

$$|t^{1-\gamma}(N(w)u)(t)| \leq \frac{|\phi(w)|}{\Gamma(\gamma)} + \frac{t^{1-\gamma}}{\Gamma(\alpha)} \int_0^t (t-s)^{\alpha-1}|f(s, u(s, w), w)|ds$$

$$\leq \frac{|\phi(w)|}{\Gamma(\gamma)} + \frac{t^{1-\gamma}}{\Gamma(\alpha)} \int_0^t (t-s)^{\alpha-1}|f(s, 0, w)|ds$$

$$+ \frac{t^{1-\gamma}}{\Gamma(\alpha)} \int_0^t (t-s)^{\alpha-1}|f(s, u(s, w), w) - f(s, 0, w)|ds$$

$$\leq \frac{|\phi(w)|}{\Gamma(\gamma)} + \frac{t^{1-\gamma}}{\Gamma(\alpha)} \int_0^t (t-s)^{\alpha-1}|f(s, 0, w)|ds$$

$$+ \frac{t^{1-\gamma}}{\Gamma(\alpha)} \int_0^t (t-s)^{\alpha-1}l(s, w)|u(s, w)|ds$$

$$\leq \frac{|\phi(w)|}{\Gamma(\gamma)} + \frac{f_p^* T^{1-\gamma}}{\Gamma(\alpha)} \int_0^t (t-s)^{\alpha-1}ds$$

$$+ \frac{l_p^* \eta_p T^{1-\gamma}}{\Gamma(\alpha)} \int_0^t (t-s)^{\alpha-1}ds$$

$$\leq \frac{\phi^*}{\Gamma(\gamma)} + \frac{(f_p^* + l_p^*\eta_p)p^{1-\gamma+\alpha}}{\Gamma(1+\alpha)}$$

$$\leq \eta_p.$$

Thus

$$\|N(w)u\|_p \leq \eta_p. \tag{5.26}$$

We shall show that the operator $N : \Omega \times B_\eta \to B_\eta$ satisfies all the assumptions of Theorem 1.8. The proof will be given in several steps.

Step 1. $N(w)$ *is a random operator on* $\Omega \times B_\eta$ *into* B_η.

Since $f(t, u, w)$ is random Carathéodory, the map $w \to f(t, u, w)$ is measurable in view of Definition 5.13. Similarly, the product $(t-s)^{\alpha-1}f(s, u(s, w), w)$ of a continuous and measurable function is again measurable. Further, the integral is a limit of a finite sum of measurable functions, therefore, the map

$$w \mapsto \frac{\phi(w)}{\Gamma(\gamma)}t^{\gamma-1} + \int_0^t \frac{(t-s)^{\alpha-1}}{\Gamma(\alpha)}f(s, u(s, w), w)ds,$$

is measurable. As a result, $N(w)$ is a random operator on $\Omega \times B_\eta$ into B_η.

Step 2. $N(w)$ *is continuous.*

Let $\{u_n\}_{n\in\mathbb{N}}$ be a sequence such that $u_n \to u$ in B_η. Then, for each $t \in I_p$, and $w \in \Omega$, we have

$$|t^{1-\gamma}(N(w)u_n)(t) - t^{1-\gamma}(N(w)u)(t)|$$

$$\leq \frac{t^{1-\gamma}}{\Gamma(\alpha)} \int_0^t (t-s)^{\alpha-1}|f(s, u_n(s,w), w) - f(s, u(s,w), w)|ds$$

$$\leq \frac{t^{1-\gamma}}{\Gamma(\alpha)} \int_0^t (t-s)^{\alpha-1}l(s,w)|u_n(s,w) - u(s,w)|ds \qquad (5.27)$$

$$\leq \frac{l_p^* T^{1-\gamma}}{\Gamma(\alpha)} \int_0^t (t-s)^{\alpha-1}|u_n(s,w) - u(s,w)|ds.$$

Since $u_n \to u$ as $n \to \infty$, then (5.27) implies

$$\|N(w)u_n - N(w)u\|_p \to 0 \quad \text{as } n \to \infty.$$

Step 3. $N(w)$ *is affine.*

For each $u, v \in B_\eta$, $t \in I_p$, and any $\lambda \in (0,1)$ and $w \in \Omega$, we have

$$N(w)(\lambda u + (1-\lambda)v) = \frac{\phi}{\Gamma(\gamma)}t^{\gamma-1} + \int_0^t (t-s)^{\alpha-1}\frac{f(s, (\lambda u + (1-\lambda)v)(s,w), w)}{\Gamma(\alpha)}ds$$

$$= \frac{\phi}{\Gamma(\gamma)}t^{\gamma-1} + \lambda\int_0^t (t-s)^{\alpha-1}\frac{f(s, u(s,w), w)}{\Gamma(\alpha)}ds$$

$$+ (1-\lambda)\int_0^t (t-s)^{\alpha-1}\frac{f(s, v(s,w), w)}{\Gamma(\alpha)}ds$$

$$= \frac{\lambda\phi}{\Gamma(\gamma)}t^{\gamma-1} + \lambda\int_0^t (t-s)^{\alpha-1}\frac{f(s, u(s,w), w)}{\Gamma(\alpha)}ds$$

$$+ \frac{(1-\lambda)\phi}{\Gamma(\gamma)}t^{\gamma-1} + (1-\lambda)\int_0^t (t-s)^{\alpha-1}\frac{f(s, v(s,w), w)}{\Gamma(\alpha)}ds$$

$$= \lambda N(w)(u) + (1-\lambda)N(w)(v).$$

Hence $N(w)$ is affine.

As a consequence of Steps 1–3, together with the Theorem 1.8, we deduce that N has a random fixed point v which is a random solution of the problem (5.22).

Step 4. *The generalized Ulam–Hyers–Rassias stability.*

Let u be a random solution of the inequality (5.8), and let us assume that v is a random solution of problem (5.22). Thus, we have

$$v(t,w) = \frac{\phi(w)}{\Gamma(\gamma)}t^{\gamma-1} + \int_0^t (t-s)^{\alpha-1}\frac{f(s,v(s,w),w)}{\Gamma(\alpha)}ds.$$

From the inequality (5.8) for each $t \in I_p$, and $w \in \Omega$, we have

$$\left| u(t,w) - \frac{\phi(w)}{\Gamma(\gamma)}t^{\gamma-1} - \int_0^t (t-s)^{\alpha-1}\frac{f(s,u(s,w),w)}{\Gamma(\alpha)}ds \right| \le (I_0^\alpha \Phi)(t,w).$$

From hypotheses (5.9.2) and (5.9.3), for each $t \in I_p$, and $w \in \Omega$, we get

$$|u(t,w) - v(t,w)| \le \left| u(t,w) - \frac{\phi(w)}{\Gamma(\gamma)}t^{\gamma-1} - \int_0^t (t-s)^{\alpha-1}\frac{f(s,u(s,w),w)}{\Gamma(\alpha)}ds \right|$$

$$+ \int_0^t (t-s)^{\alpha-1}\frac{|f(s,u(s,w),w) - f(s,v(s,w),w)|}{\Gamma(\alpha)}ds$$

$$\le (I_0^\alpha \Phi)(t,w) + \frac{l_p^*}{\Gamma(\alpha)}\int_0^t (t-s)^{\alpha-1}|u(s,w) - v(s,w)|ds.$$

From Lemma 1.14, we have

$$|u(t,w) - v(t,w)| \le \frac{\lambda_\phi}{l_p^*}\left[\Phi(t,w) + \int_0^t \left[\sum_{n=1}^\infty \frac{(l_p^*)^n}{\Gamma(n\alpha)}(t-s)^{n\alpha-1}\Phi(s,w)\right] ds \right]$$

$$\le \frac{\lambda_\phi}{l_p^*}(1 + \lambda_\phi)\Phi(t,w)$$

$$:= c_{f,\Phi}\Phi(t,w).$$

Hence, the problem (5.22) is generalized Ulam–Hyers–Rassias stable. $\quad\square$

5.4.3. *Hilfer–Hadamard fractional random differential equations*

Definition 5.22. By a random solution of the problem (5.23) we mean a measurable function $u \in C_{\gamma,\ln}$ that satisfies the condition $(^H I_1^{1-\gamma}u)(1^+,w) = \phi_0(w)$, and the equation

$$(^H D_1^{\alpha,\beta}u)(t,w) = g(t,u(t,w),w) \text{ on } [1,T] \times \Omega.$$

For each $p \in \mathbb{N}\backslash\{0,1\}$ we consider following set, $C_{p,\gamma,\ln} = C_\gamma([1,p])$, and we define in $C_{\gamma,\ln}$ the seminorms by

$$\|u\|_p = \sup_{t\in[0,p]} |(\ln t)^{1-\gamma} u(t)|.$$

Then $C_{\gamma,\ln}$ is a Fréchet space with the family of seminorms $\{\|u\|_p\}$.

Now we give (without proof) existence and Ulam stability results for problem (5.23).

The following hypotheses will be used in the sequel.

(5.10.1) The function $f : [1,p] \times \mathbb{R} \times \Omega \mapsto f(t,u,w) \in \mathbb{R}$ is random Carathéeodory on $[1,p] \times \mathbb{R} \times \Omega$, and affine with respect to u,

(5.10.2) There exists a measurable and bounded function $l' : \Omega \to L^\infty([1,p],[0,\infty))$ such that

$$|g(t,u,w) - g(t,v,w)| \le l'(t,w)|u-v|;$$

for a.e. $t \in [1,p]$, and each $u,v \in \mathbb{R}$, $w \in \Omega$,

(5.10.3) There exists $\lambda_\Phi > 0$ such that for each $t \in [1,p]$, and $w \in \Omega$, we have

$$\int_1^t \left[\sum_{n=1}^\infty \frac{(l_{p*}^*)^n}{\Gamma(n\alpha)} \left(\ln \frac{t}{s}\right)^{n\alpha-1} \Phi(s,w)\right] \frac{ds}{s} \le \lambda_\Phi \Phi(t,w),$$

where $l_{p*} = \sup_{w\in\Omega} \|l'(w)\|_{L^\infty([1,p])}$.

Theorem 5.14. *Assume that the hypotheses (5.10.1) and (5.10.2) hold. If*

$$\frac{l_{p*}(\ln p)^{1-\gamma+\alpha}}{\Gamma(1+\alpha)} < 1, \tag{5.28}$$

then problem (5.23) has at least one random solution in the space $C_{\gamma,\ln}$. Furthermore, if the hypothesis (5.43.3) holds, then problem (5.23) is generalized Ulam–Hyers–Rassias stable.

5.4.4. *Example*

Let $\Omega = (-\infty,0)$ be equipped with the usual σ-algebra consisting of Bochner measurable subsets of $(-\infty,0)$. Given a measurable function $u : \Omega \to C_{\frac{3}{4}}([0,1])$.

Consider the following problem of Hilfer fractional differential equation:

$$\begin{cases} (D_0^{\frac{1}{2},\frac{1}{2}}u)(t,w) = f(t,u(t,w),w); \ t \in [0,\infty), \\ (I_0^{\frac{1}{4}}u)(t)|_{t=0} = 1, \end{cases} \quad w \in \Omega, \quad (5.29)$$

where

$$\begin{cases} f(t,u,w) = \frac{c_p t^{\frac{-1}{4}} u \sin t}{(1+\sqrt{t})(1+w^2)}; \ t \in (0,\infty); \ u \in \mathbb{R}, \\ f(0,u,w) = 0; \quad\quad\quad\quad\quad u \in \mathbb{R}, \end{cases} \quad w \in \Omega,$$

and $0 < c_p < \frac{\sqrt{\pi}}{2}p^{-3/4}$; $p \in \mathbb{N} - \{0\}$. The hypothesis (5.9.2) is satisfied with

$$\begin{cases} l_p(t,w) = \frac{c_p t^{\frac{-1}{4}}|\sin t|}{(1+\sqrt{t})(1+w^2)}; \ t \in (0,p], \\ l_p(0,w) = 0, \end{cases} \quad w \in \Omega.$$

Also, the hypothesis (5.9.3) is satisfied with

$$\Phi(t,w) = \frac{e^3}{1+w^2}, \quad \text{and} \quad \lambda_\Phi = \sum_{n=1}^{\infty} \frac{c_p^n}{\Gamma(1+n\alpha)}.$$

A simple computation shows that conditions of Theorem 5.13 are satisfied. Hence, problem (5.29) has at least one solution defined on $[0,1]$. Moreover, problem (5.29) is generalized Ulam–Hyers–Rassias stable.

5.5. Notes and Remarks

The results of Chapter 5 are taken from [54,66,83].

Chapter 6

Nonlinear Hadamard–Pettis Fractional Integral Equations

6.1. Introduction

In this chapter, we investigate the existence of weak solutions for some classes of nonlinear Hadamard–Pettis fractional integral equations under the Pettis integrability assumption, by applying the technique of measure of weak noncompactness and convenables fixed-point theorems.

6.2. Hadamard–Pettis Fractional Integral Equations

6.2.1. *Introduction*

In this section, we investigate a class of partial integral equations via Hadamard's fractional integral, by applying the technique of measure of weak noncompactness and Mönch's fixed point theorem.

We consider the following Hadamard partial fractional integral equation:

$$u(x, y) = \mu(x, y) + \frac{1}{\Gamma(r_1)\Gamma(r_2)} \int_1^x \int_1^y \left(\ln \frac{x}{s}\right)^{r_1 - 1} \left(\ln \frac{y}{t}\right)^{r_2 - 1}$$

$$\times \frac{f(s, t, u(s, t))}{st} dt ds; \quad (x, y) \in J, \tag{6.1}$$

where $J := [1, a] \times [1, b]$, $a, b > 1$, $r_1, r_2 > 0$, $\mu : J \to E$ and $f : J \times E \to E$ are given continuous functions, $\Gamma(\cdot)$ is the Euler gamma function and E is a real (or complex) Banach space with norm $\|\cdot\|_E$ and dual E^*, such that E is the dual of a weakly compactly generated Banach space X.

This section initiates the use of the measure of weak noncompactness and Mönch's fixed-point theorem to fractional integral equations of two independent variables involving the Hadamard integral operator.

6.2.2. *Existence of weak solutions*

Let us start by defining what we mean by a solution of the integral equation (6.1).

Definition 6.1. A function $u \in C$ is said to be a solution of (6.1) if u satisfies Eq. (6.1) on J.

Further, we present conditions for the existence of a solution of Eq. (6.1).

Theorem 6.1. *Assume that the following hypotheses hold:*

(6.1.1) *For a.e. $(x, y) \in J$, the function $w \to f(x, y, w)$ is weakly sequentially continuous.*

(6.1.2) *For a.e. $w \in E$, the function $(x, y) \to f(x, y, w)$ is Pettis integrable a.e. on J.*

(6.1.3) *There exists $P \in C(J, [0, \infty))$ such that for all $\varphi \in E^*$, we have*

$$|\varphi(f(x, y, u))| \leq P(x, y)\|\varphi\|; \text{ for a.e. } (x, y) \in J, \text{ and each } u \in E.$$

(6.1.4) *For each bounded set $B \subset E$ and for each $(x, y) \in J$, we have*

$$\beta(f(x, y, B)) \leq P(x, y)\beta(B).$$

If

$$L := \frac{P^*(\ln a)^{r_1}(\ln b)^{r_2}}{\Gamma(1 + r_1)\Gamma(1 + r_2)} < 1, \tag{6.2}$$

where $P^* = \|P\|_{L^\infty}$, then the integral equation (6.1) has at least one solution defined on J.

Proof. Transform the integral equation (6.1) into a fixed-point equation. Consider the operator $N : C \to C$ defined by

$$(Nu)(x, y) = \mu(x, y) + \frac{1}{\Gamma(r_1)\Gamma(r_2)} \int_1^x \int_1^y$$

$$\left(\ln\frac{x}{s}\right)^{r_1-1}\left(\ln\frac{y}{t}\right)^{r_2-1}\frac{f(s, t, u(s, t))}{s\, t}dtds. \tag{6.3}$$

First notice that, the hypothesis (6.1.2) implies that for all $u \in C$, $f(\cdot, \cdot, u(\cdot, \cdot)) \in P(J, E)$. From (6.1.3) we have that for each $x, y \in J$, $\left(\ln\frac{x}{s}\right)^{r_1-1}\left(\ln\frac{y}{t}\right)^{r_2-1}\frac{f(s,t,u(s,t))}{st}$ is Pettis integrable and thus, the operator N makes sense.

Let $R > 0$ be such that $R > \|\mu\|_C + \frac{P^*(\ln a)^{r_1}(\ln b)^{r_2}}{\Gamma(1+r_1)\Gamma(1+r_2)}$ and consider the set

$$Q = \{u \in C : \|u\|_C \le R \text{ and } \|u(x_1, y_1) - u(x_2, y_2)\|_E \le \|\mu(x_1, y_1)$$

$$- \mu(x_2, y_2)\|_E + \frac{P^*}{\Gamma(1+r_1)\Gamma(1+r_2)}[2(\ln y_2)^{r_2}(\ln x_2 - \ln x_1)^{r_1}$$

$$+ 2(\ln x_2)^{r_1}(\ln y_2 - \ln y_1)^{r_2} + (\ln x_1)^{r_1}(\ln y_1)^{r_2}$$

$$- (\ln x_2)^{r_1}(\ln y_2)^{r_2} - 2(\ln x_2 - \ln x_1)^{r_1}(\ln y_2 - \ln y_1)^{r_2}]\}.$$

Clearly, the subset Q is closed, convex and equicontinuous. We shall show that the operator N satisfies all the assumptions of Theorem 1.7. The proof will be given in several steps.

Step 1. *N maps Q into itself.*

Let $u \in Q$, $(x, y) \in J$ and assume that $(Nu)(x, y) \ne 0$. Then there exists $\phi \in E^*$ such that $\|(Nu)(x, y)\|_E = |\phi((Nu)(x, y))|$. Thus

$$\|(Nu)(x,y)\|_E = \left| \phi \left(\mu(x,y) + \frac{1}{\Gamma(r_1)\Gamma(r_2)} \int_1^x \int_1^y \left(\ln \frac{x}{s} \right)^{r_1-1} \right.\right.$$
$$\left.\left. \times \left(\ln \frac{y}{t} \right)^{r_2-1} \frac{f(s,t,u(s,t))}{st} dt ds \right) \right|$$

$$= \left| \phi(\mu(x,y)) + \phi \left(\frac{1}{\Gamma(r_1)\Gamma(r_2)} \int_1^x \int_1^y \left(\ln \frac{x}{s} \right)^{r_1-1} \right.\right.$$
$$\left.\left. \times \left(\ln \frac{y}{t} \right)^{r_2-1} \frac{f(s,t,u(s,t))}{st} dt ds \right) \right|$$

$$\le \|\mu(x,y)\|_E + \frac{1}{\Gamma(r_1)\Gamma(r_2)} \int_1^x \int_1^y \left(\ln \frac{x}{s} \right)^{r_1-1} \left(\ln \frac{y}{t} \right)^{r_2-1}$$
$$\times \frac{P(s,t)}{st} dt ds$$

$$\le \|\mu\|_C + \frac{P^*(\ln a)^{r_1}(\ln b)^{r_2}}{\Gamma(1+r_1)\Gamma(1+r_2)}$$

$$\le R.$$

Next, let $(x_1, y_1), (x_2, y_2) \in J$ such that $x_1 < x_2$ and $y_1 < y_2$, and let $u \in Q$, with $(Nu)(x_1, y_1) - (Nu)(x_2, y_2) \ne 0$. Then there exists $\phi \in E^*$ such that

$$\|(Nu)(x_1, y_1) - (Nu)(x_2, y_2)\|_E = |\phi((Nu)(x_1, y_1) - (Nu)(x_2, y_2))|$$

and $\|\varphi\| = 1$. Then

$$\|(Nu)(x_2, y_2) - (Nu)(x_1, y_1)\|_E$$

$$= |\phi((Nu)(x_2, y_2) - (Nu)(x_1, y_1))|$$

$$\leq \|\mu(x_1, y_1) - \mu(x_2, y_2)\|_E + \frac{1}{\Gamma(r_1)\Gamma(r_2)} \int_1^{x_1} \int_1^{y_1} \left[\left| \ln \frac{x_2}{s} \right|^{r_1-1} \right.$$

$$\times \left| \ln \frac{y_2}{t} \right|^{r_2-1} - \left| \ln \frac{x_1}{s} \right|^{r_1-1} \left| \ln \frac{y_1}{t} \right|^{r_2-1} \right] \frac{|\phi(f(s,t,u(s,t)))|}{st} \, dtds$$

$$+ \frac{1}{\Gamma(r_1)\Gamma(r_2)} \int_{x_1}^{x_2} \int_{y_1}^{y_2} \left| \ln \frac{x_2}{s} \right|^{r_1-1} \left| \ln \frac{y_2}{t} \right|^{r_2-1} \frac{|\phi(f(s,t,u(s,t)))|}{st} \, dtds$$

$$+ \frac{1}{\Gamma(r_1)\Gamma(r_2)} \int_{1}^{x_1} \int_{y_1}^{y_2} \left| \ln \frac{x_2}{s} \right|^{r_1-1} \left| \ln \frac{y_2}{t} \right|^{r_2-1} \frac{|\phi(f(s,t,u(s,t)))|}{st} \, dtds$$

$$+ \frac{1}{\Gamma(r_1)\Gamma(r_2)} \int_{x_1}^{x_2} \int_{1}^{y_1} \left| \ln \frac{x_2}{s} \right|^{r_1-1} \left| \ln \frac{y_2}{t} \right|^{r_2-1} \frac{|\phi(f(s,t,u(s,t)))|}{st} \, dtds.$$

Thus

$$\|(Nu)(x_2, y_2) - (Nu)(x_1, y_1)\|_E$$

$$\leq \|\mu(x_1, y_1) - \mu(x_2, y_2)\|_E$$

$$+ \frac{1}{\Gamma(r_1)\Gamma(r_2)} \int_1^{x_1} \int_1^{y_1} \left[\left| \ln \frac{x_2}{s} \right|^{r_1-1} \left| \ln \frac{y_2}{t} \right|^{r_2-1} \right.$$

$$- \left| \ln \frac{x_1}{s} \right|^{r_1-1} \left| \ln \frac{y_1}{t} \right|^{r_2-1} \right] \frac{P^*}{st} \, dtds$$

$$+ \frac{1}{\Gamma(r_1)\Gamma(r_2)} \int_{x_1}^{x_2} \int_{y_1}^{y_2} \left| \ln \frac{x_2}{s} \right|^{r_1-1} \left| \ln \frac{y_2}{t} \right|^{r_2-1} \frac{P^*}{st} \, dtds$$

$$+ \frac{1}{\Gamma(r_1)\Gamma(r_2)} \int_{1}^{x_1} \int_{y_1}^{y_2} \left| \ln \frac{x_2}{s} \right|^{r_1-1} \left| \ln \frac{y_2}{t} \right|^{r_2-1} \frac{P^*}{st} \, dtds$$

$$+ \frac{1}{\Gamma(r_1)\Gamma(r_2)} \int_{x_1}^{x_2} \int_{1}^{y_1} \left| \ln \frac{x_2}{s} \right|^{r_1-1} \left| \ln \frac{y_2}{t} \right|^{r_2-1} \frac{P^*}{st} \, dtds$$

$$\leq \|\mu(x_1, y_1) - \mu(x_2, y_2)\|_E$$

$$+ \frac{P^*}{\Gamma(1+r_1)\Gamma(1+r_2)} [2(\ln y_2)^{r_2}(\ln x_2 - \ln x_1)^{r_1}$$

$$+ 2(\ln x_2)^{r_1}(\ln y_2 - \ln y_1)^{r_2} + (\ln x_1)^{r_1}(\ln y_1)^{r_2} - (\ln x_2)^{r_1}(\ln y_2)^{r_2}$$

$$- 2(\ln x_2 - \ln x_1)^{r_1}(\ln y_2 - \ln y_1)^{r_2}].$$

Hence $N(Q) \subset Q$.

Step 2. *N is weakly-sequentially continuous.*

Let (u_n) be a sequence in Q and let $(u_n(x,y)) \to u(x,y)$ in (E,ω) for each $(x,y) \in J$. Fix $(x,y) \in J$, since f satisfies the assumption (6.1.1), we have $f(x,y,u_n(x,y))$ converges weakly to $f(x,y,u(x,y))$. Hence the Lebesgue dominated convergence theorem for Pettis integral implies $(Nu_n)(x,y)$ converges weakly to $(Nu)(x,y)$ in (E,ω). We do it for each $(x,y) \in J$, so $N(u_n) \to N(u)$. Then $N : Q \to Q$ is weakly-sequentially continuous.

Step 3. *The implication (1.9) holds.*

Let V be a subset of Q such that $\overline{V} = \overline{\text{conv}}(N(V) \cup \{0\})$. Obviously $V(x,y) \subset \overline{\text{conv}}(NV)(x,y)) \cup \{0\})$, $\forall(x,y) \in J$. Further, as V is bounded and equicontinuous, by [172, Lemma 3] the function $(x,y) \to u(x,y) = \beta(V(x,y))$ is continuous on J. Since the functions μ is continuous on J, then the set $\{\mu(x,y); (x,y) \in J\} \subset E$ is compact. From (6.1.3), Lemma 1.7 and the properties of the measure β, for any $(x,y) \in J$, we have

$$v(x,y) \leq \beta((NV)(x,y) \cup \{0\})$$

$$\leq \beta((NV)(x,y))$$

$$\leq \frac{1}{\Gamma(r_1)\Gamma(r_2)} \int_1^x \int_1^y \left|\ln\frac{x}{s}\right|^{r_1-1} \left|\ln\frac{y}{t}\right|^{r_2-1} \frac{P(s,t)\beta(V(s,t))}{st} dt ds$$

$$\leq \frac{1}{\Gamma(r_1)\Gamma(r_2)} \int_1^x \int_1^y \left|\ln\frac{x}{s}\right|^{r_1-1} \left|\ln\frac{y}{t}\right|^{r_2-1} \frac{P(s,t)v(s,t)}{st} dt ds$$

$$\leq \frac{\|v\|_C}{\Gamma(r_1)\Gamma(r_2)} \int_1^x \int_1^y \left|\ln\frac{x}{s}\right|^{r_1-1} \left|\ln\frac{y}{t}\right|^{r_2-1} \frac{P(s,t)}{st} dt ds$$

$$\leq \frac{P^*(\ln a)^{r_1}(\ln b)^{r_2}}{\Gamma(1+r_1)\Gamma(1+r_2)} \|v\|_C.$$

Thus

$$\|v\|_C \leq L\|v\|_C.$$

From (6.2), we get $\|v\|_C = 0$, that is $v(x,y) = \beta(V(x,y)) = 0$, for each $(x,y) \in J$. and then by [290, Theorem 2], V is relatively weakly compact in C. Applying now Theorem 1.7, we conclude that N has a fixed point which is a solution of the integral equation (6.9). $\qquad\square$

6.2.3. *Example*

Let

$$E = l^1 = \left\{ u = (u_1, u_2, \ldots, u_n, \ldots), \ \sum_{n=1}^{\infty} |u_n| < \infty \right\}$$

be the Banach space with the norm

$$\|u\|_E = \sum_{n=1}^{\infty} |u_n|.$$

As an application of our results, we consider the following partial Hadamard integral equation of the form:

$$u_n(x, y) = \mu(x, y) + \frac{1}{\Gamma(r_1)\Gamma(r_2)} \int_1^x \int_1^y \left(\ln \frac{x}{s} \right)^{r_1 - 1} \left(\ln \frac{y}{t} \right)^{r_2 - 1}$$

$$\times \frac{f_n(s, t, u(s, t))}{st} dt ds, \ (x, y) \in [1, e]^2, \tag{6.4}$$

where

$$r_1, r_2 > 0, \quad \mu(x, y) = x + y^2$$

and

$$f_n(x, y, u(x, y)) = \frac{cxy^2}{1 + \|u\|_E} \left(e^{-7} + \frac{1}{e^{x+y+5}} \right) u_n(x, y),$$

with $u = (u_1, u_2, \ldots, u_n, \ldots)$ and

$$c := \frac{e^4}{8} \Gamma(1 + r_1)\Gamma(1 + r_2).$$

Set $f = (f_1, f_2, \ldots, f_n, \ldots)$. Clearly, the function f is continuous.
For each $u \in E$ and $(x, y) \in [1, e] \times [1, e]$, we have

$$\|f(x, y, u(x, y))\|_E \leq cxy^2 \left(e^{-7} + \frac{1}{e^{x+y+5}} \right).$$

Hence, the hypothesis (6.2.3) is satisfied with $P^* = 2ce^{-4}$. We shall show that condition (6.2) holds with $a = b = e$. Indeed,

$$\frac{P^*(\ln a)^{r_1}(\ln b)^{r_2}}{\Gamma(1 + r_1)\Gamma(1 + r_2)} = \frac{2c}{e^4\Gamma(1 + r_1)\Gamma(1 + r_2)} = \frac{1}{4} < 1.$$

A simple computations show that all conditions of Theorem 6.1 are satisfied. It follows that the integral equation (6.4) has at least one solution on $[1, e] \times [1, e]$.

6.3. Partial Hadamard–Pettis Fractional Integral Inclusions

6.3.1. *Introduction*

This section deals with the existence of solutions to the following Hadamard partial fractional integral inclusion of the form:

$$u(x,y) \in \mu(x,y) + (^H I_\sigma^r F)(x,y,u(x,y)) \quad \text{if } (x,y) \in J, \qquad (6.5)$$

where $J := [1,a] \times [1,b]$, $a,b > 1$, $r_1, r_2 > 0$, $F : J \times E \to \mathcal{P}(E)$ is a compact valued multivalued map, $^H I_\sigma^r F$ is the definite Hadamard integral for the set-valued function F of order $r = (r_1, r_2) \in (0,\infty) \times (0,\infty)$, $\mu : J \to E$ is a given continuous functions, $\Gamma(\cdot)$ is the Euler gamma function, E is a real (or complex) Banach space with norm $\|\cdot\|_E$ and dual E^*, and $\mathcal{P}(E)$ is the family of all nonempty subsets of E.

6.3.2. *Existence of weak solutions*

Let us start by defining what we mean by a solution of the integral inclusion (6.5).

Definition 6.2. A function $u \in C$ is said to be a solution of (6.5) if there exists a function $v \in L^1(J,E)$ with $v(x,y) \in S_{F\circ u}$ such that u satisfies the equation

$$u(x,y) = \mu(x,y) + (^H I_\sigma^r v)(x,y),$$

on J.

Further, we present conditions for the existence of a solution of the inclusion (6.5).

Theorem 6.2. *Assume that the following hypotheses hold:*

(6.2.1) $F : J \times E \to P_{cp,cl,cv}(E)$ *has weakly-sequentially closed graph.*

(6.2.2) *For each continuous* $u : J \to E$, *there exists a measurable function* $v : J \to E$ *with* $v(x,y) \in F(x,y,u(x,y))$ *a.e. on* J *and* v *is Pettis integrable on* J.

(6.2.3) *There exist* $p \in L^\infty(J,[0,\infty))$ *with*

$$(s,t) \to \left(\ln\frac{x}{s}\right)^{r_1-1} \left(\ln\frac{y}{t}\right)^{r_2-1} \frac{p(s,t)}{st} \in L^\infty(J,[0,\infty)),$$

and a continuous nondecreasing function $\psi : [0,\infty) \to (0,\infty)$ *such that:*

$$\|F(x,y,u)\|_{\mathcal{P}} = \sup\{\|v\| : v \in F(x,y,u)\} \le p(x,y)\psi(\|u\|),$$

(6.2.4) *Let $r_0 > 0$ be arbitrary (but fixed). For any $\epsilon > 0$ and for any subset $X \subset B_{r_0}$, there exists a closed subset $I_\epsilon \subset J$ such that $\mu(J \backslash I_\epsilon) < \epsilon$ and*

$$\beta(F(T \times X)) \leq \sup_{(x,y) \in T} p(x,y)\beta(X),$$

for each closed subset T of I_ϵ, where μ denotes the Bochner measure in $\mathbb{R}^2$.

If

$$L := \frac{2p^*(\ln a)^{r_1}(\ln b)^{r_2}}{\Gamma(1+r_1)\Gamma(1+r_2)} < 1, \tag{6.6}$$

where $p^* = \|p\|_{L^\infty}$, then the integral inclusion (6.5) has at least one solution defined on J.

Proof. To transform the integral inclusion (6.5) into a fixed-point equation, we define the multivalued map $N : C \to P_{cl}(C)$ by

$$N(u) = \{h \in C : h(x,y) = \mu(x,y) + (^H I_\sigma^r v)(x,y); \; v \in S_{Fou}\}. \tag{6.7}$$

First notice that, the hypothesis (6.2.2) implies that for all $u \in C$, there exists $v \in P(J,E) : v \in S_{Fou}$. From (6.2.3) and since

$$\left(\ln \frac{x}{s}\right)^{r_1-1}\left(\ln \frac{y}{t}\right)^{r_2-1}\frac{v(s,t)}{st}$$

$$\leq \left(\ln \frac{x}{s}\right)^{r_1-1}\left(\ln \frac{y}{t}\right)^{r_2-1}\frac{p(s,t)\psi(\|u(x,y)\|_E)}{st};$$

for each $(s,t) \in [1,x] \times [1,y]$, and all $(x,y) \in J$, then we have that

$$\left(\ln \frac{x}{s}\right)^{r_1-1}\left(\ln \frac{y}{t}\right)^{r_2-1}\frac{v(s,t)}{st}; \text{ for all } (x,y) \in J,$$

is Pettis integrable and thus N is well defined.

Let $R > 0$ be such that

$$R > \|\mu\|_C + \frac{p^*\psi(R)(\ln a)^{r_1}(\ln b)^{r_2}}{\Gamma(1+r_1)\Gamma(1+r_2)},$$

and consider the set

$$Q = \Big\{ u \in C : \|u\|_C \leq R \text{ and } \|u(x_1,y_1) - u(x_2,y_2)\|_E$$

$$\leq \|\mu(x_1,y_1) - \mu(x_2,y_2)\|_E$$

$$+ \frac{p^*\psi(R)}{\Gamma(1+r_1)\Gamma(1+r_2)}[2(\ln y_2)^{r_2}(\ln x_2 - \ln x_1)^{r_1}$$

$$+ 2(\ln x_2)^{r_1}(\ln y_2 - \ln y_1)^{r_2}$$

$$+ (\ln x_1)^{r_1} (\ln y_1)^{r_2} - (\ln x_2)^{r_1} (\ln y_2)^{r_2}$$
$$- 2(\ln x_2 - \ln x_1)^{r_1} (\ln y_2 - \ln y_1)^{r_2} \Big] \Big\}.$$

Clearly, the subset Q is closed, convex, bounded and equicontinuous. The remainder of the proof will be given in four steps.

Step 1. $N(u)$ *is convex for each $u \in Q$.*

Let $h_1, h_2 \in N(u)$. Then there exist $v_1, v_2 \in S_{Fou}$ such that, for each $(x, y) \in J$, we have

$$h_i(x, y) = \mu(x, y) + ({}^H I_\sigma^r v_i)(x, y); \quad \text{for } i = 1, 2.$$

Let $0 \le \lambda \le 1$. Then, for each $(x, y) \in J$, we have

$$[\lambda h_1 + (1 - \lambda)h_2](x, y) = \mu(x, y) + ({}^H I_\sigma^r (\lambda v_1 + (1 - \lambda)v_2))(x, y).$$

Since S_{Fou} is convex (because F has convex values), it follows that $\lambda h_1 + (1 - \lambda)h_2 \in N(u)$.

Step 2. N *maps Q into itself.*

Let $h \in N(Q)$. Then there exist $u \in Q$ with $h \in \Omega(u)$ and a Pettis integrable function $v : J \to E$ with $v(x, y) \in S_{Fou}$ for a.e. $(x, y) \in J$. We can consider that $h(x, y) \ne 0$. Then, there exists $\phi \in E^*$ with $\|\phi\| = 1$ such that $|\phi(h(x, y))| = \|h(x, y)\|$. Thus, we obtain

$$\|(hu)(x, y)\|_E = \left| \phi \left(\mu(x, y) + \frac{1}{\Gamma(r_1)\Gamma(r_2)} \int_1^x \int_1^y \left(\ln \frac{x}{s}\right)^{r_1-1} \left(\ln \frac{y}{t}\right)^{r_2-1} \right. \right.$$
$$\left. \left. \times \frac{v(s,t)}{st} dt ds \right) \right| = |\phi(\mu(x,y))| + \left| \phi \left(\frac{1}{\Gamma(r_1)\Gamma(r_2)} \int_1^x \int_1^y \right. \right.$$
$$\left. \left. \left(\ln \frac{x}{s}\right)^{r_1-1} \left(\ln \frac{y}{t}\right)^{r_2-1} \right| \frac{v(s,t)}{st} dt ds \right)$$
$$\le \|\mu(x,y)\|_E + \frac{1}{\Gamma(r_1)\Gamma(r_2)} \int_1^x \int_1^y \left(\ln \frac{x}{s}\right)^{r_1-1} \left(\ln \frac{y}{t}\right)^{r_2-1}$$
$$\times \frac{p(s,t)\psi(\|u(s,t)\|_E)}{st} dt ds$$
$$\le \|\mu\|_C + \frac{p^*\psi(R)(\ln a)^{r_1}(\ln b)^{r_2}}{\Gamma(1+r_1)\Gamma(1+r_2)} \le R.$$

Next, let $(x_1, y_1), (x_2, y_2) \in J$ such that $x_1 < x_2$ and $y_1 < y_2$, and let $h \in N(u)$, so $h(x_1, y_1) - h(x_2, y_2) \neq 0$. Then there exists $\phi \in E^*$ such that $\|h(x_1, y_1) - h(x_2, y_2)\| = |\varphi(h(x_1, y_1) - h(x_2, y_2))|$ and $\|\varphi\| = 1$. Then

$$\|(hu)(x_2, y_2) - (hu)(x_1, y_1)\|_E$$

$$= |\phi((hu)(x_2, y_2) - (hu)(x_1, y_1))|$$

$$\leq \|\mu(x_1, y_1) - \mu(x_2, y_2)\|_E$$

$$+ \frac{1}{\Gamma(r_1)\Gamma(r_2)} \int_1^{x_1} \int_1^{y_1} \left[\left| \ln \frac{x_2}{s} \right|^{r_1-1} \left| \ln \frac{y_2}{t} \right|^{r_2-1} \right.$$

$$\left. - \left| \ln \frac{x_1}{s} \right|^{r_1-1} \left| \ln \frac{y_1}{t} \right|^{r_2-1} \right] \frac{\|v(s,t)\|_E}{st} dt ds$$

$$+ \frac{1}{\Gamma(r_1)\Gamma(r_2)} \int_{x_1}^{x_2} \int_{y_1}^{y_2} \left| \ln \frac{x_2}{s} \right|^{r_1-1} \left| \ln \frac{y_2}{t} \right|^{r_2-1} \frac{\|v(s,t)\|_E}{st} dt ds$$

$$+ \frac{1}{\Gamma(r_1)\Gamma(r_2)} \int_1^{x_1} \int_{y_1}^{y_2} \left| \ln \frac{x_2}{s} \right|^{r_1-1} \left| \ln \frac{y_2}{t} \right|^{r_2-1} \frac{\|v(s,t)\|_E}{st} dt ds$$

$$+ \frac{1}{\Gamma(r_1)\Gamma(r_2)} \int_{x_1}^{x_2} \int_1^{y_1} \left| \ln \frac{x_2}{s} \right|^{r_1-1} \left| \ln \frac{y_2}{t} \right|^{r_2-1} \frac{\|v(s,t)\|_E}{st} dt ds.$$

Thus

$$\|(hu)(x_2, y_2) - (hu)(x_1, y_1)\|_E$$

$$\leq \|\mu(x_1, y_1) - \mu(x_2, y_2)\|_E$$

$$+ \frac{1}{\Gamma(r_1)\Gamma(r_2)} \int_1^{x_1} \int_1^{y_1} \left[\left| \ln \frac{x_2}{s} \right|^{r_1-1} \left| \ln \frac{y_2}{t} \right|^{r_2-1} \right.$$

$$\left. - \left| \ln \frac{x_1}{s} \right|^{r_1-1} \left| \ln \frac{y_1}{t} \right|^{r_2-1} \right] \frac{p^*\psi(R)}{st} dt ds$$

$$+ \frac{1}{\Gamma(r_1)\Gamma(r_2)} \int_{x_1}^{x_2} \int_{y_1}^{y_2} \left| \ln \frac{x_2}{s} \right|^{r_1-1} \left| \ln \frac{y_2}{t} \right|^{r_2-1} \frac{p^*\psi(R)}{st} dt ds$$

$$+ \frac{1}{\Gamma(r_1)\Gamma(r_2)} \int_1^{x_1} \int_{y_1}^{y_2} \left| \ln \frac{x_2}{s} \right|^{r_1-1} \left| \ln \frac{y_2}{t} \right|^{r_2-1} \frac{p^*\psi(R)}{st} dt ds$$

$$+ \frac{1}{\Gamma(r_1)\Gamma(r_2)} \int_{x_1}^{x_2} \int_1^{y_1} \left| \ln \frac{x_2}{s} \right|^{r_1-1} \left| \ln \frac{y_2}{t} \right|^{r_2-1} \frac{p^*\psi(R)}{st} dt ds.$$

This gives

$$\|(hu)(x_2, y_2) - (hu)(x_1, y_1)\|_E$$

$$\leq \|\mu(x_1, y_1) - \mu(x_2, y_2)\|_E$$

$$+ \frac{p^* \psi(R)}{\Gamma(1 + r_1)\Gamma(1 + r_2)}$$

$$\times [2(\ln y_2)^{r_2} (\ln x_2 - \ln x_1)^{r_1} + 2(\ln x_2)^{r_1} (\ln y_2 - \ln y_1)^{r_2}$$

$$+ (\ln x_1)^{r_1} (\ln y_1)^{r_2} - (\ln x_2)^{r_1} (\ln y_2)^{r_2}$$

$$- 2(\ln x_2 - \ln x_1)^{r_1} (\ln y_2 - \ln y_1)^{r_2}].$$

Hence $h \in Q$. So $N(Q) \subset Q$.

Step 3. *N has weakly sequentially closed graph.*

Let (u_n, w_n) be a sequence in $Q \times Q$, with $u_n(x, y) \to u(x, y)$ in (E, ω) for each $(x, y) \in J$, $w_n(x, y) \to w(x, y)$ in (E, ω) for each $(x, y) \in J$, and $w_n \in \Omega(u_n)$ for $n \in \{1, 2, \ldots\}$. We show that $w \in \Omega(u)$. Since $w_n \in \Omega(u_n)$, then there exists $v_n \in S_{F \circ u_n}$ such that

$$w_n(x, y) = \mu(x, y) + ({}^H I^r_\sigma v_n)(x, y).$$

We show that there exists $v \in S_{F \circ u}$ such that, for each $(x, y) \in J$, we have

$$w(x, y) = \mu(x, y) + ({}^H I^r_\sigma v)(x, y).$$

Since $F(\cdot, \cdot, \cdot)$ has compact values, then there exists a subsequence v_{n_m} such that v_{n_m} is Pettis integrable,

$$v_{n_m}(x, y) \in F(x, y, u_n(x, y)) \text{ a.e. } (x, y) \in J,$$

and

$$v_{n_m}(\cdot, \cdot) \to v(\cdot, \cdot) \text{ in } (E, \omega) \text{ as } m \to \infty.$$

As $F(x, y, \cdot)$ has weakly-sequentially closed graph, $v(x, y) \in F(x, y, u(x, y))$. Then the Lebesgue dominated convergence theorem for the Pettis integral implies that

$$\phi(w_n(x, y)) \to \phi(\mu(x, y) + ({}^H I^r_\sigma v)(x, y)),$$

i.e., $w_n(x, y) \to (Nu)(x, y)$ in (E, ω). Since this holds, for each $(x, y) \in J$, we get $w \in N(u)$.

Step 4. *The implication (1.9) holds.*

Let V be a subset of Q such that $\overline{V} = \overline{\text{conv}}(N(V) \cup \{0\})$. Obviously $V(x,y) \subset \overline{\text{conv}}(N(V(x,y)) \cup \{0\})$ for each $(x,y) \in J$. Further, as V is bounded and equicontinuous, the function $(x,y) \to v(x,y) = \beta(V(x,y))$ is continuous on J. Since μ is continuous on J, then the set $\{\mu(x,y); (x,y) \in J\} \subset E$ is compact.

From (6.4.4), Lemma 2.3 and the properties of the measure β, for any $(x,y) \in J$ we have

$$v(x,y) \leq \beta((NV)(x,y) \cup \{0\})$$

$$\leq \beta((NV)(x,y))$$

$$\leq \beta\{(Nu)(x,y) : u \in V\}$$

$$\leq \beta\{(^H I_\sigma^r v)(x,y) : v \in S_{Fou}, \ u \in V\}$$

$$\leq \beta\{(^H I_\sigma^r F)(x,y,V(x,y))\}$$

$$\leq \frac{2}{\Gamma(r_1)\Gamma(r_2)} \int_1^x \int_1^y \left|\ln \frac{x}{s}\right|^{r_1-1} \left|\ln \frac{y}{t}\right|^{r_2-1}$$
$$\times \frac{p(s,t)\beta(V(s,t))}{st} dt\,ds$$

$$\leq \frac{2}{\Gamma(r_1)\Gamma(r_2)} \int_1^x \int_1^y \left|\ln \frac{x}{s}\right|^{r_1-1} \left|\ln \frac{y}{t}\right|^{r_2-1}$$
$$\times \frac{p(s,t)v(s,t)}{st} dt\,ds$$

$$\leq \frac{2\|v\|_C}{\Gamma(r_1)\Gamma(r_2)} \int_1^x \int_1^y \left|\ln \frac{x}{s}\right|^{r_1-1} \left|\ln \frac{y}{t}\right|^{r_2-1}$$
$$\times \frac{p(s,t)}{st} dt\,ds$$

$$\leq \frac{2p^*(\ln a)^{r_1}(\ln b)^{r_2}}{\Gamma(1+r_1)\Gamma(1+r_2)} \|v\|_C.$$

Thus

$$\|v\|_C \leq L\|v\|_C.$$

From (6.6), we get $\|v\|_C = 0$, that is $v(x,y) = \beta(V(x,y)) = 0$, for each $(x,y) \in J$. and then by [290, Theorem 2], V is relatively-weakly compact in C. Applying Theorem 1.7, we conclude that N has a fixed point which is a solution of the integral inclusion (6.5). $\qquad \square$

6.3.3. *Example*

Let

$$E = l^1 = \left\{ w = (w_1, w_2, \ldots, w_n, \ldots) : \sum_{n=1}^{\infty} |w_n| < \infty \right\}$$

be the Banach space with norm

$$\|w\|_E = \sum_{n=1}^{\infty} |w_n|,$$

and consider the following partial functional fractional order integral inclusion:

$$u(x, y) - \mu(x, y) \in (^H I_\sigma^r F)(x, y, u(x, y)); \text{ a.e. } (x, y) \in [1, e] \times [1, e], \quad (6.8)$$

where $r = (r_1, r_2)$, $r_1, r_2 \in (0, \infty)$,

$$u = (u_1, u_2, \ldots, u_n, \ldots), \quad \mu(x, y) = (x + e^{-y}, 0, \ldots, 0, \ldots),$$

and

$$F(x, y, u(x, y)) = \{v \in C([1, e] \times [1, e], \mathbb{R}) : \|f_1(x, y, u(x, y))\|_E \le \|v\|_E$$
$$\le \|f_2(x, y, u(x, y))\|_E\};$$
$$(x, y) \in [1, e] \times [1, e], \text{ where } f_1, f_2 : [1, e] \times [1, e] \times E \to E,$$

$$f_k = (f_{k,1}, f_{k,2}, \ldots, f_{k,n}, \ldots); \; k \in \{1, 2\}, \; n \in \mathbb{N},$$

$$f_{1,n}(x, y, u_n(x, y)) = \frac{xy^2 u_n}{(1 + \|u_n\|_E) e^{10+x+y}}; \; n \in \mathbb{N},$$

and

$$f_{2,n}(x, y, u_n(x, y)) = \frac{xy^2 u_n}{e^{10+x+y}}; \quad n \in \mathbb{N}.$$

We assume that F is closed and convex valued. We can see that the solutions of the inclusion(6.8) are solutions of the fixed-point inclusion $u \in A(u)$ where $A : C([1, e] \times [1, e], \mathbb{R}) \to \mathcal{P}(C([1, e] \times [1, e], \mathbb{R}))$ is the multifunction operator defined by

$$(Au)(x, y) = \{\mu(x, y) + (^H I_\sigma^r f)(x, y); \; f \in S_{F \circ u}\}; \; (x, y) \in [1, e] \times [1, e].$$

For each $(x, y) \in [1, e] \times [1, e]$ and all $z_1, z_2 \in E$, we have

$$\|f_2(x, y, z_2) - f_1(x, y, z_1)\|_E \leq xy^2 e^{-10-x-y} \|z_2 - z_1\|_E.$$

A simple computations show that all conditions of Theorem 6.2 are satisfied. Indeed, $p(x, y) = xy^2 e^{-10-x-y}$, and the condition (6.6) holds with $a = b = e$. So,

$$p^* = e^{-9}, \ \Gamma(1 + r_i) > \frac{1}{2}; \ i = 1, 2,$$

and a simple computation shows that

$$L := \frac{2p^* (\ln a)^{r_1} (\ln b)^{r_2}}{\Gamma(1 + r_1)\Gamma(1 + r_2)} < 8e^{-9} < 1.$$

Hence, it follows from Theorem 6.2 that the integral inclusion(6.8) has at least one solution on $[1, e] \times [1, e]$.

6.4. Fredholm-Type Partial Hadamard–Pettis Fractional Integral Equations

6.4.1. *Introduction*

This section deals with the existence of solutions to the following Pettis–Hadamard partial fractional integral equation:

$$u(x, y) = \mu(x, y) + \int_1^a \int_1^b \left(\ln \frac{a}{s}\right)^{r_1 - 1} \left(\ln \frac{b}{t}\right)^{r_2 - 1}$$
$$\times \frac{f(x, y, s, t, u(s, t), (^H D_\sigma^r u)(s, t))}{\Gamma(r_1)\Gamma(r_2)st} dt ds \quad \text{if} \quad (x, y) \in J, \ (6.9)$$

where $J := [1, a] \times [1, b]$, $a, b > 1$, $^H D_\theta^r$ is the standard Hadamard's fractional derivative of order $r = (r_1, r_2) \in (0, 1] \times (0, 1]$, $\mu : J \to E$, $f : J \times J \times E \times E \to E$ are given continuous functions, $\Gamma(\cdot)$ is the Euler gamma function, and E is a real (or complex) Banach space with norm $\| \cdot \|_E$ and dual E^*, such that E is the dual of a weakly-compactly-generated Banach space X.

6.4.2. *Existence of weak solutions*

Define the space $X = X(J, E)$ as follows:

$$X := \{w \in C(J) : \ ^H D_\sigma^r w \text{ exists and } ^H D_\sigma^r w \in C(J)\}.$$

For $w \in X$, we define

$$\|w(x,y)\|_1 = \|w(x,y)\|_E + \|{}^H D_\sigma^r w(x,y)\|_E.$$

In the space X we define the norm

$$\|w\|_X = \sup_{(x,y)\in J} \|w(x,y)\|_1.$$

Lemma 6.1. $(X, \|\cdot\|_X)$ *is a Banach space.*

Proof. Let $\{u_n\}_{n=0}^\infty$ be a Cauchy sequence in the space $(X, \|\cdot\|_X)$. Then,

$$\forall \epsilon > 0, \ \exists N > 0 \text{ such that for all } n, m > N, \text{ we have } \|u_n - u_m\|_X < \epsilon.$$

Thus, $\{u_n(x,y)\}_{n=0}^\infty$ and $\{({}^H D_\sigma^r u_n)(x,y)\}_{n=0}^\infty$ are Cauchy sequences in E. Then $\{u_n(x,y)\}_{n=0}^\infty$ converges to some $u(x,y)$ in E, and $\{{}^H D_\sigma^r u_n\}_{n=0}^\infty$ converges uniformly to some $v(x,y) \in X$. Next, we need to prove that $u \in X$ and $v = {}^H D_\sigma^r u$. According to the uniform convergence of $\{({}^H D_\sigma^r u_n)(x,y)\}_{n=0}^\infty$ and the dominated convergence theorem, we obtain

$$v(x,y) = \lim_{n\to\infty} ({}^H D_\sigma^r u_n)(x,y).$$

Thus $\{{}^H D_\sigma^r u_n\}_{n=0}^\infty$ converges uniformly to ${}^H D_\sigma^r u$ in X. Hence, $u \in X$ and

$$v(x,y) = ({}^H D_\sigma^r u)(x,y). \qquad \square$$

Definition 6.3. A function $w \in X$ is said to be a solution of (6.9) if w satisfies the integral equation (6.9) on J.

Further, we present conditions for the existence of a solution of equation (6.9).

Theorem 6.3. *Assume that the following hypotheses hold:*

(6.3.1) *For a.e. $(s,t) \in J$, the function $(v,w) \to f(x,y,s,t,v,w)$; $(x,y) \in J$ is jointly-weakly-sequentially continuous.*

(6.3.2) *For a.e. $v, w \in E$, the function $(s,t) \to f(x,y,s,t,v,w)$; $(x,y) \in J$ is Pettis integrable a.e. on J.*

(6.3.3) *There exist $0 < r_3 < \min\{r_1, r_2\}$, functions $\rho_1 : J \times J \to \mathbb{R}^+$, $\varphi : J \to \mathbb{R}^+$, with $\rho_1(x, y, \cdot, \cdot), \varphi \in L^{\frac{1}{r_3}}(J)$ and a nondecreasing function $\psi : [0, \infty) \to (0, \infty)$ such that for all $\phi \in E^*$, we have*

$$|\phi(f(x, y, s, t, u, v))| \le \rho_1(x, y, s, t)\|\phi\|, \tag{6.10}$$

and

$$|\phi(f(x_1, y_1, s, t, u, v) - f(x_2, y_2, s, t, u, v))|$$
$$\le \frac{\varphi(s, t)\|\phi\|}{1 + \|\phi\|}(|x_1 - x_2| + |y_1 - y_2|)\psi(\|u\|_E + \|v\|_E), \tag{6.11}$$

for each $(x, y), (s, t), (x_1, y_1), (x_2, y_2) \in J$ and $u, v \in E$.

(6.3.4) *There exist nonnegative constants α, β_1, β_2 such that, for $(x, y) \in J$, we have*

$$\begin{cases} \|\mu(x, y)\|_1 \le \alpha, \\ \int_1^a \int_1^b \rho_1^{\frac{1}{r_3}}(x, y, s, t)dtds \le \beta_1^{\frac{1}{r_3}}, \\ \int_1^a \int_1^b \rho_2^{\frac{1}{r_3}}(x, y, s, t)dtds \le \beta_2^{\frac{1}{r_3}}, \end{cases} \tag{6.12}$$

where

$$\rho_2(x, y, \cdot, \cdot) \in L^{\frac{1}{r_3}}(J) \text{ and } \rho_2(x, y, s, t) = (^H D_\sigma^r \rho_1)(x, y, s, t).$$

(6.3.5) *For each measurable and bounded set $B \subset E$ and for each $(x, y) \in J$, we have*

$$\beta(f(x, y, s, t, B, {}^H D_\sigma^r B)) \le \rho_1(x, y, s, t)\beta(B),$$

where ${}^H D_\sigma^r B := \{{}^H D_\sigma^r u : u \in B\}$.

If

$$\ell := \frac{(\beta_1 + \beta_2)(\ln a)^{(\omega_1+1)(1-r_3)}(\ln b)^{(\omega_2+1)(1-r_3)}}{(\omega_1 + 1)^{(1-r_3)}(\omega_2 + 1)^{(1-r_3)}\Gamma(r_1)\Gamma(r_2)} < 1, \tag{6.13}$$

where $\omega_1 = \frac{r_1-1}{1-r_3}$, $\omega_2 = \frac{r_2-1}{1-r_3}$, then the integral equation (6.9) has at least one solution defined on J.

Remark 6.1. It is clear that the condition (6.10) implies

$$|\phi(^H D_\sigma^r f)(x, y, s, t, u, v)| \le \frac{\rho_2(x, y, s, t)\|\phi\|}{1 + \|\phi\| + \|u\|_E + \|v\|_E}. \tag{6.14}$$

Proof of Theorem 6.3. Let $u \in X$ and define the operator $N : X \to X$ by

$$(Nu)(x,y) = \mu(x,y) + \int_1^a \int_1^b \left(\ln \frac{a}{s}\right)^{r_1-1} \left(\ln \frac{b}{t}\right)^{r_2-1}$$

$$\times \frac{f(x,y,s,t,u(s,t),(^H D_\sigma^r u)(s,t))}{st\Gamma(r_1)\Gamma(r_2)} dt ds. \tag{6.15}$$

Differentiating both sides of (6.15) by applying the Hadamard fractional derivative, we get

$$^H D_\sigma^r (Nu)(x,y) = {}^H D_\sigma^r \mu(x,y) + \int_1^a \int_1^b \left(\ln \frac{a}{s}\right)^{r_1-1} \left(\ln \frac{b}{t}\right)^{r_2-1}$$

$$\times \frac{{}^H D_\sigma^r f(x,y,s,t,u(s,t),(^H D_\sigma^r u)(s,t))}{st\Gamma(r_1)\Gamma(r_2)} dt ds. \tag{6.16}$$

First notice that, the hypothesis (6.3.2) implies that

$$\forall u \in C : f(x,y,\cdot,\cdot,u(\cdot,\cdot),(^H D_\sigma^r u)(\cdot,\cdot)) \in P(J,E).$$

From (6.3.3)–(6.3.5) we have that

$$\left(\ln \frac{a}{s}\right)^{r_1-1} \left(\ln \frac{b}{t}\right)^{r_2-1} \frac{f(x,y,s,t,u(s,t),(^H D_\sigma^r u)(s,t))}{st}$$

is Pettis integrable on J and thus, the operator N is well defined. Let $R > 0$ be such that

$$R > \frac{\alpha}{1 - \ell},$$

and consider the set

$$Q = \left\{ u \in X : \|u\|_X \leq R \text{ and } \|u(x_1,y_1) - u(x_2,y_2)\|_X \right.$$

$$\leq \|\mu(x_1,y_1) - \mu(x_2,y_2)\|_X$$

$$+ \frac{(\ln a)^{(\omega_1+1)(1-r_3)}(\ln b)^{(\omega_2+1)(1-r_3)}}{(\omega_1+1)^{(1-r_3)}(\omega_2+1)^{(1-r_3)}\Gamma(r_1)\Gamma(r_2)}$$

$$\left. \times [(\|\varphi\|_{L^{\frac{1}{r_3}}})^{r_3} + (\|^H D_\sigma^r \varphi\|_{L^{\frac{1}{r_3}}})^{r_3}]\psi(R)(|x_1 - x_2| + |y_1 - y_2|) \right\}.$$

Clearly, the subset Q is closed, convex and equicontinuous. We shall show that the operator N satisfies all the assumptions of Theorem 1.8. The proof will be given in several steps.

Step 1. N *maps* Q *into itself.*

Let $u \in Q$, $(x,y) \in J$ and assume that $(Nu)(x,y) \neq 0$. Then there exists $\phi \in X^*$ such that $\|(Nu)(x,y)\|_1 = \phi((Nu)(x,y))$. Then

$$
\|(Nu)(x,y)\|_1 = \left| \phi\left(\mu(x,y) + \frac{1}{\Gamma(r_1)\Gamma(r_2)} \int_1^a \int_1^b \left(\ln\frac{a}{s}\right)^{r_1-1} \left(\ln\frac{b}{t}\right)^{r_2-1} \right. \right.
$$
$$
\left. \left. \times \frac{f(x,y,s,t,u(s,t),(^H D_\sigma^r u)(s,t))}{st} dt\,ds \right) \right| = |\phi(\mu(x,y))|
$$
$$
+ \left| \phi\left(\frac{1}{\Gamma(r_1)\Gamma(r_2)} \int_1^a \int_1^b \left(\ln\frac{a}{s}\right)^{r_1-1} \left(\ln\frac{b}{t}\right)^{r_2-1} \right. \right.
$$
$$
\left. \left. \times \frac{f(x,y,s,t,u(s,t),(^H D_\sigma^r u)(s,t))}{st} dt\,ds \right) \right|
$$

Thus
$$
\|(Nu)(x,y)\|_1 \leq \|\mu(x,y)\|_1 + \frac{1}{\Gamma(r_1)\Gamma(r_2)} \int_1^a \int_1^b \left|\ln\frac{a}{s}\right|^{r_1-1} \left|\ln\frac{b}{t}\right|^{r_2-1}
$$
$$
\times \left| \frac{\phi(f(x,y,s,t,u(s,t),(^H D_\sigma^r u)(s,t)))}{st} \right| dt\,ds
$$
$$
+ \frac{1}{\Gamma(r_1)\Gamma(r_2)} \int_1^a \int_1^b \left(\ln\frac{a}{s}\right)^{r_1-1} \left(\ln\frac{b}{t}\right)^{r_2-1}
$$
$$
\times \left| \frac{|\phi(^H D_\sigma^r f(x,y,s,t,u(s,t),(^H D_\sigma^r u)(s,t)))|}{st} \right| dt\,ds
$$
$$
\leq \|\mu(x,y)\|_1 + \frac{1}{\Gamma(r_1)\Gamma(r_2)} \left(\int_1^a \int_1^b \frac{1}{st} \left|\ln\frac{a}{s}\right|^{\frac{r_1-1}{1-r_3}} \right.
$$
$$
\times \left. \left|\ln\frac{b}{t}\right|^{\frac{r_2-1}{1-r_3}} dt\,ds \right)^{1-r_3} \left(\int_1^a \int_1^b |\phi(f(x,y,s,t,u(s,t), \right.
$$
$$
\times \left. (^H D_\sigma^r u)(s,t)))|^{\frac{1}{r_3}} dt\,ds \right)^{r_3}
$$
$$
+ \frac{1}{\Gamma(r_1)\Gamma(r_2)} \left(\int_1^a \int_1^b \frac{1}{st} \left|\ln\frac{a}{s}\right|^{\frac{r_1-1}{1-r_3}} \left|\ln\frac{b}{t}\right|^{\frac{r_2-1}{1-r_3}} dt\,ds \right)^{1-r_3}
$$
$$
\times \left(\int_1^a \int_1^b |\phi(^H D_\sigma^r f(x,y,s,t,u(s,t), \right.
$$
$$
\times \left. (^H D_\sigma^r u)(s,t)))|^{\frac{1}{r_3}} dt\,ds \right)^{r_3}.
$$

Hence, for each $(x, y) \in J$, we obtain

$$\|(Nu)(x,y)\|_1 \leq \|\mu(x,y)\|_1 + \frac{(\ln a)^{(\omega_1+1)(1-r_3)}(\ln b)^{(\omega_2+1)(1-r_3)}}{(\omega_1+1)^{(1-r_3)}(\omega_2+1)^{(1-r_3)}\Gamma(r_1)\Gamma(r_2)}$$

$$\times \left[\left(\int_1^a \int_1^b \rho_1^{\frac{1}{r_3}}(x,y,s,t)\|u(s,t)\|_1^{\frac{1}{r_3}}\,dtds\right)^{r_3}\right.$$

$$\left. + \left(\int_1^a \int_1^b \rho_2^{\frac{1}{r_3}}(x,y,s,t)\|u(s,t)\|_1^{\frac{1}{r_3}}\,dtds\right)^{r_3}\right]$$

$$\leq \alpha + \frac{(\ln a)^{(\omega_1+1)(1-r_3)}(\ln b)^{(\omega_2+1)(1-r_3)}}{(\omega_1+1)^{(1-r_3)}(\omega_2+1)^{(1-r_3)}\Gamma(r_1)\Gamma(r_2)}$$

$$\times \left[\|u\|_X\left(\int_1^a \int_1^b \rho_1^{\frac{1}{r_3}}(x,y,s,t)\,dtds\right)^{r_3}\right.$$

$$\left. + \|u\|_X\left(\int_0^a \int_0^b \rho_2^{\frac{1}{r_3}}(x,y,s,t)\,dtds\right)^{r_3}\right]$$

$$\leq \alpha + \frac{(R\beta_1 + R\beta_2)(\ln a)^{(\omega_1+1)(1-r_3)}(\ln b)^{(\omega_2+1)(1-r_3)}}{(\omega_1+1)^{(1-r_3)}(\omega_2+1)^{(1-r_3)}\Gamma(r_1)\Gamma(r_2)}$$

$$= \alpha + \frac{1}{2}R\ell$$

$$\leq \alpha + R\ell.$$

From (6.13) and the definition of R, we get

$$\|N(u)\|_X \leq R.$$

Next, let $(x_1, y_1), (x_2, y_2) \in J$ such that $x_1 < x_2$ and $y_1 < y_2$, and let $u \in Q$, with $(Nu)(x_1, y_1) - (Nu)(x_2, y_2) \neq 0$. Then there exists $\phi \in E^*$ such that

$$\|(Nu)(x_1,y_1) - (Nu)(x_2,y_2)\|_1 = |\phi((Nu)(x_1,y_1) - (Nu)(x_2,y_2))|$$

and $\|\phi\| = 1$. Then

$$\|(Nu)(x_2,y_2) - (Nu)(x_1,y_1)\|_1 = |\phi((Nu)(x_2,y_2) - (Nu)(x_1,y_1))|$$

$$\leq \|\mu(x_1,y_1) - \mu(x_2,y_2)\|_1$$

$$+ \frac{1}{\Gamma(r_1)\Gamma(r_2)}\int_1^a \int_1^b \frac{1}{st}\left|\ln\frac{a}{s}\right|^{r_1-1}\left|\ln\frac{b}{t}\right|^{r_2-1}$$

$$\times \, |\phi(f(x_2, y_2, s, t, u(s,t), ({}^H D_\sigma^r u)(s,t))$$

$$- f(x_1, y_1, s, t, u(s,t), ({}^H D_\sigma^r u)(s,t)))| dt ds$$

$$+ \frac{1}{\Gamma(r_1)\Gamma(r_2)} \int_1^a \int_1^b \frac{1}{st} \left|\ln \frac{a}{s}\right|^{r_1 - 1} \left|\ln \frac{b}{t}\right|^{r_2 - 1}$$

$$\times \, |\phi({}^H D_\sigma^r f(x_2, y_2, s, t, u(s,t), ({}^H D_\sigma^r u)(s,t))$$

$$- {}^H D_\sigma^r f(x_1, y_1, s, t, u(s,t), ({}^H D_\sigma^r u)(s,t)))| dt ds.$$

Thus

$$\|(Nu)(x_2, y_2) - (Nu)(x_1, y_1)\|_1 \le \|\mu(x_1, y_1) - \mu(x_2, y_2)\|_1$$

$$+ \frac{(\ln a)^{(\omega_1 + 1)(1 - r_3)} (\ln b)^{(\omega_2 + 1)(1 - r_3)}}{(\omega_1 + 1)^{(1 - r_3)} (\omega_2 + 1)^{(1 - r_3)} \Gamma(r_1)\Gamma(r_2)}$$

$$\times \left[\left(\int_1^a \int_1^b |\phi(f(x_2, y_2, s, t, u(s,t), ({}^H D_\sigma^r u)(s,t)) \right. \right.$$

$$\left. - f(x_1, y_1, s, t, u(s,t), ({}^H D_\sigma^r u)(s,t)))|^{\frac{1}{r_3}} dt ds \right)^{r_3}$$

$$+ \left(\int_1^a \int_1^b |\phi({}^H D_\sigma^r f(x_2, y_2, s, t, u(s,t), ({}^H D_\sigma^r u)(s,t)) \right.$$

$$\left. \left. - {}^H D_\sigma^r f(x_1, y_1, s, t, u(s,t), ({}^H D_\sigma^r u)(s,t)))|^{\frac{1}{r_3}} dt ds \right)^{r_3} \right].$$

Hence

$$\|(Nu)(x_2, y_2) - (Nu)(x_1, y_1)\|_1 \le \|\mu(x_1, y_1) - \mu(x_2, y_2)\|_1$$

$$+ \frac{(\ln a)^{(\omega_1 + 1)(1 - r_3)} (\ln b)^{(\omega_2 + 1)(1 - r_3)}}{(\omega_1 + 1)^{(1 - r_3)} (\omega_2 + 1)^{(1 - r_3)} \Gamma(r_1)\Gamma(r_2)}$$

$$\times (|x_1 - x_2| + |y_1 - y_2|)\psi(\|u\|_1)$$

$$\times \left[\left(\int_1^a \int_1^b \|\varphi(s,t)\|^{\frac{1}{r_3}} dt ds \right)^{r_3} \right.$$

$$\left. + \left(\int_1^a \int_1^b \|({}^H D_\sigma^r \varphi)(s,t)\|^{\frac{1}{r_3}} dt ds \right)^{r_3} \right]$$

$$\le \|\mu(x_1, y_1) - \mu(x_2, y_2)\|_1$$

$$+ \frac{(\ln a)^{(\omega_1 + 1)(1 - r_3)} (\ln b)^{(\omega_2 + 1)(1 - r_3)}}{(\omega_1 + 1)^{(1 - r_3)} (\omega_2 + 1)^{(1 - r_3)} \Gamma(r_1)\Gamma(r_2)}$$

$$\times [(\|\varphi\|_{L^{\frac{1}{r_3}}})^{r_3} + (\|^H D_\sigma^r \varphi\|_{L^{\frac{1}{r_3}}})^{r_3}]\psi(R)$$

$$\times (|x_1 - x_2| + |y_1 - y_2|).$$

Therefore $N(Q) \subset Q$.

Step 2. *N is weakly-sequentially continuous.*

Let (u_n) be a sequence in Q and let $(u_n(x,y)) \to u(x,y)$ in (X,ω) for each $(x,y) \in J$. Fix $(x,y) \in J$, since f satisfies the assumption (6.3.1), we have that $f(x,y,s,t,u_n(x,y),(^H D_\sigma^r u_n)(s,t))$ converges weakly to

$$f(x,y,s,t,u(x,y),(^H D_\sigma^r u)(s,t)).$$

Hence the Lebesgue dominated convergence theorem for Pettis integral implies $(Nu_n)(x,y)$ converges weakly to $(Nu)(x,y)$ in (X,ω). We do it for each $(x,y) \in J$, so $N(u_n) \to N(u)$. Then $N : Q \to Q$ is weakly-sequentially continuous.

Step 3. *The implication* (1.9) *holds.*

Let V be a subset of Q such that $\overline{V} = \overline{\text{conv}}(N(V) \cup \{0\})$. Obviously $V(x,y) \subset \overline{\text{conv}}(NV)(x,y)) \cup \{0\})$; $(x,y) \in J$. Further, as V is bounded and equicontinuous, by [172, Lemma 3], the function $(x,y) \to u(x,y) = \beta(V(x,y))$ is continuous on J. Since the functions μ and $^H D_\sigma^r \mu$ are continuous on J, the set $\{\mu(x,y); (x,y) \in J\} \subset X$ is compact. From (6.3.3) − (6.3.5), Lemma 1.7 and the properties of the measure β, for any $(x,y) \in J$, we have

$$\|v(x,y)\|_1 \leq \beta((NV)(x,y) \cup \{0\})$$

$$\leq \beta((NV)(x,y))$$

$$\leq \frac{1}{\Gamma(r_1)\Gamma(r_2)} \int_1^a \int_1^b \left|\ln\frac{a}{s}\right|^{r_1-1} \left|\ln\frac{b}{t}\right|^{r_2-1}$$

$$\times \frac{\rho_1(x,y,s,t)\beta(V(s,t))}{st} dt ds$$

$$+ \frac{1}{\Gamma(r_1)\Gamma(r_2)} \int_1^a \int_1^b \left|\ln\frac{a}{s}\right|^{r_1-1} \left|\ln\frac{b}{t}\right|^{r_2-1}$$

$$\times \frac{\rho_2(x,y,s,t)\beta(V(s,t))}{st} dt ds$$

$$\leq \frac{(\ln a)^{(\omega_1+1)(1-r_3)}(\ln b)^{(\omega_2+1)(1-r_3)}}{(\omega_1+1)^{(1-r_3)}(\omega_2+1)^{(1-r_3)}\Gamma(r_1)\Gamma(r_2)}$$

$$\times \left[\left(\int_1^a \int_1^b \rho_1^{\frac{1}{r_3}}(x,y,s,t)\|v(s,t)\|_1^{\frac{1}{r_3}}\,dtds \right)^{r_3} \right.$$

$$\left. + \left(\int_1^a \int_1^b \rho_2^{\frac{1}{r_3}}(x,y,s,t)\|v(s,t)\|_1^{\frac{1}{r_3}}\,dtds \right)^{r_3} \right]$$

$$\leq \frac{(\ln a)^{(\omega_1+1)(1-r_3)}(\ln b)^{(\omega_2+1)(1-r_3)}}{(\omega_1+1)^{(1-r_3)}(\omega_2+1)^{(1-r_3)}\Gamma(r_1)\Gamma(r_2)}$$

$$\times \left[\|v\|_X \left(\int_1^a \int_1^b \rho_1^{\frac{1}{r_3}}(x,y,s,t)dtds \right)^{r_3} \right.$$

$$\left. + \|v\|_X \left(\int_0^a \int_0^b \rho_2^{\frac{1}{r_3}}(x,y,s,t)dtds \right)^{r_3} \right]$$

$$\leq \frac{(\beta_1+R\beta_2)(\ln a)^{(\omega_1+1)(1-r_3)}(\ln b)^{(\omega_2+1)(1-r_3)}}{(\omega_1+1)^{(1-r_3)}(\omega_2+1)^{(1-r_3)}\Gamma(r_1)\Gamma(r_2)}\|v\|_X$$

$$\leq \ell\|v\|_X.$$

Thus

$$\|v\|_X \leq \ell\|v\|_X.$$

From (6.13), we get $\|v\|_X = 0$, that is $v(x,y) = \beta(V(x,y)) = 0$, for each $(x,y) \in J$. and then by [290, Theorem 2], V is relatively weakly compact in X. Applying now Theorem 1.7, we conclude that N has a fixed point which is a solution of the integral equation (6.9). $\qquad\square$

Define now the Banach space $\Lambda = \Lambda(J,E)$ with the norm

$$\|w\|_\Lambda = \sup_{(x,y)\in J} \|w(x,y)\|_1,$$

as follows:

$$\Lambda := \{w \in C(J) : {}^H D_{1,x}^{r_1}w, {}^H D_{1,y}^{r_2}w \text{ exist and } {}^H D_{1,x}^{r_1}w, {}^H D_{1,y}^{r_2}w \in C(J)\},$$

and

$$\|w(x,y)\|_1 = \|w(x,y)\|_E + \|{}^H D_{1,x}^{r_1}w(x,y)\|_E + \|{}^H D_{1,y}^{r_2}w(x,y)\|_E.$$

Corollary 6.1. *Consider the following Fredholm type Hadamard integral equation:*

$$u(x,y) = \mu(x,y) + \frac{1}{\Gamma(r_1)\Gamma(r_2)} \int_1^a \int_1^b \left(\ln\frac{x}{s}\right)^{r_1-1} \left(\ln\frac{y}{t}\right)^{r_2-1}$$

$$\times f(x,y,s,t,u(s,t),(^H D_{1,s}^{r_1}u)(s,t),(^H D_{1,t}^{r_2}u)(s,t))dtds \tag{6.17}$$

$$if \ (x,y) \in J := [1,a] \times [1,b].$$

Assume (6.3.1), (6.3.2), and the following hypotheses hold:

(6.4.1) *There exist $0 < r_3 < \min\{r_1,r_2\}$, functions $\rho_1 : J \times J \to \mathbb{R}^+$, $\varphi : J \to \mathbb{R}^+$, with $\rho_1(x,y,\cdot,\cdot), \varphi \in L^{\frac{1}{r_3}}(J)$ and a nondecreasing function $\psi : [0,\infty) \to (0,\infty)$ such that for all $\phi \in E^*$, we have*

$$|\phi(f(x,y,s,t,u,v,w))| \le \rho_1(x,y,s,t)\|\phi\|, \tag{6.18}$$

and

$$|\phi(f(x_1,y_1,s,t,u,v,w) - f(x_2,y_2,s,t,u,v,w))|$$

$$\le \frac{\varphi(s,t)}{1+\|\phi\|}(|x_1-x_2|+|y_1-y_2|)\psi(\|u\|_E + \|v\|_E + \|w\|_E), \tag{6.19}$$

for each $(x,y),(s,t),(x_1,y_1),(x_2,y_2) \in J$ and $u,v,w \in E$.

(6.4.2) *There exist nonnegative constants $\alpha, \beta_1, \beta_2, \beta_3$ such that, for $(x,y) \in J$, we have*

$$\begin{cases} \|\mu(x,y)\|_1 \le \alpha, \\ \int_1^a \int_1^b \rho_1^{\frac{1}{r_3}}(x,y,s,t)dtds \le \beta_1^{\frac{1}{r_3}}, \\ \int_1^a \int_1^b \rho_2^{\frac{1}{r_3}}(x,y,s,t)dtds \le \beta_2^{\frac{1}{r_3}}, \\ \int_1^a \int_1^b \rho_3^{\frac{1}{r_3}}(x,y,s,t)dtds \le \beta_3^{\frac{1}{r_3}}, \end{cases} \tag{6.20}$$

where

$$\rho_2(x,y,s,t) = (^H D_{1,x}^r \rho_1)(x,y,s,t), \ and \ \rho_3(x,y,s,t)$$
$$= (^H D_{1,y}^r \rho_1)(x,y,s,t).$$

(6.4.3) *For each bounded set $B \subset X$ and for each $(x, y) \in J$, we have*

$$\beta(f(x, y, s, t, B, {}^H D^{r_1}_{1,s} B, {}^H D^{r_2}_{1,t} B) \leq p_1(x, y, s, t) \beta(B),$$

where ${}^H D^r_{1,.} B := \{ {}^H D^r_{1,.} u : u \in B \}$.

If

$$\frac{(\beta_1 + \beta_2 + \beta_3) a^{(\omega_1+1)(1-r_3)} b^{(\omega_2+1)(1-r_3)}}{(\omega_1 + 1)^{(1-r_3)}(\omega_2 + 1)^{(1-r_3)} \Gamma(r_1) \Gamma(r_2)} < 1, \qquad (6.21)$$

where $\omega_1 = \frac{r_1-1}{1-r_3}$, $\omega_2 = \frac{r_2-1}{1-r_3}$, then Eq. (6.17) has at least one solution on J in Λ.

6.4.3. Example

Let

$$E = l^1 = \left\{ u = (u_1, u_2, \ldots, u_n, \ldots), \sum_{n=1}^{\infty} |u_n| < \infty \right\}$$

be the Banach space with the norm

$$\|u\|_E = \sum_{n=1}^{\infty} |u_n|.$$

As an application of our results, we consider the following Fredholm partial Hadamard integral equation:

$$u_n(x, y) = \mu(x, y) + \int_1^e \int_1^e (1 - \ln s)^{r_1 - 1} (1 - \ln t)^{r_2 - 1}$$

$$\times \frac{f_n(x, y, s, t, u(s, t), {}^H (D^r_\sigma u)(s, t))}{st \Gamma(r_1) \Gamma(r_2)} dt ds \qquad (6.22)$$

for $(x, y) \in [1, e] \times [1, e]$, where $r_1, r_2 \in (0, 1]$, $\mu(x, y) = x + y^2$; $(x, y) \in [1, e] \times [1, e]$, and

$$f_n(x, y, s, t, u(x, y), v(x, y)) = c(x + y) st^2 \frac{1}{(1 + |u(x, y)| + |v(x, y)|) e^{x+y+5}};$$

$(x, y) \in [1, e] \times [1, e]$, with

$$c := \frac{(\omega_1 + 1)^{(1-r_3)}(\omega_2 + 1)^{(1-r_3)} \Gamma(r_1) \Gamma(r_2)}{\frac{4e^{-3}}{(\Gamma(1+r_1))^{r_3}(\Gamma(1+r_2))^{r_3}} \left(1 + \frac{1}{\Gamma(1-r_1)\Gamma(1-r_2)}\right)},$$

$$0 < r_3 < \min\{r_1, r_2\}, \quad \omega_1 = \frac{r_1 - 1}{1 - r_3}, \text{ and } \omega_2 = \frac{r_2 - 1}{1 - r_3},$$

$$u = (u_1, u_2, \ldots, u_n, \ldots).$$

Set $f = (f_1, f_2, \ldots, f_n, \ldots)$.

A simple computations show that all conditions of Theorem 6.3 are satisfied. Specially, for each $u, v \in \mathbb{R}$ and $(x, y) \in [1, e] \times [1, e]$ we have

$$\|f(x, y, u, v)\|_1 \le 2ce^{-3},$$

and for each $(x, y), (s, t), (x_1, y_1), (x_2, y_2) \in [1, e] \times [1, e]$, and $u, v \in \mathbb{R}$, we have

$$|f(x_1, y_1, s, t, u, v) - f(x_2, y_2, s, t, u, v)| \le 2ce^{-3}(|x_1 - x_2| + |y_1 - y_2|)(|u| + |v|).$$

Hence condition (6.3.3) is satisfied with

$$\rho_1 = ce^{-3}, \quad \rho_2 = \frac{ce^{-3}}{\Gamma(1 - r_1)\Gamma(1 - r_2)}, \quad \varphi(s, t) = 2ce^{-3}, \quad \psi(x) = 1.$$

Also, (6.3.4) is satisfied with

$$\alpha = (e + e^2)\left(1 + \frac{1}{\Gamma(1 - r_1)\Gamma(1 - r_2)}\right),$$

$$\beta_1 = c\frac{e^{-3}}{(\Gamma(1 + r_1))^{r_3}(\Gamma(1 + r_2))^{r_3}}$$

and

$$\beta_2 = c\frac{e^{-3}}{\Gamma(1 - r_1)\Gamma(1 - r_2)\Gamma(1 + r_1))^{r_3}(\Gamma(1 + r_2))^{r_3}}.$$

We shall show that condition (6.13) holds with $a = b = e$. Indeed,

$$\begin{aligned}
\ell &= \frac{(\beta_1 + \beta_2)(\ln a)^{(\omega_1 + 1)(1 - r_3)}(\ln b)^{(\omega_2 + 1)(1 - r_3)}}{(\omega_1 + 1)^{(1 - r_3)}(\omega_2 + 1)^{(1 - r_3)}\Gamma(r_1)\Gamma(r_2)} \\
&= \frac{c\frac{e^{-3}}{(\Gamma(1 + r_1))^{r_3}(\Gamma(1 + r_2))^{r_3}}\left(1 + \frac{1}{\Gamma(1 - r_1)\Gamma(1 - r_2)}\right)}{(\omega_1 + 1)^{(1 - r_3)}(\omega_2 + 1)^{(1 - r_3)}\Gamma(r_1)\Gamma(r_2)} \\
&= \frac{1}{4} < 1.
\end{aligned}$$

Consequently, Theorem 6.3 implies the Fredholm–Hadamard integral equation (6.22) has at least one solution on $[1, e] \times [1, e]$.

6.5. Partial Hadamard–Stieltjes–Pettis Fractional Integral Equations

6.5.1. *Introduction*

This section deals with the existence of solutions to the following Hadamard–Stieltjes partial fractional integral equation.

$$
\begin{aligned}
u(x,y) = \mu(x,y) + \int_1^x \int_1^y &\left(\ln\frac{x}{s}\right)^{r_1-1}\left(\ln\frac{y}{t}\right)^{r_2-1} \\
&\times \frac{f(s,t,u(s,t))}{\Gamma(r_1)\Gamma(r_2)st}\, d_t g_2(y,t)\, d_s g_1(x,s); \quad \text{if}(x,y)\in J,
\end{aligned}
$$

(6.23)

where $J := [1,a]\times[1,b]$, $a,b > 1$, $r_1, r_2 > 0$, $\mu : J \to E$, $f : J \times E \to E$, $g_1 : [1,a]^2 \to \mathbb{R}$, $g_2 : [1,b]^2 \to \mathbb{R}$ are given continuous functions, $\Gamma(\cdot)$ is the Euler gamma function, and E is a real (or complex) Banach space with norm $\|\cdot\|_E$ and dual E^*.

6.5.2. *Existence of weak solutions*

Definition 6.4. A function $u \in C$ is said to be a solution of (6.23) if u satisfies Eq. (6.23) on J.

Further, we present conditions for the existence of a solution of Eq. (6.23). Set

$$
g^* = \sup_{(x,y)\in J} \bigvee_{k_2=1}^{y} g_2(y,k_2)\bigvee_{k_1=1}^{x} g_1(x,k_1).
$$

Theorem 6.4. *Assume that the following hypotheses hold:*

(6.5.1) *For a.e. $(x,y) \in J$, the function $w \to f(x,y,w)$ is weakly-sequentially continuous.*

(6.5.2) *For a.e. $w \in E$, the function $(x,y) \to f(x,y,w)$ is Pettis integrable a.e. on J.*

(6.5.3) *There exists $p \in C(J,[0,\infty))$ such that*

$$
\|f(x,y,u)\|_E \le p(x,y) \quad \text{for a.e. } (x,y) \in J, \text{ and each } u \in E,
$$

with

$$p^* = \sup_{(x,y)\in J} \sup_{(s,t)\in[1,x]\times[1,y]} \left|\ln\frac{x}{s}\right|^{r_1-1} \left|\ln\frac{y}{t}\right|^{r_2-1} \frac{p(s,t)}{st\Gamma(r_1)\Gamma(r_2)};$$

$i = 1, 2.$

(6.5.4) *For each bounded set $B \subset E$ and for each $(x,y) \in J$, we have*

$$\beta(f(x,y,B) \le p(x,y)\beta(B).$$

(6.5.5) *For all $x_1, x_2 \in [1,a]$ such that $x_1 < x_2$, the function $s \mapsto g(x_2,s) - g(x_1,s)$ is nondecreasing on $[1,a]$. Also, for all $y_1, y_2 \in [1,b]$ such that $y_1 < y_2$, the function $s \mapsto g(y_2,t) - g(y_1,t)$ is nondecreasing on $[1,b]$.*

(6.5.6) *The functions $s \mapsto g_1(0,s)$ and $t \mapsto g_2(0,t)$ are nondecreasing on $[1,a]$ or $[1,b]$, respectively.*

(6.5.7) *The functions $s \mapsto g_1(x,s)$ and $x \mapsto g_1(x,s)$ are continuous on $[1,a]$ for each fixed $x \in [1,a]$ or $s \in [1,a]$, respectively. Also, the functions $t \mapsto g_2(y,t)$ and $y \mapsto g_2(y,t)$ are continuous on $[1,b]$ for each fixed $y \in [1,b]$ or $t \in [1,b]$, respectively.*

If

$$L := 2g^*p^* < 1, \tag{6.24}$$

then the integral equation (6.23) has at least one solution defined on J.

Proof. Transform the integral equation (6.23) into a fixed-point equation. Consider the operator $N : C \to C$ defined by

$$(Nu)(x,y) = \mu(x,y) + \int_1^x \int_1^y \left(\ln\frac{x}{s}\right)^{r_1-1} \left(\ln\frac{y}{t}\right)^{r_2-1}$$

$$\times \frac{f(s,t,u(s,t))}{st\Gamma(r_1)\Gamma(r_2)} d_t g_2(y,t) d_s g_1(x,s). \tag{6.25}$$

First notice that, the hypotheses (6.5.2), (6.5.5)–(6.5.7) imply that $\forall u \in C : f(\cdot,\cdot,u(\cdot,\cdot)) \in P(J,E)$. From (6.11.3) we have that

$$\left(\ln\frac{x}{s}\right)^{r_1-1} \left(\ln\frac{y}{t}\right)^{r_2-1} \frac{f(s,t,u(s,t))}{st} \quad \text{for all } (x,y) \in J$$

is Pettis integrable and thus, the operator N is well defined. Let $R > 0$ be such that

$$R > \|\mu\|_C + g^* p^*,$$

and consider the set

$$Q = \left\{ u \in C : \|u\|_C \leq R \text{ and } \|u(x_1, y_1) - u(x_2, y_2)\|_E \right.$$

$$\leq \|\mu(x_1, y_1) - \mu(x_2, y_2)\|_E$$

$$+ p^* \left| \bigvee_{k_2=1}^{y_1} g_2(y_2, k_2) \bigvee_{k_1=1}^{x_1} g_1(x_2, k_1) - \bigvee_{k_2=1}^{y_1} g_2(y_1, k_2) \bigvee_{k_1=1}^{x_1} g_1(x_1, k_1) \right|$$

$$+ p^* \bigvee_{k_2=y_1}^{y_2} g_2(y_2, k_2) \bigvee_{k_1=x_1}^{x_2} g_1(x_2, k_1) + p^* \bigvee_{k_2=y_1}^{y_2} g_2(y_2, k_2) \bigvee_{k_1=1}^{x_2} g_1(x_2, k_1)$$

$$\left. + p^* \bigvee_{k_2=1}^{y_2} g_2(y_2, k_2) \bigvee_{k_1=x_1}^{x_2} g_1(x_2, k_1) \right\}.$$

Clearly, the subset Q is closed, convex and equicontinuous. We shall show that the operator N satisfies all the assumptions of Theorem 1.7. The proof will be given in several steps.

Step 1. N *maps Q into itself.*

Let $u \in Q$, $(x, y) \in J$ and assume that $(Nu)(x, y) \neq 0$. Then there exists $\phi \in E^*$ such that $\|(Nu)(x, y)\|_E = |\phi((Nu)(x, y))|$. Thus

$$\|(Nu)(x,y)\|_E = \left| \phi(\mu(x,y) + \frac{1}{\Gamma(r_1)\Gamma(r_2)} \int_1^x \int_1^y \left(\ln \frac{x}{s}\right)^{r_1-1} \left(\ln \frac{y}{t}\right)^{r_2-1} \right.$$

$$\left. \times \frac{f(s,t,u(s,t))}{st} d_t g_2(y,t) d_s g_1(x,s) \right) \right|$$

$$= |\phi(\mu(x,y))| + \left| \phi \left(\frac{1}{\Gamma(r_1)\Gamma(r_2)} \int_1^x \int_1^y \left(\ln \frac{x}{s}\right)^{r_1-1} \right.\right.$$

$$\left.\left. \times \left(\ln \frac{y}{t}\right)^{r_2-1} \frac{f(s,t,u(s,t))}{st} d_t g_2(y,t) d_s g_1(x,s) \right) \right|$$

$$\leq \|\mu(x,y)\|_E + \frac{1}{\Gamma(r_1)\Gamma(r_2)} \int_1^x \int_1^y \left(\ln \frac{x}{s}\right)^{r_1-1} \left(\ln \frac{y}{t}\right)^{r_2-1}$$

$$\times \frac{p(s,t)}{st} |d_t g_2(y,t) d_s g_1(x,s)|$$

$$\leq \|\mu(x,y)\|_E + \frac{1}{\Gamma(r_1)\Gamma(r_2)} \int_1^x \int_1^y \left(\ln\frac{x}{s}\right)^{r_1-1} \left(\ln\frac{y}{t}\right)^{r_2-1}$$

$$\times \frac{p(s,t)}{st} d_t \bigvee_{k_2=1}^t g_2(y,k_2) d_s \bigvee_{k_1=1}^s g_1(x,k_1)$$

$$\leq \|\mu(x,y)\|_E + P^* \int_1^x \int_1^y d_t \bigvee_{k_2=1}^t g_2(y,k_2) d_s \bigvee_{k_1=1}^s g_1(x,k_1)$$

$$\leq \|\mu\|_C + g^* p^* \leq R.$$

Next, let $(x_1,y_1), (x_2,y_2) \in J$ such that $x_1 < x_2$ and $y_1 < y_2$, and let $u \in Q$, with $(Nu)(x_1,y_1) - (Nu)(x_2,y_2) \neq 0$. Then there exists $\phi \in E^*$ such that

$$\|(Nu)(x_1,y_1) - (Nu)(x_2,y_2)\|_E = |\varphi((Nu)(x_1,y_1) - (Nu)(x_2,y_2))|$$

and $\|\varphi\| = 1$. Then

$$\|(Nu)(x_2,y_2) - (Nu)(x_1,y_1)\|_E = |\phi((Nu)(x_2,y_2) - (Nu)(x_1,y_1))|$$

$$\leq \|\mu(x_1,y_1) - \mu(x_2,y_2)\|_E$$

$$+ \frac{1}{\Gamma(r_1)\Gamma(r_2)} \int_1^{x_1} \int_1^{y_1} \left[\left|\ln\frac{x_2}{s}\right|^{r_1-1} \left|\ln\frac{y_2}{t}\right|^{r_2-1} - \left|\ln\frac{x_1}{s}\right|^{r_1-1} \left|\ln\frac{y_1}{t}\right|^{r_2-1} \right]$$

$$\times \frac{\|f(s,t,u(s,t))\|_E}{st} d_t g_2(y_2,t) d_s g_1(x_2,s)$$

$$+ \frac{1}{\Gamma(r_1)\Gamma(r_2)} \int_{x_1}^{x_2} \int_{y_1}^{y_2} \left|\ln\frac{x_2}{s}\right|^{r_1-1} \left|\ln\frac{y_2}{t}\right|^{r_2-1}$$

$$\times \frac{\|f(s,t,u(s,t))\|_E}{st} d_t g_2(y_2,t) d_s g_1(x_2,s)$$

$$+ \frac{1}{\Gamma(r_1)\Gamma(r_2)} \int_1^{x_1} \int_{y_1}^{y_2} \left|\ln\frac{x_2}{s}\right|^{r_1-1} \left|\ln\frac{y_2}{t}\right|^{r_2-1}$$

$$\times \frac{\|f(s,t,u(s,t))\|_E}{st} d_t g_2(y_2,t) d_s g_1(x_2,s)$$

$$+ \frac{1}{\Gamma(r_1)\Gamma(r_2)} \int_{x_1}^{x_2} \int_1^{y_1} \left|\ln\frac{x_2}{s}\right|^{r_1-1} \left|\ln\frac{y_2}{t}\right|^{r_2-1}$$

$$\times \frac{\|f(s,t,u(s,t))\|_E}{st} d_t g_2(y_2,t) d_s g_1(x_2,s).$$

Then, we obtain

$$|(Nu)(x_2, y_2) - (Nu)(x_1, y_1)|$$

$$\leq \|\mu(x_1, y_1) - \mu(x_2, y_2)\|_E$$

$$+ p^* \int_1^{x_1} \int_1^{y_1} \left| d_t \bigvee_{k_2=1}^{t} g_2(y_2, k_2) d_s \bigvee_{k_1=1}^{s} g_1(x_2, k_1) \right.$$

$$\left. - d_t \bigvee_{k_2=1}^{t} g_2(y_1, k_2) d_s \bigvee_{k_1=1}^{s} g_1(x_1, k_1) \right|$$

$$+ p^* \int_{x_1}^{x_2} \int_{y_1}^{y_2} d_t \bigvee_{k_2=1}^{t} g_2(y_2, k_2) d_s \bigvee_{k_1=1}^{s} g_1(x_2, k_1)$$

$$+ p^* \int_1^{x_1} \int_{y_1}^{y_2} d_t \bigvee_{k_2=1}^{t} g_2(y_2, k_2) d_s \bigvee_{k_1=1}^{s} g_1(x_2, k_1)$$

$$+ p^* \int_{x_1}^{x_2} \int_1^{y_1} d_t \bigvee_{k_2=1}^{t} g_2(y_2, k_2) d_s \bigvee_{k_1=1}^{s} g_1(x_2, k_1).$$

Thus, we get

$$|(Nu)(x_2, y_2) - (Nu)(x_1, y_1)|$$

$$\leq \|\mu(x_1, y_1) - \mu(x_2, y_2)\|_E$$

$$+ p^* \left| \bigvee_{k_2=1}^{y_1} g_2(y_2, k_2) \bigvee_{k_1=1}^{x_1} g_1(x_2, k_1) - \bigvee_{k_2=1}^{y_1} g_2(y_1, k_2) \bigvee_{k_1=1}^{x_1} g_1(x_1, k_1) \right|$$

$$+ p^* \bigvee_{k_2=y_1}^{y_2} g_2(y_2, k_2) \bigvee_{k_1=x_1}^{x_2} g_1(x_2, k_1) + p^* \bigvee_{k_2=y_1}^{y_2} g_2(y_2, k_2) \bigvee_{k_1=1}^{x_2} g_1(x_2, k_1)$$

$$+ p^* \bigvee_{k_2=1}^{y_2} g_2(y_2, k_2) \bigvee_{k_1=x_1}^{x_2} g_1(x_2, k_1).$$

Hence $N(Q) \subset Q$.

Step 2. *N is weakly-sequentially continuous.*

Let (u_n) be a sequence in Q and let $(u_n(x, y)) \to u(x, y)$ in (E, ω) for each $(x, y) \in J$. Fix $(x, y) \in J$, since f satisfies the assumption (6.5.1), we have $f(x, y, u_n(x, y))$ converges weakly to $f(x, y, u(x, y))$. Hence the Lebesgue dominated convergence theorem for Pettis integral implies $(Nu_n)(x, y)$ converges weakly to $(Nu)(x, y)$ in (E, ω), for each $(x, y) \in J$. Thus, $N(u_n) \to N(u)$. Hence, $N : Q \to Q$ is weakly-sequentially continuous.

Step 3. *The implication (1.9) holds.*

Let V be a subset of Q such that $\overline{V} = \overline{\text{conv}}(N(V) \cup \{0\})$. Obviously $V(x, y) \subset \overline{\text{conv}}(NV)(x, y)) \cup \{0\}$, $\forall (x, y) \in J$. Further, as V is bounded and equicontinuous, by [172, Lemma 3], the function $(x, y) \to u(x, y) = \beta(V(x, y))$ is continuous on J. Since the functions μ is continuous on J, the set $\{\mu(x, y); \ (x, y) \in J\} \subset E$ is compact. From (6.5.3), (6.5.5) − (6.5.7), Lemma 2.3 and the properties of the measure β, for any $(x, y) \in J$, we have

$$v(x, y) \le \beta((NV)(x, y) \cup \{0\})$$

$$\le \beta((NV)(x, y))$$

$$\le \frac{2}{\Gamma(r_1)\Gamma(r_2)} \int_1^x \int_1^y \left| \ln \frac{x}{s} \right|^{r_1-1} \left| \ln \frac{y}{t} \right|^{r_2-1}$$

$$\times \frac{p(s,t)\beta(V(s,t))}{st} |d_t g_2(y_2, t) d_s g_1(x_2, s)|$$

$$\le \frac{2}{\Gamma(r_1)\Gamma(r_2)} \int_1^x \int_1^y \left| \ln \frac{x}{s} \right|^{r_1-1} \left| \ln \frac{y}{t} \right|^{r_2-1}$$

$$\times \frac{p(s,t)v(s,t)}{st} d_t \bigvee_{k_2=1}^t g_2(y, k_2) d_s \bigvee_{k_1=1}^s g_1(x, k_1)$$

$$\le 2p^* \|v\|_C \int_1^x \int_1^y d_t \bigvee_{k_2=1}^t g_2(y, k_2) d_s \bigvee_{k_1=1}^s g_1(x, k_1)$$

$$\le 2g^* p^* \|v\|_C.$$

Thus

$$\|v\|_C \le L\|v\|_C.$$

From (6.24), we get $\|v\|_C = 0$, that is $v(x, y) = \beta(V(x, y)) = 0$, for each $(x, y) \in J$ and then by [290, Theorem 2], V is relatively-weakly compact in C. Applying now Theorem 1.7, we conclude that N has a fixed point which is a solution of the integral equation (6.23). $\qquad \square$

6.5.3. *Example*

Let

$$E = l^1 = \left\{ u = (u_1, u_2, \ldots, u_n, \ldots), \sum_{n=1}^{\infty} |u_n| < \infty \right\}$$

be the Banach space with the norm

$$\|u\|_E = \sum_{n=1}^{\infty} |u_n|.$$

As an application of our results, we consider the following partial Hadamard integral equation:

$$u_n(x, y) = \mu(x, y) + \int_1^x \int_1^y \left(\ln \frac{x}{s} \right)^{r_1 - 1} \left(\ln \frac{y}{t} \right)^{r_2 - 1}$$

$$\times \frac{f_n(s, t, u(s, t))}{st\Gamma(r_1)\Gamma(r_2)} d_t g_2(y, t) d_s g_1(x, s); \quad (x, y) \in [1, e] \times [1, e],$$

$$(6.26)$$

where

$$r_1, r_2 > 0, \ \mu(x, y) = x + y^2; \ (x, y) \in [1, e] \times [1, e],$$

$$g_1(x, s) = s, \ g_2(y, t) = t; \ x, s, y, t \in [1, e],$$

and

$$f_n(x, y, u(x, y)) = \frac{cxy^2}{1 + \|u\|_E} \left(e^{-7} + \frac{1}{e^{x+y+5}} \right) u_n(x, y);$$

$$(x, y) \in [1, e] \times [1, e],$$

with

$$u = (u_1, u_2, \ldots, u_n, \ldots), \quad \text{and} \quad c := \frac{e^4}{8} \Gamma(1 + r_1) \Gamma(1 + r_2).$$

Set $f = (f_1, f_2, \ldots, f_n, \ldots)$. Clearly, the function f is continuous. For each $u \in E$ and $(x, y) \in [1, e] \times [1, e]$, we have

$$\|f(x, y, u(x, y))\|_E \le cxy^2 \left(e^{-7} + \frac{1}{e^{x+y+5}} \right).$$

Hence, the hypothesis (6.5.3) is satisfied with $p^* = 2ce^{-4}$.

We shall show that condition (6.24) holds with $a = b = e$. Indeed,

$$\frac{2p^*(\ln a)^{r_1}(\ln b)^{r_2}}{\Gamma(1 + r_1)\Gamma(1 + r_2)} = \frac{4c}{e^4 \Gamma(1 + r_1)\Gamma(1 + r_2)} = \frac{1}{2} < 1.$$

A simple computations show that all conditions of Theorem 6.4 are satisfied. It follows that the integral equation (6.26) has at least one solution on $[1, e] \times [1, e]$.

6.6. Partial Random Hadamard–Pettis Fractional Integral Equations

6.6.1. *Introduction*

In this section, we apply Mönch's and Engl's fixed-point theorems associated with the technique of measure of weak noncompactness, to investigate the existence of random solutions for a class of partial random integral equations via Hadamard's fractional integral, under the Pettis integrability assumption.

This section deals with the existence of solutions to the following random Hadamard partial fractional integral equation:

$$u(x, y, w) = \mu(x, y, w) + \int_1^x \int_1^y \left(\ln \frac{x}{s} \right)^{r_1 - 1} \left(\ln \frac{y}{t} \right)^{r_2 - 1}$$

$$\times \frac{f(s, t, u(s, t, w), w)}{st \Gamma(r_1) \Gamma(r_2)} dt ds; \quad \text{if } (x, y) \in J, \ w \in \Omega, \quad (6.27)$$

where $J := [1, a] \times [1, b]$, $a, b > 1$, $r_1, r_2 > 0$, $\mu : J \times \Omega \to E$, $f : J \times E \times \Omega \to E$ are given continuous functions, $(\Omega, \mathcal{A}, \nu)$ is a measurable space, $\Gamma(\cdot)$ is the Euler gamma function, and E is a real (or complex) Banach space with norm $\| \cdot \|_E$ and dual E^*.

6.6.2. *Existence of weak solutions*

Definition 6.5. A function $u \in C$ is said to be a random solution of (6.27) if u satisfies Eq. (6.27) on J.

Further, we present conditions for the existence of a solution of Eq. (6.27).

Theorem 6.5. *Assume that the following hypotheses hold:*

(6.6.1) *The function $w \mapsto \mu(x, y, w)$ is measurable and bounded for a.e. $(x, y) \in J$.*

(6.6.2) *The function f is random Carathéodory on $J \times E \times \Omega$.*

(6.6.3) *For a.e. $(x, y) \in J$, and all $w \in \Omega$, the function $u \to f(x, y, u, w)$ is weakly-sequentially continuous.*

(6.6.4) *For a.e. $u \in E$, and all $w \in \Omega$, the function $(x, y) \to f(x, y, u, w)$ is Pettis integrable a.e. on J.*

(6.6.5) *There exists a function $p : J \times \Omega \to [0, \infty)$ with $p(w) \in L^\infty(J, [0, \infty))$ for each $w \in \Omega$ such that*

$$\|f(x, y, u, w)\|_E \leq p(x, y, w) \ for \ a.e. \ (x, y) \in J, \ and \ each \ u \in E.$$

(6.6.6) *For each bounded set $B \subset E$ and for each $(x, y) \in J$, and $w \in \Omega$, we have*

$$\beta(f(x, y, B, w)) \leq p(x, y, w)\beta(B).$$

(6.6.7) *There exists a random function $R : \Omega \to (0, \infty)$ such that*

$$R(w) > \mu^*(w) + \frac{p^*(w)(\ln a)^{r_1}(\ln b)^{r_2}}{\Gamma(1 + r_1)\Gamma(1 + r_2)},$$

where

$$\mu^*(w) = \sup_{(x,y)\in J} |\mu(x, y, w)|, \quad p^*(w) = \sup_{(x,y)\in J} p(x, y, w).$$

If

$$\ell := \frac{2p^*(\ln a)^{r_1}(\ln b)^{r_2}}{\Gamma(1 + r_1)\Gamma(1 + r_2)} < 1, \tag{6.28}$$

where $p^* = \sup ess_{w\in\Omega} p^*(w)$, then the integral equation (6.27) has at least one random solution defined on J.

Proof. Transform the integral equation (6.27) into a fixed-point equation. Consider the operator $N : \Omega \times C \to C$ defined by

$$(N(w)u)(x, y) = \mu(x, y, w) + \int_1^x \int_1^y \left(\ln \frac{x}{s}\right)^{r_1-1} \left(\ln \frac{y}{t}\right)^{r_2-1}$$

$$\times \frac{f(s, t, u(s, t, w), w)}{st\Gamma(r_1)\Gamma(r_2)} dt ds. \tag{6.29}$$

From the hypotheses (6.6.2)–(6.6.4), for each $w \in \Omega$ and almost all $(x, y) \in J$, we have that $f(x, y, u(x, y, w), w)$ is in $P(J, E)$. From (6.6.5) we have that

$$\left(\ln \frac{x}{s}\right)^{r_1-1} \left(\ln \frac{y}{t}\right)^{r_2-1} \frac{f(s, t, u(s, t, w), w)}{st}; \quad \text{for all } (x, y) \in J$$

is Pettis integrable for all $w \in \Omega$. Again, as the map μ is continuous for all $w \in \Omega$ and the indefinite integral is continuous on J, then $N(w)$ defines a mapping $N : \Omega \times C \to C$. Hence u is a solution for the integral equation (6.27) if and only if $u = (N(w))u$. We shall show that the operator N satisfies all the assumptions of Theorem 1.9. The proof will be given in several steps.

Step 1. $N(w)$ *is a random operator with stochastic domain on* C.

Since $f(x, y, u, w)$ is random Carathéodory, the map $w \to f(x, y, u, w)$ is measurable in view of Definition 5.13. Similarly, the product

$$\left(\ln \frac{x}{s} \right)^{r_1 - 1} \left(\ln \frac{y}{t} \right)^{r_2 - 1} \frac{f(s, t, u(s, t, w), w)}{st}$$

of a continuous and a measurable function is again measurable. Further, the integral is a limit of a finite sum of measurable functions, therefore, the map

$$w \mapsto \mu(x, y, w) + \int_1^x \int_1^y \left(\ln \frac{x}{s} \right)^{r_1 - 1} \left(\ln \frac{y}{t} \right)^{r_2 - 1} \frac{f(s, t, u(s, t, w), w)}{st \Gamma(r_1) \Gamma(r_2)} dt ds$$

is measurable. As a result, N is a random operator on $\Omega \times C$ into C.

Let $W : \Omega \to \mathcal{P}(C)$ be defined by

$$W(w) = \left\{ u \in C : \|u\|_C \leq R(w) \text{ and } \|u(x_1, y_1, w) \right.$$

$$- u(x_2, y_2, w)\|_E$$

$$\leq \|\mu(x_1, y_1, w) - \mu(x_2, y_2, w)\|_E$$

$$+ \frac{p^*(w)}{\Gamma(1 + r_1)\Gamma(1 + r_2)} [2(\ln y_2)^{r_2} (\ln x_2 - \ln x_1)^{r_1}$$

$$+ 2(\ln x_2)^{r_1} (\ln y_2 - \ln y_1)^{r_2} + (\ln x_1)^{r_1} (\ln y_1)^{r_2}$$

$$- (\ln x_2)^{r_1} (\ln y_2)^{r_2}$$

$$\left. - 2(\ln x_2 - \ln x_1)^{r_1} (\ln y_2 - \ln y_1)^{r_2}] \right\}.$$

Clearly, the subset $W(w)$ is closed, convex and equicontinuous for all $w \in \Omega$. Then W is measurable by [208, Lemma 17]. Therefore, N is a random operator with stochastic domain W.

Step 2. $N(w)$ *is continuous.*

Let $\{u_n\}$ be a sequence such that $u_n \to u$ in $\mathcal{C}$. Then, for each $(x, y) \in J$ and $w \in \Omega$, we have

$$\|(N(w)u_n)(x, y) - (N(w)u)(x, y)\|_E$$

$$\leq \int_1^x \int_1^y \left| \ln \frac{x}{s} \right|^{r_1 - 1} \left| \ln \frac{y}{t} \right|^{r_2 - 1}$$

$$\times \frac{\|f(s, t, u_n(s, t, w), w) - f(s, t, u(s, t, w), w)\|_E}{\Gamma(r_1)\Gamma(r_2)} dt ds.$$

Using the Lebesgue dominated convergence theorem, we get

$$\|N(w)u_n - N(w)u\|_{\mathcal{C}} \to 0 \text{ as } n \to \infty.$$

As a consequence of Steps 1 and 2, we can conclude that $N(w) : W(w) \to N(w)$ is a continuous random operator with stochastic domain W.

Step 3. *For every* $w \in \Omega$, $\{u \in W(w) : N(w)u = u\} \neq \emptyset$.

For this we apply Theorem 1.7. The proof will be given in several claims.

Claim 1. $N(w)$ *maps* $W(w)$ *into itself.*

Let $w \in \Omega$ be fixed, and let $u \in W(w)$, $(x, y) \in J$. Assume that $(N(w)u)(x, y) \neq 0$. Then there exists $\phi \in E^*$ such that $\|(N(w)u)(x, y)\|_E = |\phi((N(w)u)(x, y))|$. Then, we get

$$\|(N(w)u)(x, y)\|_E$$

$$= \left| \phi \left(\mu(x, y, w) + \frac{1}{\Gamma(r_1)\Gamma(r_2)} \int_1^x \int_1^y \left(\ln \frac{x}{s} \right)^{r_1 - 1} \right. \right.$$

$$\left. \left. \times \left(\ln \frac{y}{t} \right)^{r_2 - 1} \frac{f(s, t, u(s, t, w), w)}{st} dt ds \right) \right|$$

$$= |\phi(\mu(x, y, w))|$$

$$+ \left| \phi \left(\frac{1}{\Gamma(r_1)\Gamma(r_2)} \int_1^x \int_1^y \left(\ln \frac{x}{s} \right)^{r_1 - 1} \left(\ln \frac{y}{t} \right)^{r_2 - 1} \right. \right.$$

$$\left. \left. \times \frac{f(s, t, u(s, t, w), w)}{st} dt ds \right) \right|$$

$$\leq \mu^*(w) + \frac{1}{\Gamma(r_1)\Gamma(r_2)} \int_1^x \int_1^y \left(\ln \frac{x}{s} \right)^{r_1 - 1} \left(\ln \frac{y}{t} \right)^{r_2 - 1} \frac{p(s, t, w)}{st} dt ds$$

$$\leq \mu^*(w) + \frac{p^*(w)(\ln a)^{r_1}(\ln b)^{r_2}}{\Gamma(1 + r_1)\Gamma(1 + r_2)} \leq R(w).$$

Next, for any fixed $w \in \Omega$, let $(x_1, y_1), (x_2, y_2) \in J$ such that $x_1 < x_2$ and $y_1 < y_2$, and let $u \in W(w)$, with $(N(w)u)(x_1, y_1) - (N(w)u)(x_2, y_2) \neq 0$. Then there exists $\phi \in E^*$ such that

$$\|(N(w)u)(x_1, y_1) - (N(w)u)(x_2, y_2)\|_E = |\varphi((N(w)u)(x_1, y_1) \\ - (N(w)u)(x_2, y_2))|$$

and $\|\varphi\| = 1$. Thus, we have

$$\|(N(w)u)(x_2, y_2) - (N(w)u)(x_1, y_1)\|_E$$
$$= |\phi((N(w)u)(x_2, y_2) - (N(w)u)(x_1, y_1))|$$
$$\leq \|\mu(x_1, y_1, w) - \mu(x_2, y_2, w)\|_E + \frac{1}{\Gamma(r_1)\Gamma(r_2)}$$
$$\int_1^{x_1} \int_1^{y_1} \left[\left|\ln \frac{x_2}{s}\right|^{r_1-1} \left|\ln \frac{y_2}{t}\right|^{r_2-1} - \left|\ln \frac{x_1}{s}\right|^{r_1-1} \left|\ln \frac{y_1}{t}\right|^{r_2-1} \right]$$
$$\times \frac{\|f(s, t, u(s, t, w), w)\|_E}{st} dt ds + \frac{1}{\Gamma(r_1)\Gamma(r_2)} \int_{x_1}^{x_2} \int_{y_1}^{y_2}$$
$$\times \left|\ln \frac{x_2}{s}\right|^{r_1-1} \left|\ln \frac{y_2}{t}\right|^{r_2-1} \frac{\|f(s, t, u(s, t, w), w)\|_E}{st} dt ds$$
$$+ \frac{1}{\Gamma(r_1)\Gamma(r_2)} \int_1^{x_1} \int_{y_1}^{y_2} \left|\ln \frac{x_2}{s}\right|^{r_1-1} \left|\ln \frac{y_2}{t}\right|^{r_2-1}$$
$$\times \frac{\|f(s, t, u(s, t, w), w)\|_E}{st} dt ds$$
$$+ \frac{1}{\Gamma(r_1)\Gamma(r_2)} \int_{x_1}^{x_2} \int_1^{y_1} \left|\ln \frac{x_2}{s}\right|^{r_1-1} \left|\ln \frac{y_2}{t}\right|^{r_2-1}$$
$$\times \frac{\|f(s, t, u(s, t, w), w)\|_E}{st} dt ds.$$

Then, we obtain

$$\|(N(w)u)(x_2, y_2) - (N(w)u)(x_1, y_1)\|_E$$
$$\leq \|\mu(x_1, y_1, w) - \mu(x_2, y_2, w)\|_E$$
$$+ \frac{1}{\Gamma(r_1)\Gamma(r_2)} \int_1^{x_1} \int_1^{y_1} \left[\left|\ln \frac{x_2}{s}\right|^{r_1-1} \left|\ln \frac{y_2}{t}\right|^{r_2-1} \right.$$
$$\left. - \left|\ln \frac{x_1}{s}\right|^{r_1-1} \left|\ln \frac{y_1}{t}\right|^{r_2-1} \right] \frac{p^*(w)}{st} dt ds$$

$$+ \frac{1}{\Gamma(r_1)\Gamma(r_2)} \int_{x_1}^{x_2} \int_{y_1}^{y_2} \left| \ln \frac{x_2}{s} \right|^{r_1-1} \left| \ln \frac{y_2}{t} \right|^{r_2-1} \frac{p^*(w)}{st} dt ds$$

$$+ \frac{1}{\Gamma(r_1)\Gamma(r_2)} \int_{1}^{x_1} \int_{y_1}^{y_2} \left| \ln \frac{x_2}{s} \right|^{r_1-1} \left| \ln \frac{y_2}{t} \right|^{r_2-1} \frac{p^*(w)}{st} dt ds$$

$$+ \frac{1}{\Gamma(r_1)\Gamma(r_2)} \int_{x_1}^{x_2} \int_{1}^{y_1} \left| \ln \frac{x_2}{s} \right|^{r_1-1} \left| \ln \frac{y_2}{t} \right|^{r_2-1} \frac{p^*(w)}{st} dt ds$$

$$\leq \| \mu(x_1, y_1, w) - \mu(x_2, y_2, w) \|_E + \frac{p^*(w)}{\Gamma(1+r_1)\Gamma(1+r_2)}$$

$$\times [2(\ln y_2)^{r_2}(\ln x_2 - \ln x_1)^{r_1} + 2(\ln x_2)^{r_1}(\ln y_2 - \ln y_1)^{r_2}$$

$$+ (\ln x_1)^{r_1}(\ln y_1)^{r_2} - (\ln x_2)^{r_1}(\ln y_2)^{r_2}$$

$$- 2(\ln x_2 - \ln x_1)^{r_1}(\ln y_2 - \ln y_1)^{r_2}].$$

Hence $N(W(w)) \subset W(w)$. Therefore, $N(w) : W(w) \to N(w)$ maps $W(w)$ into itself.

Claim 2. $N(w)$ *is weakly-sequentially continuous.*
Let (u_n) be a sequence in $W(w)$ and let $(u_n(x, y, w)) \to u(x, y, w)$ in (E, ω) for any $w \in \Omega$, and each $(x, y) \in J$. Fix $(x, y) \in J$, since f satisfies the assumption (6.6.3), we have $f(x, y, u_n(x, y, w), w)$ converges weakly to $f(x, y, u(x, y, w), w)$. Hence the Lebesgue dominated convergence theorem for Pettis integral implies $(Nu_n)(x, y, w)$ converges weakly to $(N(w)u)(x, y)$ in (E, ω). We do it for any $w \in \Omega$, and each $(x, y) \in J$, so $N(w)(u_n) \to N(w)(u)$. Then $N : W(w) \to W(w)$ is weakly-sequentially continuous.

Claim 3. *The implication (1.9) holds.*
Let V be a subset of $W(w)$ such that $\overline{V} = \overline{\text{conv}}(N(w)(V) \cup \{0\})$. Obviously $V(x, y, w) \subset \overline{\text{conv}}(N(w)V)(x, y)) \cup \{0\})$. Further, as V is bounded and equicontinuous, by [172, Lemma 3], the function $(x, y, w) \to u(x, y, w) = \beta(V(x, y, w))$ is continuous on $J \times \Omega$. Since the function μ is continuous on $J \times \Omega$, then the set $\{\mu(x, y, w); \ (x, y) \in J, \ w \in \Omega\} \subset E$ is compact. From (6.6.5), Lemma 2.3 and the properties of the measure β, for any $w \in \Omega$, and each $(x, y) \in J$, we have

$$v(x, y, w) \leq \beta((N(w)V)(x, y) \cup \{0\})$$

$$\leq \beta((N(w)V)(x, y))$$

$$\leq \frac{2}{\Gamma(r_1)\Gamma(r_2)} \int_1^x \int_1^y \left| \ln \frac{x}{s} \right|^{r_1-1} \left| \ln \frac{y}{t} \right|^{r_2-1}$$
$$\times \frac{p(s,t,w)\beta(V(s,t,w))}{st} dtds$$

$$\leq \frac{2}{\Gamma(r_1)\Gamma(r_2)} \int_1^x \int_1^y \left| \ln \frac{x}{s} \right|^{r_1-1} \left| \ln \frac{y}{t} \right|^{r_2-1}$$
$$\times \frac{p(s,t,w)v(s,t,w)}{st} dtds$$

$$\leq \frac{2\|v\|_C}{\Gamma(r_1)\Gamma(r_2)} \int_1^x \int_1^y \left| \ln \frac{x}{s} \right|^{r_1-1} \left| \ln \frac{y}{t} \right|^{r_2-1}$$
$$\times \frac{p(s,t,w)}{st} dtds$$

$$\leq \frac{2p^*(\ln a)^{r_1}(\ln b)^{r_2}}{\Gamma(1+r_1)\Gamma(1+r_2)} \|v\|_C.$$

Thus

$$\|v\|_C \leq \ell \|v\|_C.$$

From (6.28), we get $\|v\|_C = 0$, that is $v(x,y,w) = \beta(V(x,y,w)) = 0$, for any $w \in \Omega$, and each $(x,y) \in J$. Hence, [290, Theorem 2] shows that V is relatively-weakly compact in C.

As consequence of claims 1–3, and from Theorem 1.8 it follows that for every $w \in \Omega$, $\{u \in W(w) : N(w)u = u\} \neq \emptyset$. Apply now Theorem 1.9, Steps 1–3 show that for each $w \in \Omega$, N has at least one fixed point in W. Since $\bigcap_{w\in\Omega} \text{int } W(w) \neq \emptyset$ the hypothesis that a measurable selector of int W exists, then N has a stochastic fixed point, i.e., the integral equation (6.27) has at least one random solution on C. $\quad\square$

6.6.3. *Example*

Let $E = \mathbb{R}$, $\Omega = (-\infty, 0)$ be equipped with the usual σ-algebra consisting of Bochner measurable subsets of $(-\infty, 0)$. Given a measurable function $u : \Omega \to C([1,e] \times [1,e])$, consider the following partial random Hadamard integral equation:

$$u(x,y,w) = \mu(x,y,w) + \int_1^x \int_1^y \left(\ln \frac{x}{s} \right)^{r_1-1} \left(\ln \frac{y}{t} \right)^{r_2-1}$$
$$\times \frac{f(s,t,u(s,t,w),w)}{st\Gamma(r_1)\Gamma(r_2)} dtds \tag{6.30}$$

for $(x, y) \in [1, e] \times [1, e]$, $w \in \Omega$, where

$$r_1, r_2 > 0, \ \mu(x, y, w) = x \sin w + y^2 \cos w; \ (x, y) \in [1, e] \times [1, e],$$

and

$$f(x, y, u(x, y)) = \frac{w^2 x y^2}{(1 + w^2 + |u(x, y, w)|) e^{x+y+3}},$$

$$(x, y) \in [1, e] \times [1, e], w \in \Omega.$$

The function $w \mapsto \mu(x, y, w) = x \sin w + y^2 \cos w$ is measurable and bounded with

$$|\mu(x, y, w)| \leq e + e^2,$$

hence, the conditions (6.13.1) is satisfied.

The map $(x, y, w) \mapsto f(x, y, u, w)$ is jointly continuous for all $u \in \mathbb{R}$ and hence jointly measurable for all $u \in \mathbb{R}$. Also the map $u \mapsto f(x, y, u, w)$ is continuous for all $(x, y) \in [1, e] \times [1, e]$ and $w \in \Omega$. So the function f is Carathéodory on $[1, e] \times [1, e] \times \mathbb{R} \times \Omega$.

For each $u \in \mathbb{R}$, $(x, y) \in [1, e] \times [1, e]$ and $w \in \Omega$, we have

$$|f(x, y, u, w)| \leq \frac{w^2 x y^2}{e^3}.$$

Hence the condition (6.6.5) is satisfied with $p^* = e^{-3}$.

We shall show that condition $\ell < 1$ holds with $a = b = e$. Indeed, for each $r_1, r_2 > 0$ we get

$$\ell = \frac{4 p^* (\ln a)^{r_1} (\ln b)^{r_2}}{\Gamma(1 + r_1) \Gamma(1 + r_2)}$$

$$\leq \frac{4}{e^3 \Gamma(1 + r_1) \Gamma(1 + r_2)}$$

$$< 1.$$

A simple computations show that all conditions of Theorem 6.5 are satisfied. It follows that the random integral equation (6.30) has at least one random solution on $[1, e] \times [1, e]$.

6.7. Partial Random Hadamard–Pettis Fractional Integral Equations with Multiple Delay

6.7.1. Introduction

In this section, we present some results concerning the existence of weak solutions for some integral equations of Hadamard fractional order with random effects and multiple delay, by applying Mönch's and Engl's fixed-point theorems associated with the technique of measure of weak noncompactness.

We discuss the existence of random solutions for the following partial Hadamard fractional integral equation:

$$u(x, y, w) = \begin{cases} \sum_{i=1}^{m} b_i(x, y, w) u(x - \xi_i, y - \mu_i, w) \\ + f\left(x, y, {}^H I_\sigma^r u(x, y, w), u(x, y, w), w\right) & \text{if } (x, y) \in J, \ w \in \Omega, \\ \Phi(x, y, w) \text{ if } (x, y) \in \tilde{J}, \ w \in \Omega, \end{cases}$$

(6.31)

where $J := [1, a] \times [1, b]$, $\tilde{J} := [-\xi, a] \times [-\mu, b] \backslash (1, a] \times (1, b]$, $a, b > 1$, $\xi_i, \mu_i \geq 1$ $(i = 1, \ldots, m)$, $\xi = \max_{i=1,\ldots,m}\{\xi_i\}$, $\mu = \max_{i=1,\ldots,m}\{\mu_i\}$, $\sigma = (1, 1)$, $r = (r_1, r_2) \in (0, \infty) \times (0, \infty)$, $b_i : J \times \Omega \to \mathbb{R}$ $(i = 1, \ldots, m)$, $f : J \times E \times E \times \Omega \to E$ are given continuous functions, $(\Omega, \mathcal{A}, \nu)$ is a measurable space, and E is a real (or complex) Banach space with norm $\| \cdot \|_E$ and dual E^*, such that E is the dual of a weakly-compactly-generated Banach space X, ${}^H I_\sigma^r$ is the left-sided mixed Hadamard integral of order r, and $\Phi : \tilde{J} \times \Omega \to E$ is a given continuous and measurable function such that

$$\Phi(x, 1, w) = \sum_{i=1}^{m} b_i(x, 1, w) \Phi(x - \xi_i, 1 - \mu_i, w); \quad x \in [1, a], \ w \in \Omega,$$

and

$$\Phi(1, y, w) = \sum_{i=1}^{m} b_i(1, y, w) \Phi(1 - \xi_i, y - \mu_i, w); \quad y \in [1, b], \ w \in \Omega.$$

6.7.2. *Existence of weak solutions*

Definition 6.6. By a random solution of the problem (6.31) we mean a measurable function $u : \Omega \to C([-\xi, a] \times [-\mu, b])$ that satisfies the integral equation

$$u(x, y, w) = \sum_{i=1}^{m} b_i(x, y, w) u(x - \xi_i, y - \mu_i, w)$$

$$+ f\left(x, y, {}^H I_\sigma^r u(x, y, w), u(x, y, w), w\right) \text{ on } J \times \Omega,$$

$$\text{and } u(x, y, w) = |\phi(x, y, w)| \text{ on } \tilde{J} \times \Omega.$$

The following hypotheses will be used in the sequel.

(6.7.1) The functions $w \mapsto b_i(x, y, w)$; $i = 1, \ldots, m$ are bounded for a.e. $(x, y) \in J$, and $b_i(\cdot, \cdot, w) \in L^\infty(J, \mathbb{R})$.

(6.7.2) The function f is random Carathéodory on $J \times E \times E \times \Omega$ for each $w \in \Omega$.

(6.7.3) For a.e. $(x, y) \in J$, and all $w \in \Omega$, the function $u \to f(x, y, {}^H I_\sigma u, u, w)$ is weakly-sequentially continuous.

(6.7.4) There exist functions $p_1, p_2, p_3 : J \times \Omega \to [0, \infty)$ with $p_i(\cdot, w) \in L^\infty(J, [0, \infty))$; $i = 1, 2, 3$, for each $w \in \Omega$, such that for all $\varphi \in E^*$, we have

$$|\varphi(f(x, y, u, v, w))|$$
$$\leq \frac{p_1(x, y, w)\|\varphi\| + p_2(x, y, w)\|u\|_E + p_3(x, y, w)\|v\|_E}{1 + \|\varphi\|}$$

for all $u, v \in E$ and a.e. $(x, y) \in J$.

(6.7.5) For all $u \in E$, there exists a continuous function $\psi : [0, \infty) \to [0, \infty)$ with $\psi(0) = 0$, such that for each $\varphi \in E^*$, $(x_1, y_1), (x_2, y_2) \in J$ and any $w \in \Omega$, we have

$$\sum_{i=1}^{m} |b_i(x_1, y_1, w)u(x_1 - \xi_i, y_1 - \mu_i, w)$$

$$- b_i(x_2, y_2, w)u(x_2 - \xi_i, y_2 - \mu_i, w)|$$

$$+ |\varphi(f(x_1, y_1, u(x_1, y_1), v(x_1, y_1, w)$$

$$- f(x_2, y_2, u(x_2, y_2), v(x_2, y_2, w)))|$$

$$\leq \frac{\psi(|x_1 - x_2| + |y_1 - y_2|)\|\varphi\|}{1 + \|\varphi\| + \|u\|_E + \|v\|_E}.$$

(6.7.6) There exists a function $q : J \times \Omega \to [0, \infty)$ with $q(\cdot, \cdot, w) \in L^\infty(J, [0, \infty))$ for each $w \in \Omega$ such that for any bounded $B \subset E$,

$$\alpha(f(x, y, {}^H I_\sigma B, B, w)) \leq q(x, y, w)\alpha(B), \ for \ a.e. \ (x, y) \in J,$$

where ${}^H I_\sigma B := \{{}^H I_\sigma u(x, y) : u(x, y) \in B; \ (x, y) \in J\}$.

(6.7.7) There exists a random function $R : \Omega \to (0, \infty)$ such that

$$p_1^*(w) + (mb^* + p_3^*(w))R(w) + \frac{p_2^*(w)R(w)(\ln a)^{r_1}(\ln b)^{r_2}}{\Gamma(1 + r_1)\Gamma(1 + r_2)} \leq R(w),$$

where

$$b^* = \max_{i=1,\ldots,m} \left\{ \operatorname*{ess\,sup}_{(x,y,w)\in J\times\Omega} |b_i(x,y)| \right\},$$

and

$$p_i^*(w) = \operatorname*{ess\,sup}_{(x,y)\in J} p_i(x,y,w); \ i = 1 , \ 2, \ 3.$$

Set

$$q^* = \operatorname*{ess\,sup}_{(x,y,w)\in J\times\Omega} q(x,y,w).$$

Theorem 6.6. *Assume that hypotheses (6.7.1)–(6.7.7) hold. If*

$$\ell := mb^* + q^* < 1, \tag{6.32}$$

then the problem (6.31) has a random solution defined on $[-\xi, a] \times [-\mu, b]$.

Proof. Define the operator $N : \Omega \times \mathcal{C} \to \mathcal{C}$ by

$$(N(w)u)(x,y) = \begin{cases} \sum_{i=1}^{m} b_i(x,y,w)u(x - \xi_i, y - \mu_i, w) \\ \quad + f\left(x,y,{}^H I_\sigma^r u(x,y,w), u(x,y,w), w\right); & (x,y) \in J, \\ \Phi(x,y,w); & (x,y) \in \tilde{J}. \end{cases}$$

The functions Φ, b_i $(i = 1, \ldots, m)$ are continuous for all $w \in \Omega$. Again, as the function f is continuous on J, then $N(w)$ defines a mapping $N : \Omega \times \mathcal{C} \to \mathcal{C}$. Thus u is a solution for the problem (6.31) if and only if $u = (N(w))u$. We shall show that the operator N satisfies all conditions of Theorem 1.7. The proof will be given in several steps.

Step 1. $N(w)$ *is a random operator with stochastic domain on* $\mathcal{C}$.

Since $f(x,y,u,v,w)$ is random Carathéodory, the map $w \to f(x,y,u,v,w)$ is measurable in view of Definition 5.13. Therefore, the map

$$w \mapsto \sum_{i=1}^{m} b_i(x,y,w)u(x - \xi_i, y - \mu_i, w)$$

$$+ f(x,y,{}^H I_\sigma^r u(x,y,w), u(x,y,w), w)$$

is measurable. As a result, N is a random operator on $\Omega \times \mathcal{C} \times \mathcal{C}$ into $\mathcal{C}$.

Let $W : \Omega \to \mathcal{P}(C)$ be defined by

$$W(w) = \{u \in C : \|u\|_C \leq R(w) \text{ and } \|u(x_1, y_1, w) - u(x_2, y_2, w)\|_E$$

$$\leq \psi(|x_1 - x_2| + |y_1 - y_2|)\}.$$

Clearly, the subset $W(w)$ is closed, convex and equicontinuous for all $w \in \Omega$. Then W is measurable by [208, Lemma 17]. Therefore, N is a random operator with stochastic domain W.

Step 2. $N(w)$ *is continuous.*

Let $\{u_n\}$ be a sequence such that $u_n \to u$ in $\mathcal{C}$. Then, there exists $\phi \in E^*$ such that $\|(N(w)u_n)(x,y)\|_E = |\phi((N(w)u_n)(x,y))|$, and $\|(N(w)u)(x,y)\|_E = |\phi((N(w)u)(x,y))|$.
Thus, for each $(x,y) \in J$ and $w \in \Omega$, we have

$$\|(N(w)u_n)(x,y) - (N(w)u)(x,y)\|_E$$
$$= |\phi((N(w)u_n)(x,y) - (N(w)u)(x,y))|$$
$$\leq \sum_{i=1}^{m} |b_i(x,y,w)| \|u_n(x-\xi_i, y-\mu_i, w) - u(x-\xi_i, y-\mu_i, w)\|_E$$
$$+ |\phi(f(x,y,{}^H I_\sigma^r u_n(x,y,w), u_n(x,y,w), w)$$
$$- f(x,y,{}^H I_\sigma^r u(x,y,w), u(x,y,w), w))|.$$

Using the Lebesgue dominated convergence theorem, we get

$$\|N(w)u_n - N(w)u\|_\infty \to 0 \quad \text{as} \quad n \to \infty.$$

As a consequence of Steps 1 and 2, we can conclude that $N(w) : W(w) \to N(w)$ is a continuous random operator with stochastic domain W, and $N(w)(W(w))$ is bounded.

Step 3. *For every* $w \in \Omega$, $\{u \in W(w) : N(w)u = u\} \neq \emptyset$.

For this, we apply Theorem 1.7. The proof will be given in several claims.

Claim 1. $N(w)$ *maps* $W(w)$ *into itself.*
Let $w \in \Omega$ be fixed, and let $u \in W(w)$, $(x,y) \in J$. Assume that $(N(w)u)(x,y) \neq 0$. Then there exists $\phi \in E^*$ such that $\|(N(w)u)(x,y)\|_E = |\phi((N(w)u)(x,y))|$. Thus, we get

$$\|(N(w)u)(x,y)\|_E \leq \left\| \sum_{i=1}^{m} b_i(x,y,w)u(x-\xi_i, y-\mu_i, w) \right\|_E$$
$$+ |\phi(f(x,y,{}^H I_\sigma^r u(x,y,w), u(x,y,w), w))|$$

$$\leq \sum_{i=1}^{m} |b_i(x,y,w)| \|u(x - \xi_i, y - \mu_i, w)\|_E$$

$$+ p_1(x,y,w) + p_2(x,y,w)\|^H I_\sigma^r u(x,y,w)\|_E + p_3(x,y,w)\|u(x,y,w)\|_E$$

$$\leq mb^* \|u\|_\infty + p_1^*(w) + \frac{p_2^*(w)}{\Gamma(r_1)\Gamma(r_2)} \int_1^x \int_1^y \left(\ln \frac{x}{s}\right)^{r_1 - 1}$$

$$\left(\ln \frac{y}{t}\right)^{r_2 - 1} \|u(s,t,w)\|_E dt ds + p_3^*(w)R(w)$$

$$\leq p_1^*(w) + (mb^* + p_3^*(w))R(w)$$

$$+ \frac{p_2^*(w)R(w)(\ln a)^{r_1}(\ln b)^{r_2}}{\Gamma(1 + r_1)\Gamma(1 + r_2)} \leq R(w).$$

Next, for any fixed $w \in \Omega$, let $(x_1, y_1), (x_2, y_2) \in J$ such that $x_1 < x_2$ and $y_1 < y_2$, and let $u \in W(w)$, with $(N(w)u)(x_1, y_1) - (N(w)u)(x_2, y_2) \neq 0$. Then there exists $\phi \in E^*$ such that $\|(N(w)u)(x_1, y_1) - (N(w)u)(x_2, y_2)\|_E = |\varphi((N(w)u)(x_1, y_1) - (N(w)u)(x_2, y_2))|$ and $\|\varphi\| = 1$. Thus, we have

$$\|(N(w)u)(x_2, y_2) - (N(w)u)(x_1, y_1)\|_E$$

$$= |\phi((N(w)u)(x_2, y_2) - (N(w)u)(x_1, y_1))|$$

$$\leq \psi(|x_1 - x_2| + |y_1 - y_2|).$$

Hence $N(W(w)) \subset W(w)$. Therefore, $N(w) : W(w) \to N(w)$ maps $W(w)$ into itself.

Claim 2. $N(w)$ *is weakly-sequentially continuous.*
Let (u_n) be a sequence in $W(w)$ and let $(u_n(x,y,w)) \to u(x,y,w)$ in (E,ω) for any $w \in \Omega$, and each $(x,y) \in J$. Fix $(x,y) \in J$, since f satisfies the assumption (6.7.6), we have $f(x, y, {}^H I_\sigma u_n(x,y,w), u_n(x,y,w), w)$ converges weakly to

$$f(x, y, {}^H I_\sigma u(x,y,w), u(x,y,w), w).$$

Hence the Lebesgue dominated convergence theorem for Pettis integral implies $(Nu_n)(x,y,w)$ converges weakly to $(N(w)u)(x,y)$ in (E,ω). We do it for any $w \in \Omega$, and each $(x,y) \in J$, so $N(w)(u_n) \to N(w)(u)$. Then $N : W(w) \to W(w)$ is weakly-sequentially continuous.

Claim 3. *The implication* (1.9) *holds.*
Let V be a subset of $W(w)$ such that $\overline{V} = \overline{\text{conv}}(N(w)(V) \cup \{0\})$. Obviously $V(x,y,w) \subset \overline{\text{conv}}(N(w)V)(x,y)) \cup \{0\})$. Further, as V is bounded and

equicontinuous, by [172, Lemma 3] the function $(x, y, w) \to u(x, y, w) = \beta(V(x, y, w))$ is continuous on $J \times \Omega$. Since the functions μ is continuous on $J \times \Omega$, then the set $\{\mu(x, y, w); (x, y) \in J, w \in \Omega\} \subset E$ is compact. From Lemma 2.3 and the properties of the measure β, for any $w \in \Omega$, and each $(x, y) \in J$, we have

$$v(x, y, w) \leq \beta((N(w)V)(x, y) \cup \{0\})$$

$$\leq \beta((N(w)V)(x, y))$$

$$= \beta\left(\sum_{i=1}^{m} b_i(x, y, w)u(x - \xi_i, y - \mu_i, w) \right.$$

$$\left. + f(x, y, {}^H I_\sigma^r u(x, y, w), u(x, y, w), w) \right)$$

$$\leq \sum_{i=1}^{m} |b_i(x, y, w)|\beta(V(x, y, w))$$

$$+ q(x, y, w)\beta(V(x, y, w))$$

$$\leq mb^* v(x, y, w) + q^* v(x, y, w)$$

$$\leq (mb^* + q^*)\|v\|_C.$$

Thus

$$\|v\|_C \leq \ell\|v\|_C.$$

From (6.32), we get $\|v\|_C = 0$, that is $v(x, y, w) = \beta(V(x, y, w)) = 0$, for any $w \in \Omega$, and each $(x, y) \in J$. Hence, Theorem 2] in [290] shows that V is relatively-weakly compact in C.

As consequence of claims 1–3, and from Theorem 1.8 it follows that for every $w \in \Omega$, $\{u \in W(w) : N(w)u = u\} \neq \emptyset$. Apply now Theorem 1.9, Steps 1–3 show that for each $w \in \Omega$, N has at least one fixed point in W. Since $\bigcap_{w \in \Omega} \text{int } W(w) \neq \emptyset$ and the hypothesis that a measurable selector of int W exists, then N has a stochastic fixed point, i.e., the problem (6.31) has at least one random solution defined on $[-\xi, a] \times [-\mu, b]$.   □

6.7.3. *Example*

Let

$$E = l^1 = \left\{ w = (w_1, w_2, \ldots, w_n, \ldots) : \sum_{n=1}^{\infty} |w_n| < \infty \right\}$$

be the Banach space with norm $\|w\|_E = \sum_{n=1}^{\infty} |w_n|$, and $\Omega = (-\infty, 0)$ be equipped with the usual σ-algebra consisting of Bochner measurable subsets of $(-\infty, 0)$. Given a measurable function $u : \Omega \to C([-\frac{7}{2}, e] \times [-5, e])$, consider the following functional random integral problem:

$$u(x, y, w) = \begin{cases} \dfrac{x^3 e^{-y}}{17 + w^2} u\left(x - 3, y - \dfrac{4}{3}, w\right) + \dfrac{xy^2}{10 + w^2} u(x - 2, y - 6, w) \\[2ex] + \dfrac{1}{11 + w^2 + x^2 + y^2} u\left(x - \dfrac{9}{2}, y - \dfrac{5}{4}, w\right) \\[2ex] + \dfrac{w^2 e^{-x-y-3}}{1 + w^2 + |u(x, y, w)| + |^H I_\sigma^r u(x, y, w)|}; \\[2ex] (x, y) \in J = [1, e] \times [1, e], \ w \in \Omega, \\[1ex] \Phi(x, y, w); \ (x, y) \in \tilde{J}, \ w \in \Omega, \end{cases}$$

(6.33)

where

$$\tilde{J} := \left[-\dfrac{7}{2}, e\right] \times [-5, e] \backslash (1, e] \times (1, e],$$

$m = 3$, $r = (r_1, r_2)$; $r_1, r_2 \in (0, \infty)$, $u = (u_1, u_2, \ldots, u_n, \ldots)$, $f = (f_1, f_2, \ldots, f_n, \ldots)$,

$$f_n(x, y, {}^H I_\sigma^r u, u, w) = \dfrac{w^2 e^{-x-y-3}}{1 + w^2 + |u_n(x, y, w)| + |^H I_\sigma^r u_n(x, y, w)|};$$

$(x, y) \in J = [0, 1] \times [0, 1]$, $w \in \Omega$, $n \in \mathbb{N}$, and $\Phi : \tilde{J} \to E$ is a continuous and measurable function such that $\phi = (\phi_1, \phi_2, \ldots, \phi_n, \ldots)$, with

$$\phi_n(x, 1, w) = \dfrac{x^3}{17 + w^2} \phi_n\left(x - 3, -\dfrac{1}{3}, w\right)$$

$$+ \dfrac{1}{11 + w^2 + x^2} \phi_n\left(x - \dfrac{9}{2}, -\dfrac{1}{4}, w\right); \ x \in [0, 1], \ n \in \mathbb{N},$$

and

$$\phi_n(1, y, w) = \dfrac{1}{11 + w^2 + y^2} \phi_n\left(-\dfrac{7}{2}, y - \dfrac{5}{4}, w\right); \quad y \in [0, 1], \ n \in \mathbb{N}.$$

Set

$$b_1(x, y, w) = \dfrac{x^3 e^{-y}}{17}, \quad b_2(x, y) = \dfrac{xy^2}{10}, \quad b_3(x, y, w) = \dfrac{1}{11 + x^2 + y^2}.$$

Then, $b^* = \frac{1}{10}$. For each $u, v \in E$, $(x, y) \in [0, 1] \times [0, 1]$ and $w \in \Omega$, we have

$$\|f(x, y, u, v, w)\|_E \leq 1 + \frac{1}{e^3}(\|u\|_E + \|v\|_E).$$

Hence the conditions (6.7.4) is satisfied with $p_1^* = 1$ and $p_2^* = p_3^* = \frac{1}{e^3}$.

A simple computation shows that all other conditions of Theorem 6.6 are satisfied. Consequently, Theorem 6.6 implies that the problem (6.33) has a random solution defined on $[-\frac{7}{2}, e] \times [-5, e]$.

6.8. Notes and Remarks

The results of Chapter 6 are taken from [15,17,18,37,38,50].

Chapter 7

Nonlinear Implicit Hadamard–Pettis Fractional Differential Equations

7.1. Introduction

In this chapter, we present some results concerning the existence of weak solutions for some implicit differential equations of Hadamard fractional derivative. The main results are proved by applying Mönch's fixed-point theorem associated with the technique of measure of weak noncompactness.

7.2. Implicit Hadamard–Pettis Fractional Differential Equations

7.2.1. *Introduction*

Recently, considerable attention has been given to the existence of solutions of initial and boundary value problem for fractional differential equations with Hadamard fractional derivative; see [9,53,58]. In this section, we discuss the existence of weak solutions for the following implicit Hadamard fractional differential equation:

$$
\begin{cases}
(^H D_1^r u)(t) = f(t, u(t), (^H D_1^r u)(t)); & t \in I := [1, T], \\
(^H I_1^{1-r} u)(t)|_{t=1} = \phi,
\end{cases}
\tag{7.1}
$$

where $T > 1$, $\phi \in E$, $f : I \times E \times E \to E$ is a given continuous function, E is a real (or complex) Banach space with norm $\| \cdot \|_E$ and dual E^*, such that E is the dual of a weakly compactly generated Banach space X, $^H I_1^r$ is the left-sided mixed Hadamard integral of order $r \in (0, 1]$, and $^H D_1^r$ is the Hadamard fractional derivative of order r.

7.2.2. Existence of weak solutions

Definition 7.1. By a weak solution of the problem (7.1) we mean a measurable function $u \in C(I)$ that satisfies the condition $({}^H I_1^{1-r} u)(t)|_{t=1} = \phi$, and the equation $({}^H D_1^r u)(t) = f(t, u(t), ({}^H D_1^r u)(t))$ on I.

The following hypotheses will be used in the sequel.

(7.1.1) For a.e. $t \in I$, the functions $v \to f(t, v, \cdot)$ and $w \to f(t, \cdot, w)$ are weakly sequentially continuous.

(7.1.2) For a.e. $v, w \in E$, the function $t \to f(t, v, w)$ is Pettis integrable a.e. on I.

(7.1.3) There exists $p \in C(I, [0, \infty))$ such that for all $\varphi \in E^*$, we have

$$|\varphi(f(t, u, v))| \le p(t)\|\varphi\|; \quad \text{for a.e. } t \in I, \text{ and each } u, v \in E.$$

(7.1.4) For each bounded set $B \subset E$ and for each $t \in I$, we have

$$\beta(f(t, B, {}^H D_1^r B) \le (\ln t)^{1-r} p(t) \beta(B),$$

where ${}^H D_1^r B = \{{}^H D_1^r w : w \in B\}$.

Set

$$p^* = \sup_{t \in I} p(t),$$

Theorem 7.1. *Assume that the hypotheses (7.1.1)–(7.1.4) hold. If*

$$L := \frac{p^* \ln T}{\Gamma(1+r)} < 1, \tag{7.2}$$

then the problem (7.1) has at least one solution defined on I.

Proof. Transform the problem (7.1) into a fixed-point equation. Consider the operator $N : \mathcal{C} \to \mathcal{C}$ defined by

$$(Nu)(t) = \frac{\phi}{\Gamma(r)}(\ln t)^{r-1} + \int_1^t \left(\ln \frac{t}{s}\right)^{r-1} \frac{g(s)}{s\Gamma(r)} ds. \tag{7.3}$$

where $g \in C$ with

$$g(t) = f\left(t, \frac{\phi}{\Gamma(r)}(\ln t)^{r-1} + ({}^H I_1^r g)(t), g(t)\right).$$

First notice that, the hypotheses imply that $\left(\ln \frac{t}{s}\right)^{r-1}\frac{g(s)}{s}$ for all $t \in I$ is Pettis integrable, and for each $u \in C$, the function

$$t \mapsto f\left(t, \frac{\phi}{\Gamma(r)}(\ln t)^{r-1} + (^H I_1^r g)(t), g(t)\right)$$

is Pettis integrable over I. Thus, the operator N is well defined. Let $R > 0$ be such that

$$R > \frac{p^* \ln T}{\Gamma(1+r)},$$

and consider the set

$$Q = \left\{ u \in C : \|u\|_C \leq R \text{ and } \|(\ln t_2)^{1-r}u(t_2) - (\ln t_1)^{1-r}u(t_1)\|_E \right.$$

$$\leq \frac{p^*}{\Gamma(1+r)}(\ln T)^{1-r}\left|\ln \frac{t_2}{t_1}\right|^r$$

$$\left. + \frac{p^*}{\Gamma(r)}\int_1^{t_1}\left|(\ln t_2)^{1-r}\left(\ln \frac{t_2}{s}\right)^{r-1} - (\ln t_1)^{1-r}\left(\ln \frac{t_1}{s}\right)^{r-1}\right|ds \right\}.$$

Clearly, the subset Q is closed, convex and equicontinuous. We shall show that the operator N satisfies all the assumptions of Theorem 1.7. The proof will be given in several steps.

Step 1. *N maps Q into itself.*

Let $u \in Q$, $t \in I$ and assume that $(Nu)(t) \neq 0$. Then there exists $\varphi \in E^*$ such that $\|(\ln t)^{1-r}(Nu)(t)\|_E = |\varphi((\ln t)^{1-r}(Nu)(t))|$. Thus

$$\|(\ln t)^{1-r}(Nu)(t)\|_E = \left\|\varphi\left(\frac{\phi}{\Gamma(r)} + \frac{(\ln t)^{1-r}}{\Gamma(r)}\int_1^t\left(\ln \frac{t}{s}\right)^{r-1}\frac{g(s)}{s}ds\right)\right\|,$$

where $g \in C$ with

$$g(t) = f\left(t, \frac{\phi}{\Gamma(r)}(\ln t)^{r-1} + (^H I_1^r g)(t), g(t)\right).$$

Then

$$\|(\ln t)^{1-r}(Nu)(t)\|_E \leq \frac{(\ln t)^{1-r}}{\Gamma(r)}\int_1^t\left(\ln \frac{t}{s}\right)^{r-1}\frac{|\varphi(g(s))|}{s}ds$$

$$\leq \frac{p^*(\ln T)^{1-r}}{\Gamma(r)}\int_1^t\left(\ln \frac{t}{s}\right)^{r-1}\frac{ds}{s}$$

$$\le \frac{p^* \ln T}{\Gamma(1+r)}$$

$$\le R.$$

Next, let $t_1, t_2 \in I$ such that $t_1 < t_2$ and let $u \in Q$, with

$$(\ln t_2)^{1-r}(Nu)(t_2) - (\ln t_1)^{1-r}(Nu)(t_1) \ne 0.$$

Then there exists $\varphi \in E^*$ such that

$$\|(\ln t_2)^{1-r}(Nu)(t_2) - (\ln t_1)^{1-r}(Nu)(t_1)\|_E$$
$$= |\varphi((\ln t_2)^{1-r}(Nu)(t_2) - (\ln t_1)^{1-r}(Nu)(t_1))|$$

and $\|\varphi\| = 1$. Then

$$\|(\ln t_2)^{1-r}(Nu)(t_2) - (\ln t_1)^{1-r}(Nu)(t_1)\|_E$$
$$= |\varphi((\ln t_2)^{1-r}(Nu)(t_2) - (\ln t_1)^{1-r}(Nu)(t_1))|$$
$$\le \left| \varphi \left((\ln t_2)^{1-r} \int_1^{t_2} \left(\ln \frac{t_2}{s} \right)^{r-1} \frac{g(s)}{s\Gamma(r)} ds - (\ln t_1)^{1-r} \int_1^{t_1} \left(\ln \frac{t_1}{s} \right)^{r-1} \right. \right.$$
$$\left. \left. \times \frac{g(s)}{s\Gamma(r)} ds \right) \right|,$$

where $g \in C$ with

$$g(t) = f\left(t, \frac{\phi}{\Gamma(r)}(\ln t)^{r-1} + (^H I_1^r g)(t), g(t) \right).$$

Then

$$\|(\ln t_2)^{1-r}(Nu)(t_2) - (\ln t_1)^{1-r}(Nu)(t_1)\|_E$$
$$\le (\ln t_2)^{1-r} \int_{t_1}^{t_2} \left| \ln \frac{t_2}{s} \right|^{r-1} \frac{|\varphi(g(s))|}{s\Gamma(r)} ds$$
$$+ \int_1^{t_1} \left| (\ln t_2)^{1-r} (\ln \frac{t_2}{s})^{r-1} - (\ln t_1)^{1-r} \left(\ln \frac{t_1}{s} \right)^{r-1} \right| \frac{|\varphi(g(s))|}{s\Gamma(r)} ds$$
$$\le (\ln t_2)^{1-r} \int_{t_1}^{t_2} \left| \ln \frac{t_2}{s} \right|^{r-1} \frac{p(s)}{\Gamma(r)} ds$$
$$+ \int_1^{t_1} \left| (\ln t_2)^{1-r} \left(\ln \frac{t_2}{s} \right)^{r-1} - (\ln t_1)^{1-r} \left(\ln \frac{t_1}{s} \right)^{r-1} \right| \frac{p(s)}{\Gamma(r)} ds.$$

Thus, we get

$$\|(\ln t_2)^{1-r}(Nu)(t_2) - (\ln t_1)^{1-r}(Nu)(t_1)\|_E$$

$$\leq \frac{p^*}{\Gamma(1+r)}(\ln T)^{1-r}\left|\ln \frac{t_2}{t_1}\right|^r$$

$$+ \frac{p^*}{\Gamma(r)}\int_1^{t_1}\left|(\ln t_2)^{1-r}\left(\ln \frac{t_2}{s}\right)^{r-1} - (\ln t_1)^{1-r}\left(\ln \frac{t_1}{s}\right)^{r-1}\right|ds.$$

Hence $N(Q) \subset Q$.

Step 2. *N is weakly-sequentially continuous.*

Let (u_n) be a sequence in Q and let $(u_n(t)) \to u(t)$ in (E, ω) for each $t \in I$. Fix $t \in I$, since f satisfies the assumption (7.1.1), we have $f(t, u_n(t), {}^H D_1 u_n(t))$ converges weakly to $f(t, u(t), {}^H D_1 u(t))$. Hence the Lebesgue dominated convergence theorem for Pettis integral implies $(Nu_n)(t)$ converges weakly to $(Nu)(t)$ in (E, ω), for each $t \in I$. Thus, $N(u_n) \to N(u)$. Hence, $N : Q \to Q$ is weakly-sequentially continuous.

Step 3. *The implication (1.9) holds.*

Let V be a subset of Q such that $\overline{V} = \overline{\text{conv}}(N(V) \cup \{0\})$. Obviously

$$V(t) \subset \overline{\text{conv}}(NV)(t)) \cup \{0\}), \ \forall t \in I.$$

Further, as V is bounded and equicontinuous, by [172, Lemma 3], the function $t \to v(t) = \beta(V(t))$ is continuous on I. From (7.1.3), (7.1.4), Lemma 1.7 and the properties of the measure β, for any $t \in I$, we have

$$(\ln t)^{1-r}v(t) \leq \beta((\ln t)^{1-r}(NV)(t) \cup \{0\})$$

$$\leq \beta((\ln t)^{1-r}(NV)(t))$$

$$\leq \frac{(\ln T)^{1-r}}{\Gamma(r)}\int_1^t\left|\ln \frac{t}{s}\right|^{r-1}\frac{p(s)\beta(V(s))}{s}ds$$

$$\leq \frac{(\ln T)^{1-r}}{\Gamma(r)}\int_1^t\left|\ln \frac{t}{s}\right|^{r-1}\frac{(\ln s)^{1-r}p(s)v(s)}{s}ds$$

$$\leq \frac{p^*\ln T}{\Gamma(1+r)}\|v\|_C.$$

Thus

$$\|v\|_C \leq L\|v\|_C.$$

From (7.2), we get $\|v\|_C = 0$, that is $v(t) = \beta(V(t)) = 0$, for each $t \in I$. and then by [290, Theorem 2], V is relatively-weakly compact in C. Applying now Theorem 1.7, we conclude that N has a fixed point which is a solution of the problem (7.1). □

7.2.3. Example

Let

$$E = l^1 = \left\{ u = (u_1, u_2, \ldots, u_n, \ldots), \sum_{n=1}^{\infty} |u_n| < \infty \right\}$$

be the Banach space with the norm

$$\|u\|_E = \sum_{n=1}^{\infty} |u_n|.$$

As an application of our results we consider the following problem of implicit Hadamard fractional differential equation

$$\begin{cases} (^H D_1^{\frac{1}{2}} u_n)(t) = f_n(t, u(t), (^H D_1^{\frac{1}{2}} u)(t)); \ t \in [1, e], \\ (^H I_1^{\frac{1}{2}} u)(t)|_{t=1} = 0, \end{cases} \tag{7.4}$$

where

$$f_n(t, u(t), (^H D_1^{\frac{1}{2}} u)(t)) = \frac{ct^2}{1 + \|u\|_E + \|^H D_1^{\frac{1}{2}} u\|_E} \left(e^{-7} + \frac{1}{e^{t+5}} \right) u_n(t);$$

$$t \in [1, e],$$

with

$$u = (u_1, u_2, \ldots, u_n, \ldots) \quad \text{and} \quad c := \frac{e^4}{8} \Gamma\left(\frac{1}{2}\right).$$

Set $f = (f_1, f_2, \ldots, f_n, \ldots)$. Clearly, the function f is continuous. For each $u \in E$ and $t \in [1, e]$, we have

$$\|f(t, u(t), (^H D_1^{\frac{1}{2}})(t))\|_E \le ct^2 \left(e^{-7} + \frac{1}{e^{t+5}} \right).$$

Hence, the hypothesis (7.2.3) is satisfied with $p^* = ce^{-4}$,

We shall show that condition (7.2) holds with $T = e$. Indeed,

$$\frac{p^* \ln T}{\Gamma(1+r)} = \frac{c}{e^4 \Gamma(\frac{3}{2})} = \frac{1}{4} < 1.$$

A simple computations show that all conditions of Theorem 7.1 are satisfied. It follows that the problem (7.4) has at least one solution on $[1, e]$.

7.3. Implicit Hadamard–Pettis Fractional Differential Equations with Delay

7.3.1. *Introduction*

In this section, we discuss the existence of weak solutions for the following delay implicit Pettis–Hadamard fractional differential equation:

$$\begin{cases} (^H D_1^r u)(t) = f(t, u_t, (^H D_1^r u)(t)); & t \in I := [1, T], \\ u(t) = \phi(t); & t \in \tilde{I} := [1 - \alpha, 1], \end{cases} \quad (7.5)$$

where $T > 1$, $f : I \times C[-\alpha, 0] \times E \to E$ is a given continuous function, $\phi \in C[1 - \alpha, 1]$ with $\phi(1) = 0$, E is a real (or complex) Banach space with norm $\|\cdot\|_E$ and dual E^*, such that E is the dual of a weakly-compactly-generated Banach space X, $C[-\alpha, 0]$ is the space of continuous functions from $[-\alpha, 0]$ to E, $^H I_1^r$ is the left-sided mixed Pettis–Hadamard integral of order $r \in (0, 1]$, and $^H D_1^r$ is the Pettis–Hadamard fractional derivative of order r.

We denote by u_t the element of $C[-\alpha, 0]$ defined by

$$u_t(s) = u(t + s); \; t \in [-\alpha, 0].$$

Here $u_t(\cdot)$ represents the history of the state from time $t - \alpha$ up to the present time t.

7.3.2. *Existence of weak solutions*

Definition 7.2. By a weak solution of the problem (7.5) we mean a measurable function $u \in C[1 - \alpha, T]$ that satisfies the condition $u(1) = \phi(t)$ on $\tilde{I}$, and the equation $(^H D_1^r u)(t) = f(t, u_t, (^H D_1^r u)(t))$ on I.

The following hypotheses will be used in the sequel.

(7.2.1) For a.e. $t \in I$, the functions $v \to f(t, v, \cdot)$ and $w \to f(t, \cdot, w)$ are weakly-sequentially continuous.

(7.2.2) For a.e. $v \in C[-\alpha, 0]$, and $w \in E$, the function $t \to f(t, v, w)$ is Pettis integrable a.e. on I.

(7.2.3) There exists $p \in C(I, [0, \infty))$ such that for all $\varphi \in E^*$, we have

$$|\varphi(f(t, u, v))| \le p(t)\|\varphi\|$$

for a.e. $t \in I$, and each $u \in C[-\alpha, 0]$, and $v \in E$,

(7.2.4) For each bounded set $B \subset E$ and for each $t \in I$, we have

$$\beta(f(t, B, {}^H D_1^r B) \leq p(t)\beta(B),$$

where ${}^H D_1^r B = \{{}^H D_1^r w : w \in B\}$.

Set

$$p^* = \sup_{t \in I} p(t),$$

Theorem 7.2. *Assume that the hypotheses (7.2.1)–(7.2.4) hold. If*

$$L := \frac{p^*(\ln T)^r}{\Gamma(1+r)} < 1, \tag{7.6}$$

then the problem (7.5) has at least one solution defined on I.

Proof. Transform the problem (7.5) into a fixed-point equation. Consider the operator $N : C[1-\alpha, 1] \to C[1-\alpha, 1]$ defined by

$$(Nu)(t) = \begin{cases} ({}^H I_1^q g)(t); & t \in I, \\ \phi(t); & t \in \tilde{I}, \end{cases} \tag{7.7}$$

where $g(t) \in C(I)$ with

$$g(t) = f\left(t, {}^H I_1^r g_t, g(t)\right).$$

First notice that, the function ϕ is continuous on $\tilde{I}$, and the hypotheses imply that for all $t \in I$, the functions $t \mapsto \left(\ln \frac{t}{s}\right)^{r-1} \frac{g(s)}{s}$, and $t \mapsto f(t, u, v)$ are Pettis integrables, over I. Thus, the operator N is well defined.

In the following we denote $\|w\|_{C[1-\alpha,T]}$ by $\|w\|_C$.

Let $R > 0$ be such that

$$R > \max\left\{\frac{p^* \ln T}{\Gamma(1+r)}, \|\phi\|_{C[1-\alpha,1]}\right\}$$

and consider the set

$$Q = \left\{u \in C : \|u\|_C \leq R \text{ and } \|u(t_2) - u(t_1)\|_E \leq \frac{p^*}{\Gamma(1+r)} \left|\ln \frac{t_2}{t_1}\right|^r \right.$$

$$\left. + \frac{p^*}{\Gamma(r)} \int_1^{t_1} \left|\left(\ln \frac{t_2}{s}\right)^{r-1} - \left(\ln \frac{t_1}{s}\right)^{r-1}\right| ds\right\}.$$

Clearly, the subset Q is closed, convex and equicontinuous. We shall show that the operator N satisfies all the assumptions of Theorem 1.7. The proof will be given in several steps.

Step 1. N *maps Q into itself.*

Let $u \in Q$; $t \in I$ and assume that $(Nu)(t) \neq 0$. Then there exists $\varphi \in E^*$ such that $\|(Nu)(t)\|_E = |\varphi((Nu)(t))|$. Thus

$$\|(Nu)(t)\|_E = \left| \varphi \left(\frac{1}{\Gamma(r)} \int_1^t \left(\ln \frac{t}{s} \right)^{r-1} \frac{g(s)}{s} ds \right) \right|,$$

where $g \in C$ with

$$g(t) = f\left(t, {}^H I_1^r g_t, g(t)\right).$$

Then

$$\|(Nu)(t)\|_E \leq \frac{1}{\Gamma(r)} \int_1^t \left(\ln \frac{t}{s} \right)^{r-1} \frac{|\varphi(g(s))|}{s} ds$$

$$\leq \frac{p^*}{\Gamma(r)} \int_1^t \left(\ln \frac{t}{s} \right)^{r-1} \frac{ds}{s}$$

$$\leq \frac{p^* (\ln T)^r}{\Gamma(1+r)}$$

$$\leq R.$$

Also, if $u \in Q$; $t \in \tilde{I}$, we have

$$\|(Nu)(t)\|_E = \|\phi(t)\|_{C[1-\alpha,1]} \leq R.$$

Next, let $t_1, t_2 \in I$ such that $t_1 < t_2$ and let $u \in Q$, with

$$(Nu)(t_2) - (Nu)(t_1) \neq 0.$$

Then there exists $\varphi \in E^*$ such that

$$\|(Nu)(t_2) - (Nu)(t_1)\|_E = |\varphi((Nu)(t_2) - (Nu)(t_1))|$$

and $\|\varphi\| = 1$. Then

$$\|(Nu)(t_2) - (Nu)(t_1)\|_E$$

$$= |\varphi((Nu)(t_2) - (Nu)(t_1))|$$

$$\leq \left| \varphi \left(\int_1^{t_2} \left(\ln \frac{t_2}{s} \right)^{r-1} \frac{g(s)}{s\Gamma(r)} ds - \int_1^{t_1} \left(\ln \frac{t_1}{s} \right)^{r-1} \frac{g(s)}{s\Gamma(r)} ds \right) \right|,$$

where $g \in C$ with

$$g(t) = f\left(t, {}^H I_1^r g_t, g(t)\right).$$

Then

$$\|(Nu)(t_2) - (Nu)(t_1)\|_E \leq \int_{t_1}^{t_2} \left| \ln \frac{t_2}{s} \right|^{r-1} \frac{|\varphi(g(s))|}{s\Gamma(r)} ds$$

$$+ \int_1^{t_1} \left| \left(\ln \frac{t_2}{s} \right)^{r-1} - \left(\ln \frac{t_1}{s} \right)^{r-1} \right| \frac{|\varphi(g(s))|}{s\Gamma(r)} ds$$

$$\leq \int_{t_1}^{t_2} \left| \ln \frac{t_2}{s} \right|^{r-1} \frac{p(s)}{\Gamma(r)} ds$$

$$+ \int_1^{t_1} \left| \left(\ln \frac{t_2}{s} \right)^{r-1} - \left(\ln \frac{t_1}{s} \right)^{r-1} \right| \frac{p(s)}{\Gamma(r)} ds.$$

Thus, we get

$$\|(Nu)(t_2) - (Nu)(t_1)\|_E \leq \frac{p^*}{\Gamma(1+r)} \left| \ln \frac{t_2}{t_1} \right|^r$$

$$+ \frac{p^*}{\Gamma(r)} \int_1^{t_1} \left| \left(\ln \frac{t_2}{s} \right)^{r-1} - \left(\ln \frac{t_1}{s} \right)^{r-1} \right| ds.$$

Hence $N(Q) \subset Q$.

Step 2. *N is weakly-sequentially continuous.*

Let (u_n) be a sequence in Q and let $(u_n(t)) \to u(t)$ in (E, ω) for each $t \in [1 - \alpha, T]$. Fix $t \in [1 - \alpha, T]$, since f satisfies the assumption (7.2.1), we have $f(t, u_{nt}, {}^H D_1 u_n(t))$ converges weakly to $f(t, u_t, {}^H D_1 u(t))$. Hence the Lebesgue dominated convergence theorem for Pettis integral implies $(Nu_n)(t)$ converges weakly to $(Nu)(t)$ in (E, ω), for each $t \in [1-\alpha, T]$. Thus, $N(u_n) \to N(u)$. Hence, $N : Q \to Q$ is weakly-sequentially continuous.

Step 3. *The implication (1.9) holds.*

Let V be a subset of Q such that $\overline{V} = \overline{\text{conv}}(N(V) \cup \{0\})$. Obviously

$$V(t) \subset \overline{conv}(NV)(t)) \cup \{0\}), \ \forall t \in [1 - \alpha, T].$$

Further, as V is bounded and equicontinuous, by [172, Lemma 3], the function $t \to v(t) = \beta(V(t))$ is continuous on $[1 - \alpha, T]$. From (7.2.3), (7.2.4), Lemma 1.7 and the properties of the measure β, for any $t \in [1 - \alpha, T]$, we have

$$v(t) \leq \beta((NV)(t) \cup \{0\})$$

$$\leq \beta((NV)(t))$$

$$\leq \frac{1}{\Gamma(r)} \int_1^t \left| \ln \frac{t}{s} \right|^{r-1} \frac{p(s)\beta(V(s))}{s} ds$$

$$\leq \frac{1}{\Gamma(r)} \int_1^t \left| \ln \frac{t}{s} \right|^{r-1} \frac{p(s)v(s)}{s} ds$$

$$\leq \frac{p^*(\ln T)^r}{\Gamma(1+r)} \|v\|_C.$$

Thus

$$\|v\|_C \leq L\|v\|_C.$$

From (7.6), we get $\|v\|_C = 0$, that is $v(t) = \beta(V(t)) = 0$, for each $t \in [1-\alpha, T]$. and then by [290, Theorem 2], V is relatively-weakly compact in $C[1-\alpha, T]$. Applying now Theorem 1.7, we conclude that N has a fixed point which is a solution of the problem (7.5). $\qquad\qquad\square$

7.3.3. *Example*

Let

$$E = l^1 = \left\{ u = (u_1, u_2, \ldots, u_n, \ldots), \sum_{n=1}^{\infty} |u_n| < \infty \right\}$$

be the Banach space with the norm

$$\|u\|_E = \sum_{n=1}^{\infty} |u_n|.$$

As an application of our results we consider the following problem of implicit Hadamard fractional differential equation

$$\begin{cases} ({}^H D_1^{\frac{1}{2}} u_n)(t) = f_n(t, u_t, ({}^H D_1^{\frac{1}{2}} u)(t)); & t \in [1, e], \\ u(t) = 1 - e^{2t}; & t \in [-2, 1], \end{cases} \tag{7.8}$$

where

$$f_n(t, u_t, ({}^H D_1^{\frac{1}{2}} u)(t)) = \frac{ct^2}{1 + \|u\|_{C[-3,0]} + \|{}^H D_1^{\frac{1}{2}} u\|_E} \left(e^{-7} + \frac{1}{e^{t+5}} \right) u_n(t);$$

$$t \in [1, e],$$

with

$$u = (u_1, u_2, \ldots, u_n, \ldots), \quad \text{and} \quad c := \frac{e^4}{8} \Gamma\left(\frac{1}{2}\right).$$

Set $f = (f_1, f_2, \ldots, f_n, \ldots)$.

Clearly, the function f is continuous.

For each $u \in E$ and $t \in [1, e]$, we have

$$\|f(t, u(t), (^H D_1^{\frac{1}{2}})(t))\|_E \leq ct^2 \left(e^{-7} + \frac{1}{e^{t+5}}\right).$$

Hence, the hypothesis (7.2.3) is satisfied with $p^* = ce^{-4}$,

We shall show that condition (7.6) holds with $T = e$. Indeed,

$$\frac{p^*(\ln T)^r}{\Gamma(1 + r)} = \frac{c}{e^4 \Gamma(\frac{3}{2})} = \frac{1}{4} < 1.$$

A simple computations show that all the conditions of Theorem 7.2 are satisfied. It follows that the problem (7.8) has at least one solution on $[-2, e]$.

7.4. Successive Approximations for Implicit Hadamard–Pettis Fractional Differential Equations

7.4.1. *Introduction*

This section deals with the global convergence of successive approximations as well as the uniqueness of solutions for a class of implicit differential equations involving the Hadamard fractional derivative. We prove a theorem on the global convergence of successive approximations to the unique solution of our problem.

Convergence of successive approximations for ordinary functional differential and integral equations is a well-established property. It has been studied by De Blasi and Myjak [202], Chen [187], Faina [210], Shin [343], and the references therein. Recently, Człapiński [194], and Abbas *et al.* [12,21,22,47] got the global convergence of successive approximations as well as the uniqueness of solutions for the Darboux problem for partial fractional differential equations. In the present section, we discuss the global convergence of successive approximations for the fractional hadamard implicit differential equation

$$(^H D_1^r u)(t) = f(t, u(t), (^H D_1^r u)(t)); \quad t \in I := [1, T], \tag{7.9}$$

with the initial Hadamard integral condition

$$(^H I_1^{1-r} u)(t)|_{t=1} = \phi, \tag{7.10}$$

where $T > 1$, $\phi \in E$, $f : I \times E \times E \to E$ is a given continuous function, E is a real (or complex) Banach space with norm $\|\cdot\|_E$ and dual E^*, such that E is the dual of a weakly-compactly generated Banach space X, $^H I_1^r$ is the left-sided mixed Hadamard integral of order $r \in (0, 1]$, and $^H D_1^r$ is the Hadamard fractional derivative of order r.

7.4.2. *Successive approximations and uniqueness results*

Now we present the main result for the global convergence of successive approximations to a unique solution of problem (7.9)–(7.10). Let $L^\infty(I)$ be the Banach space of measurable functions $u : J \to E$ which are essentially bounded, equipped with the norm

$$\|u\|_{L^\infty} = \inf\{c > 0 : \|u(t)\|_E \le c, \text{ a.e. } t \in I\}.$$

and let $(E, w) = (E, \sigma(E, E^*))$ be the Banach space E with its weak topology.

Definition 7.3. The function $f : I \times E \times E \to E$ is said to be weakly-Carathéodory if

(i) for a.e. $v, w \in E$, the function $t \to f(t, v, w)$ is Pettis integrable a.e. on I;
(ii) for a.e. $t \in I$, the functions $v \to f(t, v, \cdot)$ and $w \to f(t, \cdot, w)$ are weakly-sequentially continuous.

The function f is said to be L^∞-weakly Carathéodory if (i), (ii) and the following condition hold;

(iii) for every positive integer k and all $\varphi \in E^*$, there exists a function $h_k \in L^\infty(I, \mathbb{R}_+)$ such that

$$|\varphi(f(t, u, v))| \le h_k(t); \quad \text{for all } \|u\|_E \le k, \ \|v\|_E \le k,$$

$$\text{and almost each } t \in I.$$

From [263, Theorem 2.3], we have

$$(^H I_1^q)(^H D_1^q w)(x) = w(x) - \frac{(^H I_1^{1-q} w)(1)}{\Gamma(q)} (\ln x)^{q-1}.$$

Corollary 7.1. *Let* $h : I \to E$ *be a continuous function. A function* $u \in$ $L^1(I, E)$ *is said to be a solution of the equation*

$$(^H D_1^q w)(t) = h(t),$$

if and only if u *satisfies the following Hadamard integral equation:*

$$w(t) = \frac{(^H I_1^{1-q} u)(1)}{\Gamma(q)} (\ln t)^{q-1} + (^H I_1^q h)(t).$$

Let $C_{r,\ln}(I)$, be the weighted space of continuous functions defined by

$$\mathcal{C}_{r,\ln} := \left\{ w \in \mathcal{C} : (\ln t)^r w(t) \in \mathcal{C}, \ \|w\|_{C_{r,\ln}} := \sup_{t \in I} \|(\ln t)^r w(t)\|_E \right\}.$$

Define the space $X := X(I, E)$ as follows:

$$X := \{ w \in \mathcal{C}_{r,\ln} : \ {}^H D_1^r w \text{ exists and } {}^H D_1^r w \in \mathcal{C}_{r,\ln} \}.$$

For $w \in X$, we define

$$\|w(t)\|_1 = \|(\ln t)^r w(t)\|_E + \|(\ln t)^r (^H D_1^r w)(t)\|_E.$$

In the space X we define the norm

$$\|w\|_X = \sup_{t \in I} \|w(t)\|_1.$$

Lemma 7.1. $(X, \| \cdot \|_X)$ *is a Banach space.*

Proof. Let $\{u_n\}_{n=0}^\infty$ be a Cauchy sequence in the space $(X, \| \cdot \|_X)$. Then,

$\forall \epsilon > 0, \ \exists N > 0$ such that for all $n, m > N$, we have $\|u_n - u_m\|_X < \epsilon$.

Thus, $\{u_n(t)\}_{n=0}^\infty$ and $\{(^H D_1^r u_n)(t)\}_{n=0}^\infty$ are Cauchy sequences in E. Then $\{u_n(t)\}_{n=0}^\infty$ converges to some $u(t)$ in E, and $\{^H D_1^r u_n\}_{n=0}^\infty$ converges uniformly to some $v(t) \in X$. Next, we need to prove that $u \in X$ and $v =^H D_1^r u$. According to the uniform convergence of $\{(^H D_1^r u_n)(t)\}_{n=0}^\infty$ and the dominated convergence theorem, we obtain

$$v(t) = \lim_{n \to \infty} (^H D_1^r u_n)(t).$$

Thus $\{^H D_1^r u_n\}_{n=0}^\infty$ converges uniformly to $^H D_1^r u$ in X. Hence, $u \in X$ and

$$v(t) = (^H D_t^r u)(t). \qquad \square$$

From Corollary 7.1, we conclude the following lemma.

Lemma 7.2. *Let $r \in (0,1]$ and $f : I \times E \times E \to E$ be an L^{∞}-weakly-Carathéodory function. A function $u \in X \cap \mathcal{AC}$ is a weak solution of the fractional integral equation:*

$$u(t) = \frac{\phi}{\Gamma(r)}(\ln t)^{r-1} + \int_1^t \left(\ln \frac{t}{s}\right)^{r-1} \frac{f(s, u(s), (^H D_1^r u)(s))}{s\Gamma(r)} ds, \quad (7.11)$$

if and only if u is a solution of the problem (7.9)–(7.10).

Definition 7.4. A generalized solution of the problem (7.9)–(7.10) is an absolutely continuous function satisfying the fractional integral equation (7.11) almost everywhere on J.

Set $I_\sigma := [1, 1+\sigma(T-1)]$ for any $\sigma \in [0,1]$. Let us introduce the following hypotheses.

(7.3.1) The function $f : I \times E \times E \to E$ is L^{∞}-weakly Carathéodory.
(7.3.2) There exist a constant $\rho > 0$ and a weakly-Carathéodory function $w : I \times [0, \rho] \times [0, \rho] \to [0, \infty)$ such that $w(t, \cdot, \cdot)$ is nondecreasing for a.e. $t \in I$, and for all $\varphi \in E^*$, and the inequality

$$|\varphi(f(t, u, v) - f(t, \overline{u}, \overline{v}))| \leq w(t, (\ln t)^{1-r} \|u - \overline{u}\|_E, (\ln t)^{1-r} \|v - \overline{v}\|_E)$$
$$(7.12)$$

holds for all $t \in I$ and $u, v, \overline{u}, \overline{v} \in E$ such that $\|u - \overline{u}\|_E \leq \rho$ and $\|v - \overline{v}\|_E \leq \rho$.
(7.3.3) $v \equiv 0$ is the only function in $X(I_\lambda, [0, \rho])$ satisfying the integral inequality

$$v(t) \leq \frac{1}{\Gamma(r)} \int_1^{1+\lambda(T-1)} \left(\ln \frac{t}{s}\right)^{r-1}$$
$$\times w(s, (\ln s)^{1-r} v(s), (\ln s)^{1-r}(^H D_1^r)v(s))ds, \quad (7.13)$$

with $\sigma \leq \lambda \leq 1$.

Define the successive approximations of the problem (7.9)-(7.10) as follows:

$$u_0(t) = \frac{\phi}{\Gamma(r)}(\ln t)^{r-1}; \quad t \in I,$$

$$u_{n+1}(t) = \frac{\phi}{\Gamma(r)}(\ln t)^{r-1} + \int_1^t \left(\ln \frac{t}{s}\right)^{r-1} \frac{f(s, u_n(s), {}^H D_1^r u_n(s))}{s\Gamma(r)} ds; \quad t \in I.$$

Theorem 7.3. *Assume that the hypotheses (7.3.1)–(7.3.3) are satisfied. Then the successive approximations u_n; $n \in \mathbb{N}$ are well defined and converge to the unique solution of the problem (7.9)–(7.10) uniformly on I.*

Proof. Differentiating both sides of the successive approximations u_n; $n \in \mathbb{N}$ and applying the Hadamard fractional derivative, we get

$$({}^H D_1^r u_0)(t) = 0; \quad t \in I,$$

and

$$({}^H D_1^r u_{n+1})(t) = f(t, u_n(t), {}^H D_1^r u_n(t)); \quad t \in I.$$

From (7.3.1), and since u_n and ${}^H D_1^r u_n$ are in $C_{r,\ln}$, then the successive approximations are well defined. Next, for each $t_1, t_2 \in I$ with $t_1 < t_2$ and for all $t \in I$, there exists $\varphi \in E^*$ such that

$$\|(\ln t_2)^{1-r} u_n(t_2) - (\ln t_1)^{1-r} u_n(t_1)\|_E$$

$$= |\varphi((\ln t_2)^{1-r} u_n(t_2) - (\ln t_1)^{1-r} u_n(t_1))|$$

$$\leq (\ln t_2)^{1-r} \int_{t_1}^{t_2} \left|\ln \frac{t_2}{s}\right|^{r-1} \frac{|\varphi(f(s, u_n(s), {}^H D_1^r u_n(s)))|}{s\Gamma(r)} ds$$

$$+ \int_1^{t_1} \left|(\ln t_2)^{1-r}(\ln \frac{t_2}{s})^{r-1} - (\ln t_1)^{1-r}\left(\ln \frac{t_1}{s}\right)^{r-1}\right|$$

$$\times \frac{|\varphi(f(s, u_n(s), {}^H D_1^r u_n(s)))|}{s\Gamma(r)} ds.$$

Then, from (7.3.1), for every positive integer k there exists a function $h_k \in L^\infty(J, \mathbb{R}_+)$ such that

$$\|(\ln t_2)^{1-r} u_n(t_2) - (\ln t_1)^{1-r} u_n(t_1)\|_E$$

$$\leq (\ln t_2)^{1-r} \int_{t_1}^{t_2} \left|\ln \frac{t_2}{s}\right|^{r-1} \frac{\|h_k\|_{L^\infty}}{\Gamma(r)} ds$$

$$+ \int_1^{t_1} \left|(\ln t_2)^{1-r}\left(\ln \frac{t_2}{s}\right)^{r-1} - (\ln t_1)^{1-r}\left(\ln \frac{t_1}{s}\right)^{r-1}\right| \frac{\|h_k\|_{L^\infty}}{\Gamma(r)} ds.$$

Thus, we get

$$\|(\ln t_2)^{1-r}u_n(t_2) - (\ln t_1)^{1-r}u_n(t_1)\|_E$$

$$\leq \frac{\|h_k\|_{L^\infty}}{\Gamma(1+r)}(\ln T)^{1-r}\left|\ln\frac{t_2}{t_1}\right|^r$$

$$+ \frac{\|h_k\|_{L^\infty}}{\Gamma(r)}\int_1^{t_1}\left|(\ln t_2)^{1-r}\left(\ln\frac{t_2}{s}\right)^{r-1} - (\ln t_1)^{1-r}\left(\ln\frac{t_1}{s}\right)^{r-1}\right|ds$$

$$\to 0, \quad \text{as } t_1 \to t_2.$$

On the other hand, and since f is weakly sequentially continuous, then by using the Lebesgue dominated convergence theorem, we get

$$\|(\ln t_2)^{1-r}(^HD_1^r u_n)(t_2) - (\ln t_1)^{1-r}(^HD_1^r u_n)(t_1)\|_E$$

$$= |\varphi((\ln t_2)^{1-r}(^HD_1^r u_n)(t_2) - (\ln t_1)^{1-r}(^HD_1^r u_n)(t_1))|$$

$$\leq \|(\ln t_2)^{1-r}f(t_2, u_{n-1}(t_2), ^HD_1^r u_{n-1}(t_2))$$

$$- (\ln t_1)^{1-r}f(t_1, u_{n-1}(t_1), ^HD_1^r u_{n-1}(t_1))\|_E$$

$$\leq (\ln t_2)^{1-r}\|f(t_2, u_{n-1}(t_2), ^HD_1^r u_{n-1}(t_2))$$

$$- f(t_1, u_{n-1}(t_1), ^HD_1^r u_{n-1}(t_1))\|_E + |(\ln t_2)^{1-r} - (\ln t_1)^{1-r}|$$

$$\times \|f(t_1, u_{n-1}(t_1), ^HD_1^r u_{n-1}(t_1)) - f(t_1, u_{n-1}(t_1), ^HD_1^r u_{n-1}(t_1))\|_E$$

$$\leq (\ln T)^{1-r}\|f(t_2, u_{n-1}(t_2), ^HD_1^r u_{n-1}(t_2))$$

$$- f(t_1, u_{n-1}(t_1), ^HD_1^r u_{n-1}(t_1))\|_E$$

$$+ 2\|h_k\|_{L^\infty}|(\ln t_2)^{1-r} - (\ln t_1)^{1-r}|$$

$$\to 0, \text{ as } t_1 \to t_2.$$

Thus

$$\|u_n(t_2) - u_n(t_1)\|_1 \to 0, \quad \text{as } t_1 \to t_2.$$

Hence, the sequence $\{u_n(t); n \in \mathbb{N}\}$ is equicontinuous on I. Let

$$\tau := \sup\{\sigma \in [0,1] : \{u_n(t)\} \text{ converges uniformly on } I_\sigma\}.$$

If $\tau = 1$, then we have the global convergence of successive approximations. Suppose that $\tau < 1$, then the sequence $\{u_n(t)\}$ converges uniformly on I_τ. Since this sequence is equicontinuous, then it converges uniformly to a continuous function $\tilde{u}(t)$. If we prove that there exists $\lambda \in (\tau, 1]$ such that $\{u_n(t)\}$ converges uniformly on I_λ, this will yield a contradiction.

Put $u(t) = \tilde{u}(t)$; for $t \in I_\tau$. From (7.3.2), there exist a constant $\rho > 0$ and a weakly-Carathéodory function $w : I \times [0, \rho] \times [0, \rho] \to [0, \infty)$ satisfying inequality (7.12). Also, there exist $\lambda \in [\tau, 1]$ and $n_0 \in \mathbb{N}$, such that, for all $t \in I_\lambda$ and $n, m > n_0$, we have

$$\|u_n(t) - u_m(t)\|_E \leq \rho,$$

and

$$\|(^H D_1^r u_n)(t) - (^H D_1^r u_m)(t)\|_E \leq \rho.$$

For any $t \in I_\lambda$, put

$$v^{(n,m)}(t) = \|u_n(t) - u_m(t)\|_E,$$

$$v_k(t) = \sup_{n,m \geq k} v^{(n,m)}(t),$$

$$^H D_1^r v^{(n,m)}(t) = \|^H D_1^r u_n(t) - ^H D_1^r u_m(t)\|_E,$$

and

$$^H D_1^r v_k(t) = \sup_{n,m \geq k} (^H D_1^r v^{(n,m)}(t)).$$

Since the sequence $v_k(t)$ is nonincreasing, it is convergent to a function $v(t)$ for each $t \in I_\lambda$. From the equicontinuity of $\{v_k(t)\}$, it follows that $\lim_{k \to \infty} v_k(t) = v(t)$ uniformly on I_λ. Furthermore, for $t \in I_\lambda$ and $n, m \geq k$, there exists $\varphi \in E^*$ such that

$$v^{(n,m)}(t) = \|u_n(t) - u_m(t)\|_E$$

$$\leq \sup_{s \in [1,t]} \|u_n(s) - u_m(s)\|_E \leq \int_1^t \left(\ln \frac{t}{s}\right)^{r-1}$$

$$\times \frac{\|f(s, u_{n-1}(s), ^H D_1^r u_{n-1}(s)) - f(s, u_{m-1}(s), ^H D_1^r u_{m-1}(s))\|_E}{s\Gamma(r)} ds$$

$$\leq \frac{1}{\Gamma(r)} \int_1^{1+\lambda(T-1)} \left(\ln \frac{t}{s}\right)^{r-1} \|f(s, u_{n-1}(s), ^H D_1^r u_{n-1}(s))$$

$$- f(s, u_{m-1}(s), ^H D_1^r u_{m-1}(s))\|_E ds$$

$$\leq \frac{1}{\Gamma(r)} \int_1^{1+\lambda(T-1)} \left(\ln \frac{t}{s}\right)^{r-1} |\varphi(f(s, u_{n-1}(s), ^H D_1^r u_{n-1}(s))$$

$$- f(s, u_{m-1}(s), ^H D_1^r u_{m-1}(s)))| ds$$

Thus, by (7.12) we get

$$v^{(n,m)}(t) \le \frac{1}{\Gamma(r)} \int_1^{1+\lambda(T-1)} \left(\ln \frac{t}{s}\right)^{r-1}$$

$$\times w(s, (\ln s)^{1-r}\|u_{n-1}(s) - u_{m-1}(s)\|_E, (\ln s)^{1-r}$$

$$\times \|^H D_1^r u_{n-1}(s) -^H D_1^r u_{m-1}(s)\|_E) ds$$

$$= \frac{1}{\Gamma(r)} \int_1^{1+\lambda(T-1)} \left(\ln \frac{t}{s}\right)^{r-1}$$

$$\times w(s, (\ln s)^{1-r} v^{(n-1,m-1)}(s), (\ln s)^{1-r} (^H D_1^r) v^{(n-1,m-1)}(s)) ds.$$

Hence

$$v_k(t) \le \frac{1}{\Gamma(r)} \int_1^{1+\lambda(T-1)} \left(\ln \frac{t}{s}\right)^{r-1} w(s, (\ln s)^{1-r} v_{k-1}(s),$$

$$(\ln s)^{1-r} (^H D_1^r) v_{k-1}(s)) ds.$$

By the Lebesgue dominated convergence theorem, we get

$$v(t) \le \frac{1}{\Gamma(r)} \int_1^{1+\lambda(T-1)} \left(\ln \frac{t}{s}\right)^{r-1} w(s, (\ln s)^{1-r} v(s), (\ln s)^{1-r} (^H D_1^r) v(s)) ds.$$

Then, by (7.3.1) and (7.3.3) we get $v \equiv 0$ on I_λ, which yields that $\lim_{k\to\infty} v_k(x, y) = 0$ uniformly on J_λ. Thus $\{u_k(t)\}_{k=1}^\infty$ is a Cauchy sequence on I_λ. Consequently $\{u_k(t)\}_{k=1}^\infty$ is uniformly convergent on I_λ which yields the contradiction.

Thus $\{u_k(t)\}_{k=1}^\infty$ converges uniformly on I to a continuous function $u_*(t)$. By the weakly-Carathéodory condition (iii) and the Lebesgue dominated convergence theorem, we get

$$\lim_{k\to\infty} \int_1^t \left(\ln \frac{t}{s}\right)^{r-1} \frac{f(s, u_k(s), ^H D_1^r u_k(s))}{s\Gamma(r)} ds$$

$$= \int_1^t \left(\ln \frac{t}{s}\right)^{r-1} \frac{f(s, u_*(s), ^H D_1^r u_*(s))}{s\Gamma(r)} ds,$$

for each $t \in I$. This yields that u_* is a weak solution of the problem (7.9)–(7.10).

Finally, we show the uniqueness of solutions of the problem (7.9)–(7.10). Let u_1 and u_2 be two solutions of (7.11). As above, put

$$\tau := \sup\{\sigma \in [0,1] : u_1(t) = u_2(t) \text{ for } t \in I_\sigma\},$$

and suppose that $\tau < 1$. There exist a constant $\rho > 0$ and a comparison function $w : I_\tau \times [0,\rho] \times [0,\rho] \to [0,\infty)$ satisfying inequality (7.12). We choose $\lambda \in (\sigma, 1)$ such that

$$\|u_1(t) - u_2(t)\|_E \le \rho \text{ and } \|(^H D_1^r u_1)(t) - (^H D_1^r u_2)(t)\|_E \le \rho;$$

for $t \in I_\lambda$. Then for all $t \in I_\lambda$ there exists $\varphi \in E^*$ such that

$$\|u_1(t) - u_2(t)\|_E$$

$$\le \frac{1}{\Gamma(r)} \int_1^{1+\lambda(T-1)} \left(\ln \frac{t}{s}\right)^{r-1} \|f(s, u_1(s), {}^H D_1^r u_1(s))$$

$$- f(s, u_2(s), {}^H D_1^r u_2(s))\|_E ds$$

$$\le \frac{1}{\Gamma(r)} \int_1^{1+\lambda(T-1)} \left(\ln \frac{t}{s}\right)^{r-1} |\varphi(f(s, u_1(s), {}^H D_1^r u_1(s))$$

$$- f(s, u_2(s), {}^H D_1^r u_2(s)))| ds$$

$$\le \frac{1}{\Gamma(r)} \int_1^{1+\lambda(T-1)} \left(\ln \frac{t}{s}\right)^{r-1}$$

$$\times w(s, (\ln s)^{1-r} \|u_1(s) - u_2(s)\|_E, (\ln s)^{1-r}$$

$$\times \|^H D_1^r u_1(s) - {}^H D_1^r u_2(s)\|_E) ds.$$

Again, by (7.3.1) and (7.3.3) we get $u_1 - u_2 \equiv 0$ on I_λ. This gives $u_1 = u_2$ on I_λ, which yields a contradiction. Consequently, $\tau = 1$ and the weak solution of the problem (7.9)–(7.10) is unique on I.   □

7.4.3. Example

Let

$$E = l^1 = \left\{ w = (w_1, w_2, \ldots, w_p, \ldots) : \sum_{p=1}^\infty |w_p| < \infty \right\}$$

be the Banach space with the norm

$$\|w\|_E = \sum_{p=1}^\infty |w_p|.$$

Consider the following implicit Hadamard fractional differential equation:

$$({}^{H}D_1^r u_p)(t) = \frac{te^{t-3}}{1 + |u_p(t)| + |{}^{H}D_1^r u_p(t)|}; \quad t \in [1, e]; \; p \in \mathbb{N}^*, \qquad (7.14)$$

with the initial integral condition

$$({}^{H}I_1^{1-r} u)(t)|_{t=1} = (0, 0, \ldots), \qquad (7.15)$$

where $r \in (0, 1]$,

$$u = (u_1, u_2, \ldots, u_p, \ldots), {}^{H}D_1^r u = ({}^{H}D_1^r u_1, {}^{H}D_1^r u_2, \ldots, {}^{H}D_1^r u_p, \ldots),$$
$$f = (f_1, f_2, \ldots, f_p, \ldots).$$

For each $p \in \mathbb{N}^*$, set

$$f_p(t, u(t), v(t)) = \frac{te^{t-3}}{1 + |u_p(t)| + |v_p(t)|}; \quad t \in [1, e].$$

For each u, v, $\overline{u}$, $\overline{v} \in E$, $p \in \mathbb{N}^*$ and $t \in [1, e]$ we have

$$|f_p(t, u, v) - f_p(t, \overline{u}, \overline{v})| \le te^t(|u_p - \overline{u}_p| + |v_p - \overline{v}_p|).$$

Thus, for each u, v, $\overline{u}$, $\overline{v} \in E$ and $t \in [1, e]$, we get

$$\|f(t, u(t), v(t) - f(t, \overline{u}(t), \overline{v}(t))\|_E$$

$$= \sum_{p=1}^{\infty} |f_p(t, u(t), v(t)) - f_p(t, \overline{u}(t), \overline{v}(t))|$$

$$\le te^t \sum_{p=1}^{\infty} (|u_p - \overline{u}_p| + |v_p - \overline{v}_p|)$$

$$= te^t(\|u - \overline{u}\|_E + \|v - \overline{v}\|_E).$$

This means that condition (7.12) holds with any $t \in [1, e]$, $\rho > 0$ and a comparison function $w : [1, e] \times [1, e] \times [0, \rho] \times [0, \rho] \to [0, \infty)$ given by

$$w(t, v, w) = te^t(v + w).$$

We see that w satisfies the weakly-Carathéodory conditions with $h : [1, e] \times [[1, e] \to [0, \infty)$ given by $h(t) = \rho t e^t$.

Consequently, Theorem 7.3 implies that the successive approximations u_n; $n \in \mathbb{N}$, defined by

$$u_0(x,y) = (0,0,\ldots),$$

$$u_{n+1}(x,y) = \int_1^t \left(\ln \frac{t}{s}\right)^{r-1} \frac{f(s,u_n(s),{}^H D_1^r u_n(s))}{s\Gamma(r)} ds; \ t \in [1,e],$$

converge to the unique solution of the problem (7.14)–(7.15) uniformly on $[1,e]$.

7.5. Impulsive Implicit Hadamard–Pettis Fractional Differential Equations

7.5.1. *Introduction*

Our intention in this section is to extend the results to implicit impulsive differential equations of Hadamard fractional derivative. We discuss the existence of weak solutions for the following implicit Hadamard fractional differential equation

$$\begin{cases} ({}^H D_{t_k}^r u)(t) = f(t,u(t),({}^H D_{t_k}^r u)(t)); \ t \in J_k, \ k = 0,\ldots,m, \\ \dfrac{(\ln t)^{r-1}}{\Gamma(r)}({}^H I_{t_k}^{1-r} u)(t_k^+) = u(t_k^-) + L_k(u(t_k^-)); \ k = 1,\ldots,m, \qquad (7.16) \\ ({}^H I_1^{1-r} u)(t)|_{t=1} = \phi, \end{cases}$$

where $T > 1$, $\phi \in E$, $J_0 - [1,t_1]$, $J_k := (t_k,t_{k+1}]$; $k = 1,\ldots,m$, $1 = t_0 < t_1 < \cdots < t_m < t_{m+1} = T$, $f : J_k \times E \times E \to E$; $k = 1,\ldots,m$, $L_k : E \to E$; $k = 1,\ldots,m$ are given continuous functions, E is a real (or complex) Banach space with norm $\|\cdot\|_E$ and dual E^*, such that E is the dual of a weakly-compactly-generated Banach space X, $\ln = \log_e$, ${}^H I_{t_k}^r$ is the left-sided mixed Hadamard integral of order $r \in (0,1]$, and ${}^H D_{t_k}^r$ is the Hadamard fractional derivative of order r.

Consider the Banach space

$$PC = \big\{u : J \to E : u \in C(J_k); \ k = 0,\ldots,m, \ \text{and there exist } u(t_k^-) \\ \text{and}({}^H I_{t_k}^{1-r} u)(t_k^+); \ k = 1,\ldots,m, \ \text{with } u(t_k^-) = u(t_k)\big\},$$

with the norm

$$\|u\|_C = \sup_{t \in J} \|u(t)\|_E.$$

Also, we can define the weighted space of PC by

$$PC_{r,\ln}(I) = \left\{ w(t) : (\ln t)^r w(t) \in PC, \; \|w\|_{PC_{r,\ln}} := \sup_{t \in J} \|(\ln t)^r w(t)\|_E \right\}.$$

In the sequel we denote $\|w\|_{PC_{r,\ln}}$ by $\|w\|_{PC}$.

From [263, Theorem 2.3], we have

$$({}^H I_1^q)({}^H D_1^q w)(x) = w(x) - \frac{({}^H I_1^{1-q} w)(1)}{\Gamma(q)} (\ln x)^{q-1}.$$

Lemma 7.3. *Let* $h : J_0 \to E$ *be a continuous function. Then the Cauchy problem*

$$\begin{cases} ({}^H D_1^q u)(t) = h(t), \\ ({}^H I_1^{1-r} u)(t)|_{t=1} = \phi \end{cases} \tag{7.17}$$

has a unique solution $u \in L^1(J_0, E)$ *given by*

$$u(t) = \frac{({}^H I_1^{1-q} u)(1)}{\Gamma(q)} (\ln t)^{q-1} + ({}^H I_1^q h)(t).$$

Lemma 7.4. *Let* $h : J \to E$ *be a continuous function. A function* $u \in L^1(J, E)$ *is said to be a solution of the fractional integral equations*

$$u(t) = \begin{cases} \frac{\phi}{\Gamma(r)} (\ln t)^{r-1} + ({}^H I_1^r h)(t) & \text{if } t \in J_0, \\[2mm] \frac{\phi}{\Gamma(r)} (\ln t)^{r-1} + \sum_{i=1}^{k} L_i(u(t_i^-)) \\[2mm] + \sum_{i=1}^{k} \int_{t_{i-1}}^{t_i} \left(\ln \frac{t_i}{s}\right)^{r-1} \frac{h(s)}{s\Gamma(r)} ds \\[2mm] + \int_{t_k}^{t} \left(\ln \frac{t}{s}\right)^{r-1} \frac{h(s)}{s\Gamma(r)} ds; & \text{if } t \in J_k, \; k = 1, \ldots, m, \end{cases} \tag{7.18}$$

if and only if u *is a solution of the following problem*

$$\begin{cases} ({}^H D_{t_k}^r u)(t) = h(t); \quad t \in J_k, \; k = 0, \ldots, m, \\[2mm] \frac{(\ln t)^{r-1}}{\Gamma(r)} ({}^H I_{t_k}^{1-r} u)(t_k^+) = u(t_k^-) + L_k(u(t_k^-)); \quad k = 1, \ldots, m, \\[2mm] ({}^H I_1^{1-r} u)(t)|_{t=1} = \phi. \end{cases} \tag{7.19}$$

Proof. Assume u satisfies (7.19). If $t \in J_0$, then

$$({}^{H}D_1^r u)(t) = h(t).$$

Lemma 7.3 implies

$$u(t) = \frac{\phi}{\Gamma(r)}(\ln t)^{r-1} + ({}^{H}I_1^r h)(t).$$

If $t \in J_1$, then

$$({}^{H}D_{t_1}^r u)(t) = h(t).$$

Lemma 7.3 implies

$$u(t) = \frac{({}^{H}I_{t_1}^{1-r} u)(t_1^+)}{\Gamma(r)}(\ln t)^{r-1} + ({}^{H}I_{t_1}^r h)(t)$$

$$= L_1(u(t_1^-)) + u(t_1^-) + ({}^{H}I_{t_1}^r h)(t)$$

$$= L_1(u(t_1^-)) + \frac{\phi}{\Gamma(r)}(\ln t)^{r-1} + ({}^{H}I_1^r h)(t_1) + ({}^{H}I_{t_1}^r h)(t).$$

If $t \in J_2$, then

$$({}^{H}D_2^r u)(t) = h(t).$$

Lemma 7.3 implies

$$u(t) = \frac{({}^{H}I_{t_2}^{1-r} u)(t_2^+)}{\Gamma(r)}(\ln t)^{r-1} + ({}^{H}I_{t_2}^r h)(t)$$

$$= L_2(u(t_2^-)) + u(t_2^-) + ({}^{H}I_{t_2}^r h)(t)$$

$$= L_2(u(t_2^-)) + L_1(u(t_1^-)) + \frac{\phi}{\Gamma(r)}(\ln t)^{r-1} + ({}^{H}I_1^r h)(t_1)$$

$$+ ({}^{H}I_{t_1}^r h)(t_2) + ({}^{H}I_{t_2}^r h)(t).$$

If $t \in J_k$, then again from Lemma 7.3 we get (7.18). Conversely, assume that u satisfies the impulsive fractional integral equations (7.18). If $t \in J_0$, then $u(t) = \frac{\phi}{\Gamma(r)}(\ln t)^{r-1} + ({}^{H}I_1^r h)(t)$. Thus, $({}^{H}I_1^{1-r} u)(t)|_{t=1} = \phi$ and using the fact that ${}^{H}D_1^r$ is the left inverse of ${}^{H}I_1^r$ we get $({}^{H}D_1^r u)(t) = h(t)$.

Now, if $t \in J_k$; $k = 1, \dots, m$, we get $({}^{H}D_{t_k}^r u)(t) = h(t)$. Also, we can easily show that

$$\frac{(\ln t)^{r-1}}{\Gamma(r)}({}^{H}I_{t_k}^{1-r} u)(t_k^+) = u(t_k^-) + L_k(u(t_k^-)).$$

Hence, if u satisfies the impulsive fractional integral equations (7.18) then we get (7.19). $\qquad\Box$

From Lemma 7.4 and Lemma 5.1 in [360], we conclude the following lemma.

Lemma 7.5. *Let $f(t, u, z) : J_k \times E \times E \to E$; $k = 0, \ldots, m$ be a continuous function. Then problem (7.16) is equivalent to the problem of the solution of the equation*

$$g(t) = f\left(t, \frac{\phi}{\Gamma(r)}(\ln t)^{r-1} + ({}^H I_{t_k}^r g)(t), g(t)\right),$$

and if $g(t) \in C(J_k)$; $k = 0, \ldots, m$ is the solution of this equation, then

$$u(t) = \begin{cases} \dfrac{\phi}{\Gamma(r)}(\ln t)^{r-1} + ({}^H I_1^r g)(t) & \text{if } t \in J_0, \\[2ex] \dfrac{\phi}{\Gamma(r)}(\ln t)^{r-1} + \displaystyle\sum_{i=1}^{k}(L_i(({}^H I_{t_i}^{1-r} u)(t_i^-)) \\[2ex] \quad + \displaystyle\sum_{i=1}^{k}\int_{t_{i-1}}^{t_i}\left(\ln\dfrac{t_i}{s}\right)^{r-1}\dfrac{g(s)}{s\Gamma(r)}ds \\[2ex] \quad + \displaystyle\int_{t_k}^{t}\left(\ln\dfrac{t}{s}\right)^{r-1}\dfrac{g(s)}{s\Gamma(r)}ds; & \text{if } t \in J_k, \ k = 1, \ldots, m, \end{cases}$$

7.5.2. *Existence of weak solutions*

Let us start by defining what we mean by a weak solution of the problem (7.16).

Definition 7.5. By a weak solution of the problem (7.16) we mean a measurable function $u \in PC(J)$ that satisfies the condition $({}^H I_1^{1-r} u)(t)|_{t=1} = \phi$, and the equation $({}^H D_{t_k}^r u)(t) = f(t, u(t), ({}^H D_{t_k}^r u)(t))$ on J_k; $k = 0, \ldots, m$.

The following hypotheses will be used in the sequel.

(7.4.1) For a.e. $t \in J_k$; $k = 0, \ldots, m$, the functions $v \to f(t, v, \cdot)$ and $w \to f(t, \cdot, w)$ are weakly-sequentially continuous.

(7.4.2) For a.e. $v, w \in E$, the function $t \to f(t, v, w)$ is Pettis integrable a.e. on J_k; $k = 0, \ldots, m$.

(7.4.3) There exists $p \in C(J_k, [0, \infty))$; $k = 0, \ldots, m$ such that for all $\varphi \in E^*$, we have

$$|\varphi(f(t, u, v))| \le p(t)\|\varphi\| \quad \text{for a.e. } t \in J_k, \text{ and each } u, v \in E.$$

(7.4.4) For each bounded and measurable set $B \subset E$ and for each $t \in J_k$; $k = 0, \ldots, m$, we have

$$\beta(f(t, B, {}^H D_1^r B)) \leq (\ln t)^{1-r} p(t) \beta(B),$$

where ${}^H D_1^r B = \{{}^H D_1^r w : w \in B\}$,

(7.4.5) There exists a constant $l^* > 0$ such that for all $\varphi \in E^*$, we have

$$|\varphi(L_k(u))| \leq l^* \|\varphi\|; \text{ for a.e. } t \in J_k; \ k = 1, \ldots, m, \text{ and each } u \in E.$$

Set

$$p^* = \sup_{t \in J} p(t),$$

Theorem 7.4. *Assume that the hypotheses (7.15.1)–(7.15.5) hold. If*

$$L := ml^* (\ln T)^{1-r} + \frac{2p^* \ln T}{\Gamma(1+r)} < 1, \tag{7.20}$$

then the problem (7.16) has at least one solution defined on I.

Proof. Transform the problem (7.16) into a fixed point equation. Consider the operator $N : \mathcal{PC} \to \mathcal{PC}$ defined by:

$$(Nu)(t) - \begin{cases} \dfrac{\phi}{\Gamma(r)} (\ln t)^{r-1} + ({}^H I_1^r g)(t); & \text{if } t \in J_0, \\[4mm] \dfrac{\phi}{\Gamma(r)} (\ln t)^{r-1} + \displaystyle\sum_{i=1}^{k} L_i(u(t_i^-)) \\[4mm] \quad + \displaystyle\sum_{i=1}^{k} \int_{t_{i-1}}^{t_i} \left(\ln \dfrac{t_i}{s}\right)^{r-1} \dfrac{g(s)}{s\Gamma(r)} ds \\[4mm] \quad + \displaystyle\int_{t_k}^{t} \left(\ln \dfrac{t}{s}\right)^{r-1} \dfrac{g(s)}{s\Gamma(r)} ds; & \text{if } t \in J_k, \ k = 1, \ldots, m, \end{cases} \tag{7.21}$$

where $g \in C(J_k)$; $k = 0, \ldots, m$, with

$$g(t) = f\left(t, \frac{\phi}{\Gamma(r)} (\ln t)^{r-1} + ({}^H I_{t_k}^r g)(t), g(t)\right).$$

First notice that, the hypotheses imply that $\left(\ln \frac{t_k}{s}\right)^{r-1} \frac{g(s)}{s}$ for all $t \in J_k$, $k = 0, \ldots, m$, is Pettis integrable, and for each $u \in C$, the function

$$t \mapsto f\left(t, \frac{\phi}{\Gamma(r)} (\ln t)^{r-1} + ({}^H I_{t_k}^r g)(t), g(t)\right): \quad k = 0, \ldots, m,$$

is Pettis integrable over J_k; $k = 0, \ldots, m$. Thus, the operator N is well defined. Let $R > 0$ be such that

$$R > ml^*(\ln T)^{1-r} + \frac{2p^* \ln T}{\Gamma(1+r)},$$

and consider the set

$$Q = \left\{ u \in \mathcal{PC} : \|u\|_{\mathcal{PC}} \leq R \text{ and } \|(\ln x_2)^{1-r} u(x_2) - (\ln x_1)^{1-r} u(x_1)\|_E \right.$$

$$\leq ml^* \left| (\ln x_2)^{1-r} - (\ln x_1)^{1-r} \right| + \frac{2p^*}{\Gamma(1+r)} (\ln T)^{1-r} \left| \ln \frac{x_2}{x_1} \right|^r$$

$$\left. + \frac{2p^*}{\Gamma(r)} \int_1^{x_1} \left| (\ln x_2)^{1-r} \left(\ln \frac{x_2}{s} \right)^{r-1} - (\ln x_1)^{1-r} \left(\ln \frac{x_1}{s} \right)^{r-1} \right| ds \right\}.$$

Clearly, the subset Q is closed, convex and equicontinuous. We shall show that the operator N satisfies all the assumptions of Theorem 1.7. The proof will be given in several steps.

Step 1. *N maps Q into itself.*

Let $u \in Q$, $t \in J_0$ and assume that $(Nu)(t) \neq 0$. Then there exists $\varphi \in E^*$ such that $\|(\ln t)^{1-r}(Nu)(t)\|_E = |\varphi((\ln t)^{1-r}(Nu)(t))|$. Thus

$$\|(\ln t)^{1-r}(Nu)(t)\|_E = \left| \varphi \left(\frac{\phi}{\Gamma(r)} + \frac{(\ln t)^{1-r}}{\Gamma(r)} \int_1^t \left(\ln \frac{t}{s} \right)^{r-1} \frac{g(s)}{s} ds \right) \right|,$$

where $g \in C$ with

$$g(t) = f \left(t, \frac{\phi}{\Gamma(r)} (\ln t)^{r-1} + ({}^H I_1^r g)(t), g(t) \right).$$

Then

$$\|(\ln t)^{1-r}(Nu)(t)\|_E \leq \frac{(\ln t)^{1-r}}{\Gamma(r)} \int_1^t \left(\ln \frac{t}{s} \right)^{r-1} \frac{|\varphi(g(s))|}{s} ds$$

$$\leq \frac{p^*(\ln T)^{1-r}}{\Gamma(r)} \int_1^t \left(\ln \frac{t}{s} \right)^{r-1} \frac{ds}{s}$$

$$\leq \frac{p^* \ln T}{\Gamma(1+r)}$$

$$\leq R.$$

Also, if $u \in Q$, $t \in J_k : \ k = 1, \ldots, m$, we get

$$\|(\ln t)^{1-r}(Nu)(t)\|_E \leq \sum_{i=1}^{k} |\varphi((\ln t)^{1-r}L_i(u(t_i^-)))|$$

$$+ (\ln T)^{1-r} \sum_{i=1}^{k} \int_{t_{i-1}}^{t_i} \left(\ln \frac{t_i}{s}\right)^{r-1} \frac{|\varphi(g(s))|}{s\Gamma(r)} ds$$

$$+ (\ln T)^{1-r} \int_{t_k}^{t} \left(\ln \frac{t}{s}\right)^{r-1} \frac{|\varphi(g(s))|}{s\Gamma(r)} ds$$

$$\leq ml^*(\ln T)^{1-r} + \frac{2p^* \ln T}{\Gamma(1+r)}$$

$$\leq R.$$

Next, let $x_1, x_2 \in J_0$ such that $1 \leq x_1 < x_2 \leq t_1$ and let $u \in Q$, with

$$(\ln x_2)^{1-r}(Nu)(x_2) - (\ln x_1)^{1-r}(Nu)(x_1) \neq 0.$$

Then there exists $\varphi \in E^*$ such that

$$\|(\ln x_2)^{1-r}(Nu)(x_2) - (\ln x_1)^{1-r}(Nu)(x_1)\|_E$$
$$= |\varphi((\ln x_2)^{1-r}(Nu)(x_2) - (\ln x_1)^{1-r}(Nu)(x_1))|$$

and $\|\varphi\| = 1$. Then

$$\|(\ln x_2)^{1-r}(Nu)(x_2) - (\ln x_1)^{1-r}(Nu)(x_1)\|_E$$
$$= |\varphi((\ln x_2)^{1-r}(Nu)(x_2) - (\ln x_1)^{1-r}(Nu)(x_1))|$$
$$\leq ml^*|(\ln x_2)^{1-r} - (\ln x_1)^{1-r}|$$
$$+ \left|\varphi\left((\ln x_2)^{1-r}\int_1^{x_2}\left(\ln\frac{x_2}{s}\right)^{r-1}\frac{g(s)}{s\Gamma(r)}ds - (\ln x_1)^{1-r}\right.\right.$$
$$\left.\left.\int_1^{x_1}\left(\ln\frac{x_1}{s}\right)^{r-1}\frac{g(s)}{s\Gamma(r)}ds\right)\right|,$$

where $g \in C$ with

$$g(t) = f\left(t, \frac{\phi}{\Gamma(r)}(\ln t)^{r-1} + ({}^H I_1^r g)(t), g(t)\right).$$

Then

$$\|(\ln x_2)^{1-r}(Nu)(x_2) - (\ln x_1)^{1-r}(Nu)(x_1)\|_E$$

$$\leq ml^* \left|(\ln x_2)^{1-r} - (\ln x_1)^{1-r}\right|$$

$$+ (\ln x_2)^{1-r} \int_{x_1}^{x_2} \left|\ln \frac{x_2}{s}\right|^{r-1} \frac{|\varphi(g(s))|}{s\Gamma(r)} ds$$

$$+ \int_1^{x_1} \left|(\ln x_2)^{1-r}\left(\ln \frac{x_2}{s}\right)^{r-1} - (\ln x_1)^{1-r}\left(\ln \frac{x_1}{s}\right)^{r-1}\right| \frac{|\varphi(g(s))|}{s\Gamma(r)} ds$$

$$\leq ml^* \left|(\ln x_2)^{1-r} - (\ln x_1)^{1-r}\right|$$

$$+ (\ln x_2)^{1-r} \int_{x_1}^{x_2} \left|\ln \frac{x_2}{s}\right|^{r-1} \frac{p(s)}{\Gamma(r)} ds$$

$$+ \int_1^{x_1} \left|(\ln x_2)^{1-r}\left(\ln \frac{x_2}{s}\right)^{r-1} - (\ln x_1)^{1-r}\left(\ln \frac{x_1}{s}\right)^{r-1}\right| \frac{p(s)}{\Gamma(r)} ds.$$

Thus, we get

$$\|(\ln x_2)^{1-r}(Nu)(x_2) - (\ln x_1)^{1-r}(Nu)(x_1)\|_E$$

$$\leq ml^* \left|(\ln x_2)^{1-r} - (\ln x_1)^{1-r}\right|$$

$$+ \frac{p^*}{\Gamma(1+r)}(\ln T)^{1-r}\left|\ln \frac{x_2}{x_1}\right|^r$$

$$+ \frac{p^*}{\Gamma(r)} \int_1^{x_1} \left|(\ln x_2)^{1-r}\left(\ln \frac{x_2}{s}\right)^{r-1} - (\ln x_1)^{1-r}\left(\ln \frac{x_1}{s}\right)^{r-1}\right| ds.$$

Also, if we let $x_1, x_2 \in J_k$; $k = 1, \ldots, m$ such that $t_k \leq x_1 < x_2 \leq t_{k+1}$ and let $u \in Q$, we obtain

$$\|(\ln x_2)^{1-r}(Nu)(x_2) - (\ln x_1)^{1-r}(Nu)(x_1)\|_E$$

$$\leq ml^* \left|(\ln x_2)^{1-r} - (\ln x_1)^{1-r}\right| + \frac{2p^*}{\Gamma(1+r)}(\ln T)^{1-r}\left|\ln \frac{x_2}{x_1}\right|^r$$

$$+ \frac{2p^*}{\Gamma(r)} \int_1^{x_1} \left|(\ln x_2)^{1-r}\left(\ln \frac{x_2}{s}\right)^{r-1} - (\ln x_1)^{1-r}\left(\ln \frac{x_1}{s}\right)^{r-1}\right| ds.$$

Hence $N(Q) \subset Q$.

Step 2. N *is weakly-sequentially continuous.*

Let (u_n) be a sequence in Q and let $(u_n(t)) \to u(t)$ in (E, ω) for each $t \in J_k$; $k = 0, \ldots, m$. Fix $t \in J_k$; $k = 0, \ldots, m$, since f satisfies the assumption (7.4.1), we have $f(t, u_n(t), {}^H D_{t_k} u_n(t))$ converges weakly to

$f(t, u(t), {}^H D_{t_k} u(t))$. Hence the Lebesgue dominated convergence theorem for Pettis integral implies $(Nu_n)(t)$ converges weakly to $(Nu)(t)$ in (E, ω), for each $t \in J_k$; $k = 0, \ldots, m$. Thus, $N(u_n) \to N(u)$. Hence, $N : Q \to Q$ is weakly-sequentially continuous.

Step 3. *The implication (1.9) holds.*

Let V be a subset of Q such that $\overline{V} = \overline{\text{conv}}(N(V) \cup \{0\})$. Obviously

$$V(t) \subset \overline{\text{conv}}(NV)(t)) \cup \{0\}), \ \forall t \in J_k; \ k = 0, \ldots, m.$$

Further, as V is bounded and equicontinuous, by [172, Lemma 3] the function $t \to v(t) = \beta(V(t))$ is continuous on J_k; $k = 0, \ldots, m$. From $(7.4.3) - (7.4.5)$, Lemma 1.7 and the properties of the measure β, for any $t \in J_0$, we have

$$(\ln t)^{1-r} v(t) \leq \beta((\ln t)^{1-r}(NV)(t) \cup \{0\})$$

$$\leq \beta((\ln t)^{1-r}(NV)(t))$$

$$\leq \frac{(\ln T)^{1-r}}{\Gamma(r)} \int_1^t \left| \ln \frac{t}{s} \right|^{r-1} \frac{p(s)\beta(V(s))}{s} ds$$

$$\leq \frac{(\ln T)^{1-r}}{\Gamma(r)} \int_1^t \left| \ln \frac{t}{s} \right|^{r-1} \frac{(\ln s)^{1-r} p(s) v(s)}{s} ds$$

$$\leq \frac{p^* \ln T}{\Gamma(1+r)} \|v\|_C.$$

Thus

$$\|v\|_C \leq L \|v\|_C.$$

Also, for any $t \in J_k$; $k = 1, \ldots, m$, we get

$$(\ln t)^{1-r} v(t) \leq \beta((\ln t)^{1-r}(NV)(t) \cup \{0\})$$

$$\leq \beta((\ln t)^{1-r}(NV)(t))$$

$$\leq (\ln T)^{1-r} \sum_{i=1}^k l^* \beta(V(s))$$

$$+ (\ln T)^{1-r} \sum_{i=1}^k \int_{t_{i-1}}^{t_i} \left(\ln \frac{t_i}{s} \right)^{r-1} \frac{p(s)\beta(V(s))}{s\Gamma(r)} ds$$

$$+ (\ln T)^{1-r} \int_{t_k}^t \left(\ln \frac{t}{s} \right)^{r-1} \frac{p(s)\beta(V(s))}{s\Gamma(r)} ds$$

$$\leq l^*(\ln T)^{1-r}\sum_{i=1}^{k}(\ln t)^{1-r}v(t)$$

$$+ (\ln T)^{1-r}\sum_{i=1}^{k}\int_{t_{i-1}}^{t_i}\left(\ln\frac{t_i}{s}\right)^{r-1}\frac{(\ln s)^{1-r}p(s)v(s)}{s\Gamma(r)}ds$$

$$+ (\ln T)^{1-r}\int_{t_k}^{t}\left(\ln\frac{t}{s}\right)^{r-1}\frac{(\ln s)^{1-r}p(s)v(s)}{s\Gamma(r)}ds$$

$$\leq \left(ml^*(\ln T)^{1-r} + \frac{2p^*\ln T}{\Gamma(1+r)}\right)\|v\|_C.$$

Hence

$$\|v\|_C \leq L\|v\|_C.$$

From (7.20), we get $\|v\|_C = 0$, that is $v(t) = \beta(V(t)) = 0$, for each $t \in I$. and then by [290, Theorem 2], V is relatively-weakly compact in C. Applying now Theorem 1.7, we conclude that N has a fixed point which is a solution of the problem (7.16). □

7.5.3. *Example*

Let

$$E = l^1 = \left\{ u = (u_1, u_2, \ldots, u_n, \ldots), \sum_{n=1}^{\infty}|u_n| < \infty \right\}$$

be the Banach space with the norm

$$\|u\|_E = \sum_{n=1}^{\infty}|u_n|.$$

Consider the following problem of implicit impulsive Hadamard fractional differential equations:

$$\begin{cases} ({}^H D^r_{t_k}u)(t) = f(t, u(t), ({}^H D^r_{t_k}u)(t)); & t \in J_k, \ k = 0, \ldots, m, \\ \dfrac{(\ln t)^{r-1}}{\Gamma(r)}({}^H I^{1-r}_{t_k}u)(t_k^+) = u(t_k^-) + L_k(u(t_k^-)); & k = 1, \ldots, m, \\ ({}^H I^{1-r}_1 u)(t)|_{t=1} = 0, \end{cases} \quad (7.22)$$

where $J = [1, e]$, $r \in (0, 1]$, $u = (u_1, u_2, \ldots, u_n, \ldots)$,

$$f = (f_1, f_2, \ldots, f_n, \ldots), \; {}^H D^r_{t_k} u$$
$$= ({}^H D^r_{t_k} u_1, {}^H D^r_{t_k} u_2, \ldots, {}^H D^r_{t_k} u_n, \ldots);$$
$$k = 0, \ldots, m,$$

$$f_n(t, u(t), ({}^H D^r_{t_k} u)(t)) = \frac{ct^2}{1 + \|u(t)\|_E + \|{}^H D^r_{t_k} u(t)\|_E} \left(e^{-7} + \frac{1}{e^{t+5}} \right) u(t);$$

$$t \in [1, e],$$

$$L_k(u(t_k^-)) = \frac{1}{(3e^4)(1 + \|u(t_k^-)\|_E)}; \quad k = 1, \ldots, m.$$

Clearly, the function f is continuous.

For each $u \in E$ and $t \in [1, e]$, we have

$$\|f(t, u(t), ({}^H D^r_{t_k})(t))\|_E \le ct^2 \left(e^{-7} + \frac{1}{e^{t+5}} \right).$$

and

$$\|L_k(u)\|_E \le \frac{1}{3e^4}.$$

Hence, the hypothesis (7.4.3) is satisfied with $p^* = ce^{-4}$, and (7.4.5) is satisfied with $l^* = \frac{1}{3e^4}$.

We shall show that condition (7.20) holds with $T = e$. Indeed, if we assume, for instance, that the number of impulses $m = 3$, and $r = \frac{1}{2}$, then we have

$$ml^*(\ln T)^{1-r} + \frac{2p^* \ln T}{\Gamma(1+r)} = \frac{1}{e^4} + \frac{2c}{e^4 \Gamma(\frac{3}{2})} = \frac{9}{16} < 1.$$

A simple computations show that all conditions of Theorem 7.4 are satisfied. It follows that the problem (7.22) has at least one solution on $[1, e]$.

7.6. Implicit Hadamard–Pettis Fractional Differential Equations with Not Instantaneous Impulses

7.6.1. *Introduction*

In this section, we discuss the existence of weak solutions for the following implicit Hadamard fractional differential equation with not instantaneous impulses:

$$\begin{cases} {}^{H}D_{s_k}^{r}u(t) = f(t, u(t), {}^{H}D_{s_k}^{r}u(t)); & \text{if } t \in I_k, \ k = 0, \dots, m, \\ u(t) = g_k(t, u(t_k^-)); & \text{if } t \in J_k, \ k = 1, \dots, m, \quad (7.23) \\ ({}^{H}I_1^{1-r}u)(t)|_{t=1} = \phi_0, \end{cases}$$

where $I_0 := [1, t_1]$, $J_k := (t_k, s_k]$, $I_k := (s_k, t_{k+1}]$; $k = 1, \dots, m$, $T > 1$, $\phi_0 \in E$, $f : I_k \times E \times E \to E$, $g_k : J_k \times E \to E$ are given continuous functions such that $({}^{H}I_{s_k}^{1-r}g_k)(t, u(t_k^-))|_{t=s_k} = \phi_k \in E$; $k = 1, \dots, m$, E is a real (or complex) Banach space with norm $\|\cdot\|_E$ and dual E^*, such that E is the dual of a weakly-compactly-generated Banach space X, ${}^{H}I_{s_k}^{r}$ is the left-sided mixed Hadamard integral of order $r \in (0, 1]$, and ${}^{H}D_{s_k}^{r}$ is the Hadamard fractional derivative of order r, $1 = s_0 < t_1 \le s_1 < t_2 \le s_2 < \cdots \le s_{m-1} < t_m \le s_m < t_{m+1} = T$.

7.6.2. *Existence of weak solutions*

Let $\mathcal{C}$ be the Banach space of all continuous functions v from $I := [1, T]$ into E with the supremum (uniform) norm

$$\|v\|_\infty := \sup_{t \in I} \|v(t)\|_E.$$

Denote by

$$\mathcal{PC} = \{u : I \to E : u \in \mathcal{C}(\cup_{k=1}^m (t_k, t_{k+1})), \ u(t_k^-) = u(t_k)\},$$

the Banach space equipped with the standard supremum norm.

By $PC_{r,\ln}(I)$, we denote the weighted space of continuous functions defined by

$$PC_{r,\ln}(I) = \left\{ w(t) : (\ln t)^r w(t) \in \mathcal{PC}, \ \|w\|_{PC_{r,\ln}} := \sup_{t \in I} \|(\ln t)^r w(t)\|_E \right\}.$$

In the following we denote $\|w\|_{PC_{r,\ln}}$ by $\|w\|_{\mathcal{PC}}$.

Let $(E, w) = (E, \sigma(E, E^*))$ be the Banach space E with its weak topology. From [263, Theorem 2.3], we have

$$({}^{H}I_{s_k}^{q})({}^{H}D_{s_k}^{q}w)(x) = w(x) - \frac{({}^{H}I_{s_k}^{1-q}w)(s_k)}{\Gamma(q)}(\ln x)^{q-1}.$$

Lemma 7.6. *Let $h : I \to E$ be a continuous function. A function $u \in L^1(I_k, E)$; $k = 0, \ldots, m$ is said to be a solution of the equation*

$$({}^H D^q_{s_k} w)(t) = h(t),$$

if and only if u satisfies the following Hadamard integral equation:

$$w(t) = \frac{({}^H I^{1-q}_{s_k} u)(s_k)}{\Gamma(q)} (\ln t)^{q-1} + ({}^H I^q_{s_k} h)(t).$$

From the above lemma and Lemma 5.1 [82] we concluded the following lemma.

Lemma 7.7. *Let $f(t, u, z) : I_k \times E \times E \to E$ and $g_k(t, u) : J_k \times E \to E$ be a continuous functions. A function $u \in PC$ is a weak solution of problem (7.23), if and only if u satisfies*

$$u(t) = \begin{cases} \frac{\phi_k}{\Gamma(r)} (\ln t)^{r-1} + ({}^H I^r_{s_k} h)(t); & \text{if } t \in I_k, \ k = 0, \ldots, m, \\ g_k(t, u(t)); & \text{if } t \in J_k, \ k = 1, \ldots, m, \end{cases} \tag{7.24}$$

where $h \in C(I_k, E)$; $k = 0, \ldots, m$, such that

$$h(t) = f\left(t, \frac{\phi_k}{\Gamma(r)} (\ln t)^{r-1} + ({}^H I^r_{s_k} h)(t), h(t)\right); \quad k = 0, \ldots, m.$$

Definition 7.6. By a weak solution of the problem (7.23) we mean a measurable function $u \in PC$ that satisfies the condition $({}^H I^{1-r}_1 u)(t)|_{t=1} = \phi_0$, and the equations $({}^H D^r_{s_k} u)(t) = f(t, u(t), ({}^H D^r_{s_k} u)(t))$ on I_k; $k = 0, \ldots, m$, and $u(t) = g_k(t, u(t_k^-))$ on J_k; $k = 1, \ldots, m$.

The following hypotheses will be used in the sequel.

(7.5.1) The function $f : I_k \times E \times E \to E$; $k = 0, \ldots, m$ is weakly Carathéodory.

(7.5.2) There exist $p_k \in C(I_k, [0, \infty))$; $k = 0, \ldots, m$, such that for all $\varphi \in E^*$, we have

$$|\varphi(f(t, u, {}^H D^r_{s_k} u))| \le p_k(t) \|\varphi\|; \quad \text{for a.e. } t \in J_k; \ k = 1, \ldots, m,$$

and each $u \in E$,

and for each bounded and measurable set $B \subset E$, we have

$$\beta(f(t, B,^H D^r_{s_k} B)) \le (\ln t)^{1-r} p_k(t) \beta(B); \quad \text{for each } t \in I_k;$$

$$k = 0, \dots, m,$$

where $^H D^r_{s_k} B = \{^H D^r_{s_k} w : w \in B\}$.
(7.5.3) There exist $q_k \in C(J_k, [0, \infty))$; $k = 1, \dots, m$, such that for all $\varphi \in E^*$, we have

$$|\varphi(g_k(t, u_{t_k^-}))| \le q_k(t) \|\varphi\|; \quad \text{for a.e. } t \in J_k; \; k = 1, \dots, m,$$

Set

$$p^* = \max_{k=0,\dots,m} \sup_{t \in I_k} p_k(t), \quad q^* = \max_{k=1,\dots,m} \sup_{t \in J_k} q_k(t).$$

Theorem 7.5. *Assume that the hypotheses (7.5.1)–(7.5.3) hold. If*

$$L := \frac{p^* \ln T}{\Gamma(1+r)} < 1, \tag{7.25}$$

then the problem (7.23) has at least one weak solution defined on I.

Proof. Transform the problem (7.23) into a fixed-point equation. Consider the operator $N : \mathcal{PC} \to \mathcal{PC}$ defined by

$$(Nu)(t) = \begin{cases} \dfrac{\phi_k}{\Gamma(r)} (\ln t)^{r-1} + \displaystyle\int_{s_k}^t \left(\ln \dfrac{t}{s}\right)^{r-1} \dfrac{h(s)}{s\Gamma(r)} ds; & \text{if } t \in I_k, \; k = 0, \dots, m, \\[4mm] g_k(t, u(t_k^-)); & \text{if } t \in J_k, \; k = 1, \dots, m, \end{cases} \tag{7.26}$$

where $h \in C(I_k, E)$; $k = 0, \dots, m$, with

$$h(t) = f\left(t, \frac{\phi_k}{\Gamma(r)} (\ln t)^{r-1} + (^H I^r_{s_k} h)(t), h(t)\right).$$

First notice that, the hypotheses imply that $\left(\ln \frac{t}{s}\right)^{r-1} \frac{h(s)}{s}$ for all $t \in I_k$ is Pettis integrable, and for each $u \in \mathcal{PC}$, the function

$$t \mapsto f\left(t, \frac{\phi_k}{\Gamma(r)} (\ln t)^{r-1} + (^H I^r_{s_k} h)(t), h(t)\right)$$

is Pettis integrable over I_k. Thus, the operator N is well defined.

Let $R > 0$ be such that

$$R \geq \max \left\{ \frac{p^* \ln T}{\Gamma(1+r)}, q^*(\ln T)^{1-r} \right\},$$

and consider the set

$$Q = \left\{ u \in \mathcal{PC} : \|u\|_{\mathcal{PC}} \leq R \text{ and } \|(\ln t_2)^{1-r} u(t_2) - (\ln t_1)^{1-r} u(t_1)\|_E \right.$$

$$\leq \frac{p^*}{\Gamma(1+r)} (\ln T)^{1-r} \left| \ln \frac{t_2}{t_1} \right|^r + \frac{p^*}{\Gamma(r)}$$

$$\times \int_1^{t_1} \left| (\ln t_2)^{1-r} \left(\ln \frac{t_2}{s} \right)^{r-1} - (\ln t_1)^{1-r} (\ln \frac{t_1}{s})^{r-1} \right| ds;$$

on I_k, $k = 0, \ldots, m$, and $\|(\ln t_2)^{1-r} u(t_2) - (\ln t_1)^{1-r} u(t_1)\|_E$

$$\leq \|(\ln t_2)^{1-r} g_k(t_2, u(t_k^-)) - (\ln t_1)^{1-r} g_k(t_1, u(t_k^-))\|_E;$$

$$\left. \text{on } J_k, \ k = 1, \ldots, m \right\}.$$

Clearly, the subset Q is closed, convex and equicontinuous. We shall show that the operator N satisfies all the assumptions of Theorem 1.7. The proof will be given in several steps.

Step 1. *N maps Q into itself.*

Let $u \in Q$, $t \in I$ and assume that $(Nu)(t) \neq 0$. Then there exists $\varphi \subset E^*$ such that $\|(\ln t)^{1-r}(Nu)(t)\|_E - |\varphi((\ln t)^{1-r}(Nu)(t))|$. Thus, for each $t \in I_k$, $k = 0, \ldots, m$,

$$\|(\ln t)^{1-r}(Nu)(t)\|_E = \left| \varphi \left(\frac{\phi_k}{\Gamma(r)} + \frac{(\ln t)^{1-r}}{\Gamma(r)} \int_1^t \left(\ln \frac{t}{s} \right)^{r-1} \frac{g(s)}{s} ds \right) \right|,$$

where $g \in C(I_k)$; $k = 0, \ldots, m$, with

$$g(t) = f \left(t, \frac{\phi_k}{\Gamma(r)} (\ln t)^{r-1} + (^H I_1^r g)(t), g(t) \right).$$

Then

$$\|(\ln t)^{1-r}(Nu)(t)\|_E \leq \frac{(\ln t)^{1-r}}{\Gamma(r)} \int_1^t \left(\ln \frac{t}{s} \right)^{r-1} \frac{|\varphi(g(s))|}{s} ds$$

$$\leq \frac{p^*(\ln T)^{1-r}}{\Gamma(r)} \int_1^t \left(\ln \frac{t}{s} \right)^{r-1} \frac{ds}{s}$$

$$\leq \frac{p^* \ln T}{\Gamma(1+r)}$$

$$\leq R.$$

Also, for each $t \in J_k$; $k = 1, \ldots, m$, it is clear that

$$\|(\ln t)^{1-r}(Nu)(t)\|_E \leq q^* (\ln T)^{1-r} \leq R.$$

Hence,

$$\|N(u)\|_{\mathcal{PC}} \leq R.$$

Next, let $t_1, t_2 \in I_k$; $k = 0, \ldots, m$ such that $t_1 < t_2$ and let $u \in Q$, with

$$(\ln t_2)^{1-r}(Nu)(t_2) - (\ln t_1)^{1-r}(Nu)(t_1) \neq 0.$$

Then there exists $\varphi \in E^*$ such that

$$\|(\ln t_2)^{1-r}(Nu)(t_2) - (\ln t_1)^{1-r}(Nu)(t_1)\|_E$$
$$= |\varphi((\ln t_2)^{1-r}(Nu)(t_2) - (\ln t_1)^{1-r}(Nu)(t_1))|,$$

and $\|\varphi\| = 1$. Then

$$\|(\ln t_2)^{1-r}(Nu)(t_2) - (\ln t_1)^{1-r}(Nu)(t_1)\|_E$$
$$= |\varphi((\ln t_2)^{1-r}(Nu)(t_2) - (\ln t_1)^{1-r}(Nu)(t_1))|$$
$$\leq \left| \varphi \left((\ln t_2)^{1-r} \int_1^{t_2} \left(\ln \frac{t_2}{s} \right)^{r-1} \frac{g(s)}{s\Gamma(r)} ds - (\ln t_1)^{1-r} \right. \right.$$
$$\left. \left. \times \int_1^{t_1} \left(\ln \frac{t_1}{s} \right)^{r-1} \frac{g(s)}{s\Gamma(r)} ds \right) \right|,$$

where $g \in C(I_k)$ with

$$g(t) = f\left(t, \frac{\phi_k}{\Gamma(r)} (\ln t)^{r-1} + ({}^H I_1^r g)(t), g(t) \right).$$

Then

$$\|(\ln t_2)^{1-r}(Nu)(t_2) - (\ln t_1)^{1-r}(Nu)(t_1)\|_E$$
$$\leq (\ln t_2)^{1-r} \int_{t_1}^{t_2} \left| \ln \frac{t_2}{s} \right|^{r-1} \frac{|\varphi(g(s))|}{s\Gamma(r)} ds$$
$$+ \int_1^{t_1} \left| (\ln t_2)^{1-r} (\ln \frac{t_2}{s})^{r-1} - (\ln t_1)^{1-r} \left(\ln \frac{t_1}{s} \right)^{r-1} \right| \frac{|\varphi(g(s))|}{s\Gamma(r)} ds$$

$$\leq (\ln t_2)^{1-r} \int_{t_1}^{t_2} \left| \ln \frac{t_2}{s} \right|^{r-1} \frac{p(s)}{\Gamma(r)} ds$$

$$+ \int_1^{t_1} \left| (\ln t_2)^{1-r} \left(\ln \frac{t_2}{s} \right)^{r-1} - (\ln t_1)^{1-r} \left(\ln \frac{t_1}{s} \right)^{r-1} \right| \frac{p(s)}{\Gamma(r)} ds.$$

Thus, we get

$$\| (\ln t_2)^{1-r} (Nu)(t_2) - (\ln t_1)^{1-r} (Nu)(t_1) \|_E$$

$$\leq \frac{p^*}{\Gamma(1+r)} (\ln T)^{1-r} \left| \ln \frac{t_2}{t_1} \right|^r$$

$$+ \frac{p^*}{\Gamma(r)} \int_1^{t_1} \left| (\ln t_2)^{1-r} \left(\ln \frac{t_2}{s} \right)^{r-1} - (\ln t_1)^{1-r} \left(\ln \frac{t_1}{s} \right)^{r-1} \right| ds.$$

Also, for $t_1, t_2 \in J_k$; $k = 1, \ldots, m$, such that $t_1 < t_2$ and let $u \in Q$, with

$$(\ln t_2)^{1-r} (Nu)(t_2) - (\ln t_1)^{1-r} (Nu)(t_1) \neq 0,$$

then there exists $\varphi \in E^*$ such that

$$\| (\ln t_2)^{1-r} (Nu)(t_2) - (\ln t_1)^{1-r} (Nu)(t_1) \|_E$$

$$\leq \| (\ln t_2)^{1-r} g_k(t_2, u(t_k^-)) - (\ln t_1)^{1-r} g_k(t_1, u(t_k^-)) \|_E.$$

Hence $N(Q) \subset Q$.

Step 2. *N is weakly-sequentially continuous.*

Let (u_n) be a sequence in Q and let $(u_n(t)) \to u(t)$ in (E, ω) for each $t \in I$. Fix $t \in I$, since f satisfies the assumption (7.5.1), we have $f(t, u_n(t), {}^H D_{s_k} u_n(t))$ converges weakly to $f(t, u(t), {}^H D_{s_k} u(t))$ on I_k; $k = 0, \ldots, m$. Hence the Lebesgue dominated convergence theorem for Pettis integral implies $(Nu_n)(t)$ converges weakly to $(Nu)(t)$ in (E, ω), for each $t \in I$. Thus, $N(u_n) \to N(u)$.

Hence, $N : Q \to Q$ is weakly-sequentially continuous.

Step 3. *The implication (1.9) holds.*

Let V be a subset of Q such that $\overline{V} = \overline{\text{conv}}(N(V) \cup \{0\})$. Obviously

$$V(t) \subset \overline{\text{conv}}(NV)(t)) \cup \{0\}), \quad \forall t \in I.$$

Further, as V is bounded and equicontinuous, by [172, Lemma 3] the function $t \to v(t) = \beta(V(t))$ is continuous on I. From (7.5.2), Lemma 1.7 and the properties of the measure β, for any $t \in I_k;\;\; k = 0, \ldots, m$, we have

$$(\ln t)^{1-r} v(t) \leq \beta((\ln t)^{1-r}(NV)(t) \cup \{0\})$$

$$\leq \beta((\ln t)^{1-r}(NV)(t))$$

$$\leq \frac{(\ln T)^{1-r}}{\Gamma(r)} \int_1^t \left| \ln \frac{t}{s} \right|^{r-1} \frac{p(s)\beta(V(s))}{s} ds$$

$$\leq \frac{(\ln T)^{1-r}}{\Gamma(r)} \int_1^t \left| \ln \frac{t}{s} \right|^{r-1} \frac{(\ln s)^{1-r} p(s) v(s)}{s} ds$$

$$\leq \frac{p^* \ln T}{\Gamma(1+r)} \|v\|_{\mathcal{PC}}.$$

Thus

$$\|v\|_{\mathcal{PC}} \leq L \|v\|_{\mathcal{PC}}.$$

From (7.25), we get $\|v\|_{\mathcal{PC}} = 0$, that is $v(t) = \beta(V(t)) = 0$, for each $t \in I$ and then by [290, Theorem 2], V is relatively weakly compact in $\mathcal{PC}$. Applying now Theorem 1.7, we conclude that N has a fixed point which is a solution of the problem (7.23). $\qquad\square$

7.6.3. *Example*

Let

$$E = l^1 = \left\{ u = (u_1, u_2, \ldots, u_n, \ldots), \sum_{n=1}^{\infty} |u_n| < \infty \right\}$$

be the Banach space with the norm

$$\|u\|_E = \sum_{n=1}^{\infty} |u_n|.$$

As an application of our results we consider the following problem of implicit Hadamard fractional differential equation:

$$\begin{cases} (^H D_1^{\frac{1}{2}} u_n)(t) = f_n(t, u(t), (^H D_1^{\frac{1}{2}} u)(t)); & t \in [1, e], \\ u(t) = g(t, e^-); & t \in (e, e^2], \\ (^H D_{e^2}^{\frac{1}{2}} u_n)(t) = f_n(t, u(t), (^H D_{e^2}^{\frac{1}{2}} u)(t)); & t \in (e^2, e^3], \\ (^H I_1^{\frac{1}{2}} u)(t)|_{t=1} = 0, \end{cases} \tag{7.27}$$

where

$$f_n(t, u(t), (^H D_k^{\frac{1}{2}} u)(t)) = \frac{ct^2(e^{-7} + e^{-t-5})}{1 + \|u\|_E + \|^H D_k^{\frac{1}{2}} u\|_E} u_n(t);$$

$t \in [1, e] \cup (e^2, e^3], \ k \in \{1, e^2\},$

$$g(t, e^-) = \frac{e^{-2t}}{1 + e}; \ t \in (e, e^2],$$

with

$$u = (u_1, u_2, \ldots, u_n, \ldots), \text{ and } c := \frac{1}{48} \Gamma \left(\frac{1}{2} \right).$$

Set $f = (f_1, f_2, \ldots, f_n, \ldots)$. Clearly, the function f is continuous.

For each $u \in E$ and $t \in [1, e] \cup (e^2, e^3]$, we have

$$\|f(t, u(t), (^H D_k^{\frac{1}{2}})(t))\|_E \le ct^2 \left(e^{-7} + \frac{1}{e^{t+5}} \right); \quad k \in \{1, e^2\}.$$

Hence, the hypothesis (7.5.2) is satisfied with $p^* = 2c$,

We shall show that condition (7.25) holds with $T = e^3$. Indeed,

$$\frac{p^* \ln T}{\Gamma(1 + r)} = \frac{6c}{\Gamma(\frac{3}{2})} = \frac{12c}{\Gamma(\frac{1}{2})} = \frac{1}{4} < 1.$$

A simple computations show that all the conditions of Theorem 7.5 are satisfied. It follows that the problem (7.27) has at least one solution on $[1, e]$.

7.7. Notes and Remarks

The results of Chapter 7 are taken from [39–41,45,57,60].

Chapter 8

Hilfer–Pettis Fractional Differential Equations and Inclusions

8.1. Introduction

In this chapter, we present some results concerning the existence of weak solutions for some Hilfer differential equations. The main results are proved by applying Mönch's fixed-point theorem associated with the technique of measure of weak noncompactness.

8.2. Hilfer–Pettis Fractional Differential Equations

8.2.1. *Introduction*

In this section, we discuss the existence of weak solutions for the following problem of Hilfer fractional differential equation

$$\begin{cases} (D_0^{\alpha,\beta}u)(t) = f(t, u(t)); & t \in I := [0, T], \\ (I_0^{1-\gamma}u)(t)|_{t=0} = \phi, \end{cases} \tag{8.1}$$

where $\alpha \in (0, 1)$, $\beta \in [0, 1]$, $\gamma = \alpha + \beta - \alpha\beta$, $T > 0$, $\phi \in E$, $f : I \times E \to E$ is a given continuous function, E is a real (or complex) Banach space with norm $\| \cdot \|_E$ and dual E^*, such that E is the dual of a weakly-compactly-generated Banach space X, $I_0^{1-\gamma}$ is the left-sided mixed Riemann–Liouville integral of order $1 - \gamma$, and $D_0^{\alpha,\beta}$ is the generalized Riemann–Liouville derivative operator of order α and type β, introduced by Hilfer in [238].

8.2.2. *Existence of weak solutions*

Definition 8.1. By a weak solution of the problem (8.1) we mean a measurable function $u \in C_\gamma$ that satisfies the condition $(I_0^{1-\gamma}u)(0^+) = \phi$, and the equation $(D_0^{\alpha,\beta}u)(t) = f(t, u(t))$ on I.

The following hypotheses will be used in the sequel.

(8.1.1) For a.e. $t \in I$, the function $v \to f(t, v)$ is weakly-sequentially continuous.

(8.1.2) For each $v \in E$, the function $t \to f(t, v)$ is Pettis integrable a.e. on I.

(8.1.3) There exists $p \in C(I, [0, \infty))$ such that for all $\varphi \in E^*$, we have
$$|\varphi(f(t, u))| \le p(t)\|\varphi\| \text{ for a.e. } t \in I, \text{ and each } u \in E.$$

(8.1.4) For each bounded and measurable set $B \subset E$ and for each $t \in I$, we have
$$\beta(f(t, B)) \le t^{1-r} p(t) \beta(B).$$

Set
$$p^* = \sup_{t \in I} p(t).$$

Theorem 8.1. *Assume that the hypotheses (8.1.1)–(8.1.4) hold. If*
$$L := \frac{p^* T^{1-\gamma+\alpha}}{\Gamma(1+\alpha)} < 1, \tag{8.2}$$

then the problem (8.1) has at least one weak solution defined on I.

Proof. Consider the operator $N : C_\gamma \to C_\gamma$ defined by
$$(Nu)(t) = \frac{\phi}{\Gamma(\gamma)} t^{\gamma-1} + \int_0^t (t-s)^{\alpha-1} \frac{f(s, u(s))}{\Gamma(\alpha)} ds. \tag{8.3}$$

First notice that, the hypotheses imply that for each $u \in C_\gamma$, the function $t \mapsto (t-s)^{\alpha-1} f(s, u(s))$, $t \in I$, is Pettis integrable. Thus, the operator N is well defined. Let $R > 0$ be such that
$$R > \frac{p^* T^{1-\gamma+\alpha}}{\Gamma(1+\alpha)},$$

and consider the set
$$Q = \left\{ u \in C_\gamma : \|u\|_C \le R \text{ and } \|t_2^{1-\gamma} u(t_2) - t_1^{1-\gamma} u(t_1)\|_E \right.$$
$$\le \frac{p^* T^{1-\gamma+\alpha}}{\Gamma(1+\alpha)} (t_2 - t_1)^\alpha + \frac{p^*}{\Gamma(\alpha)}$$
$$\left. \times \int_0^{t_1} |t_2^{1-\gamma}(t_2 - s)^{\alpha-1} - t_1^{1-\gamma}(t_1 - s)^{\alpha-1}| ds \right\}.$$

Clearly, the subset Q is closed, convex and equicontinuous. We shall show that the operator N satisfies all the assumptions of Theorem 1.7. The proof will be given in several steps.

Step 1. *N maps Q into itself.*

Let $u \in Q$, $t \in I$ and assume that $(Nu)(t) \neq 0$. Then there exists $\varphi \in E^*$ such that $\|t^{1-\gamma}(Nu)(t)\|_E = |\varphi(t^{1-\gamma}(Nu)(t))|$. Thus

$$\|t^{1-\gamma}(Nu)(t)\|_E = \left| \varphi \left(\frac{\phi}{\Gamma(\gamma)} + \frac{t^{1-\gamma}}{\Gamma(\alpha)} \int_0^t (t-s)^{\alpha-1} f(s, u(s)) ds \right) \right|.$$

Then

$$\|t^{1-\gamma}(Nu)(t)\|_E \leq \frac{t^{1-\gamma}}{\Gamma(\alpha)} \int_0^t (t-s)^{\alpha-1} |\varphi(f(s, u(s)))| ds$$

$$\leq \frac{p^* T^{1-\gamma}}{\Gamma(\alpha)} \int_0^t (t-s)^{\alpha-1} ds$$

$$\leq \frac{p^* T^{1-\gamma+\alpha}}{\Gamma(1+\alpha)}$$

$$\leq R.$$

Next, let $t_1, t_2 \in I$ such that $t_1 < t_2$ and let $u \in Q$, with

$$t_2^{1-\gamma}(Nu)(t_2) - t_1^{1-\gamma}(Nu)(t_1) \neq 0.$$

Then there exists $\varphi \in E^*$ such that

$$\|t_2^{1-\gamma}(Nu)(t_2) - t_1^{1-\gamma}(Nu)(t_1)\|_E = |\varphi(t_2^{1-\gamma}(Nu)(t_2) - t_1^{1-\gamma}(Nu)(t_1))|,$$

and $\|\varphi\| = 1$. Then

$$\|t_2^{1-\gamma}(Nu)(t_2) - t_1^{1-\gamma}(Nu)(t_1)\|_E$$

$$= |\varphi(t_2^{1-\gamma}(Nu)(t_2) - t_1^{1-\gamma}(Nu)(t_1))|$$

$$\leq \left| \varphi \left(t_2^{1-\gamma} \int_0^{t_2} (t_2-s)^{\alpha-1} \frac{f(s, u(s))}{\Gamma(\alpha)} ds - t_1^{1-\gamma} \right. \right.$$

$$\left. \left. \times \int_0^{t_1} (t_1-s)^{\alpha-1} \frac{f(s, u(s))}{\Gamma(\alpha)} ds \right) \right|.$$

Then

$$\|t_2^{1-\gamma}(Nu)(t_2) - t_1^{1-\gamma}(Nu)(t_1)\|_E$$

$$\leq t_2^{1-\gamma} \int_{t_1}^{t_2} (t_2-s)^{\alpha-1} \frac{|\varphi(f(s, u(s)))|}{\Gamma(\alpha)} ds$$

$$+ \int_0^{t_1} |t_2^{1-\gamma}(t_2-s)^{\alpha-1} - t_1^{1-\gamma}(t_1-s)^{\alpha-1}| \frac{|\varphi(f(s, u(s)))|}{\Gamma(\alpha)} ds$$

$$\leq t_2^{1-\gamma} \int_{t_1}^{t_2} (t_2 - s)^{\alpha-1} \frac{p(s)}{\Gamma(\alpha)} ds$$

$$+ \int_0^{t_1} |t_2^{1-\gamma}(t_2 - s)^{\alpha-1} - t_1^{1-\gamma}(t_1 - s)^{\alpha-1}| \frac{p(s)}{\Gamma(\alpha)} ds.$$

Thus, we get

$$\|t_2^{1-\gamma}(Nu)(t_2) - t_1^{1-\gamma}(Nu)(t_1)\|_E \leq \frac{p^* T^{1-\gamma+\alpha}}{\Gamma(1+\alpha)} (t_2 - t_1)^{\alpha}$$

$$+ \frac{p^*}{\Gamma(\alpha)} \int_0^{t_1} |t_2^{1-\gamma}(t_2 - s)^{\alpha-1} - t_1^{1-\gamma}(t_1 - s)^{\alpha-1}| ds.$$

Hence $N(Q) \subset Q$.

Step 2. *N is weakly-sequentially continuous.*

Let (u_n) be a sequence in Q and let $(u_n(t)) \to u(t)$ in (E, ω) for each $t \in I$. Fix $t \in I$, since f satisfies the assumption (8.1.1), we have $f(t, u_n(t))$ converges weakly to $f(t, u(t))$. Hence the Lebesgue dominated convergence theorem for Pettis integral implies $(Nu_n)(t)$ converges weakly to $(Nu)(t)$ in (E, ω), for each $t \in I$. Thus, $N(u_n) \to N(u)$. Hence, $N : Q \to Q$ is weakly-sequentially continuous.

Step 3. *The implication (1.9) holds.*

Let V be a subset of Q such that $\overline{V} = \overline{\text{conv}}(N(V) \cup \{0\})$. Obviously

$$V(t) \subset \overline{\text{conv}}(NV)(t)) \cup \{0\}), \ \forall t \in I.$$

Further, as V is bounded and equicontinuous, by [172, Lemma 3] the function $t \to v(t) = \beta(V(t))$ is continuous on I. From (8.1.3), (8.1.4), Lemma 1.7 and the properties of the measure β, for any $t \in I$, we have

$$t^{1-\gamma}v(t) \leq \beta(t^{1-\gamma}(NV)(t) \cup \{0\})$$

$$\leq \beta(t^{1-\gamma}(NV)(t))$$

$$\leq \frac{T^{1-\gamma}}{\Gamma(\alpha)} \int_0^t |t - s|^{\alpha-1} p(s)\beta(V(s)) ds$$

$$\leq \frac{T^{1-\gamma}}{\Gamma(\alpha)} \int_0^t |t - s|^{\alpha-1} s^{1-\gamma} p(s)v(s) ds$$

$$\leq \frac{p^* T^{1-\gamma+\alpha}}{\Gamma(1+\alpha)} \|v\|_C.$$

Thus

$$\|v\|_C \leq L\|v\|_C.$$

From (8.2), we get $\|v\|_C = 0$, that is $v(t) = \beta(V(t)) = 0$, for each $t \in I$, and then by [290, Theorem 2], V is weakly-relatively compact in C. Applying now Theorem 1.7, we conclude that N has a fixed point which is a weak solution of the problem (8.1). □

8.2.3. *Example*

Let

$$E = l^1 = \left\{ u = (u_1, u_2, \ldots, u_n, \ldots), \sum_{n=1}^{\infty} |u_n| < \infty \right\}$$

be the Banach space with the norm

$$\|u\|_E = \sum_{n=1}^{\infty} |u_n|.$$

Consider the following problem of Hilfer fractional differential equation:

$$\begin{cases} (D_0^{\frac{1}{2},\frac{1}{2}} u_n)(t) = f_n(t, u(t)); & t \in [0,1], \\ (I_0^{\frac{1}{4}} u)(t)|_{t=0} = (2^{-1}, 2^{-2}, \ldots, 2^{-n}, \ldots), \end{cases} \tag{8.4}$$

where

$$f_n(t, u(t)) = \frac{ct^2}{1 + \|u(t)\|_E} \frac{u_n(t)}{e^{t+4}}, \quad t \in [0,1],$$

with

$$u = (u_1, u_2, \ldots, u_n, \ldots), \quad \text{and} \quad c := \frac{e^4}{8} \Gamma\left(\frac{1}{2}\right).$$

Set

$$f = (f_1, f_2, \ldots, f_n, \ldots).$$

Clearly, the function f is continuous.

For each $u \in E$ and $t \in [0,1]$, we have

$$\|f(t, u(t))\|_E \leq ct^2 \frac{1}{e^{t+4}}.$$

Hence, the hypothesis (8.2.3) is satisfied with $p^* = ce^{-4}$. We shall show that condition (8.2) holds with $T = 1$. Indeed,

$$\frac{p^* T^{1-\gamma+\alpha}}{\Gamma(1+\alpha)} = \frac{2ce^{-4}}{\Gamma(\frac{1}{2})} = \frac{1}{4} < 1.$$

Simple computations show that all conditions of Theorem 8.1 are satisfied. It follows that the problem (8.4) has at least one weak solution defined on $[0, 1]$.

8.3. Hilfer–Pettis Fractional Differential Inclusions

8.3.1. *Introduction*

In this section, we present some results concerning the existence of weak solutions for some functional Hilfer differential inclusions. The main results are proved by applying Mönch's fixed-point theorem associated with the technique of measure of weak noncompactness. We discuss the existence of weak solutions for the following problem of Hilfer fractional differential inclusion:

$$\begin{cases} (D_0^{\alpha,\beta} u)(t) \in F(t, u(t)); & t \in I := [0, T], \\ (I_0^{1-\gamma} u)(t)|_{t=0} = \phi, \end{cases} \tag{8.5}$$

where $\alpha \in (0, 1)$, $\beta \in [0, 1]$, $\gamma = \alpha + \beta - \alpha\beta$, $T > 0$, $\phi \in E$, $I_0^{1-\gamma}$ is the left-sided mixed Riemann–Liouville integral of order $1 - \gamma$, $D_0^{\alpha,\beta}$ is the generalized Riemann–Liouville derivative operator of order α and type β, introduced by Hilfer in [238], $F : I \times E \to \mathcal{P}(E)$ is a given multivalued map, E is a real (or complex) Banach space with norm $\|\cdot\|_E$ and dual E^*, such that E is the dual of a weakly-compactly-generated Banach space X, and $\mathcal{P}(E)$ is the family of all nonempty subsets of E.

8.3.2. *Existence of weak solutions*

Definition 8.2. By a weak solution of the problem (8.5) we mean a measurable function $u \in C_\gamma$ that satisfies the condition $(I_0^{1-\gamma} u)(0^+) = \phi$, and the equation $(D_0^{\alpha,\beta} u)(t) = h(t)$ on I, where $h \in S_{Fou}$.

Lemma 8.1. *Let $F : I \times E \to E$ be such that $S_{Fou} \subset C_\gamma(I)$ for any $u \in C_\gamma(I)$. Then problem (8.5) is equivalent to the problem of the solutions of the integral equation*

$$u(t) = \frac{\phi}{\Gamma(\gamma)} t^{\gamma-1} + (I_0^\alpha v)(t),$$

where $v \in S_{Fou}$.

The following hypotheses will be used in the sequel.

(8.2.1) $F : I \times E \to \mathcal{P}_{cp,cl,cv}(E)$ has weakly-sequentially closed graph.

(8.2.2) For each continuous $u : I \to E$, there exists a measurable function $v \in S_{Fou}$ a.e. on I and v is Pettis integrable on I.

(8.2.3) There exists $p \in C(I, [0, \infty))$ such that for all $\varphi \in E^*$, we have

$$\|F(t, u)\|_{\mathcal{P}} = \sup_{v \in S_{Fou}} |\varphi(v)| \le p(t)\|\varphi\|,$$

for a.e. $t \in I$, and each $u \in E$.

(8.2.4) For each bounded and measurable set $B \subset E$ and for each $t \in I$, we have

$$\beta(F(t, B)) \le t^{1-r} p(t) \beta(B).$$

Set

$$p^* = \sup_{t \in I} p(t).$$

Theorem 8.2. *Assume that the hypotheses (8.2.1)–(8.2.4) hold. If*

$$L := \frac{p^* T^{1-\gamma+\alpha}}{\Gamma(1+\alpha)} < 1, \tag{8.6}$$

then the problem (8.5) has at least one weak solution defined on I.

Proof. Consider the multivalued map $N : C_\gamma \to \mathcal{P}_{cl}(C_\gamma)$ defined by:

$$(Nu)(t) = \left\{ h \in C_\gamma : h(t) = \frac{\phi}{\Gamma(\gamma)} t^{\gamma-1} + \int_0^t (t-s)^{\alpha-1} \frac{v(s)}{\Gamma(\alpha)} ds; \ v \in S_{Fou} \right\}. \tag{8.7}$$

First notice that, the hypotheses imply that for each $u \in C_\gamma$, there exists a Pettis integrable function $v \in S_{Fou}$, and for each $s \in [0, t]$, the function

$$t \mapsto (t-s)^{\alpha-1} v(s) \quad \text{for a.e. } t \in I,$$

is Pettis integrable. Thus, the multifunction N is well defined. Let $R > 0$ be such that

$$R > \frac{p^* T^{1-\gamma+\alpha}}{\Gamma(1+\alpha)},$$

and consider the set

$$Q = \left\{ u \in C_\gamma : \|u\|_C \leq R \text{ and } \|t_2^{1-\gamma}u(t_2) - t_1^{1-\gamma}u(t_1)\|_E \right.$$

$$\leq \frac{p^* T^{1-\gamma+\alpha}}{\Gamma(1+\alpha)}(t_2 - t_1)^\alpha + \frac{p^*}{\Gamma(\alpha)}$$

$$\left. \times \int_0^{t_1} |t_2^{1-\gamma}(t_2 - s)^{\alpha-1} - t_1^{1-\gamma}(t_1 - s)^{\alpha-1}|ds \right\}.$$

Clearly, the subset Q is closed, convex and equicontinuous. We shall show that the operator N satisfies all the assumptions of Theorem 1.7. The proof will be given in several steps.

Step 1. *$N(u)$ is convex for each $u \in Q$.*

For that, let $h_1, h_2 \in N(u)$. Then there exist $v_1, v_2 \in S_{Fou}$ such that, for each $t \in I$, and for any $i = 1, 2$, we have

$$h_i(t) = \frac{\phi}{\Gamma(\gamma)}t^{\gamma-1} + \int_0^t (t - s)^{\alpha-1}\frac{v_i(s)}{\Gamma(\alpha)}ds.$$

Let $0 \leq \lambda \leq 1$. Then, for each $t \in I$, we have

$$[\lambda h_1 + (1 - \lambda)h_2](t) = \frac{\phi}{\Gamma(\gamma)}t^{\gamma-1} + \int_0^t (t - s)^{\alpha-1}\frac{\lambda v_1(s) + (1 - \lambda)v_2(s)}{\Gamma(\alpha)}ds.$$

Since S_{Fou} is convex (because F has convex values), it follows that

$$\lambda h_1 + (1 - \lambda)h_2 \in N(u).$$

Step 2. *N maps Q into itself.*

Take $h \in N(Q)$. Then there exists $u \in Q$ with $h \in N(u)$, and there exists a Pettis integrable $v : I \to E$ with $v(t) \in F(t, u(t))$; for a.e. $t \in I$. Assume that $h(t) \neq 0$, then there exists $\varphi \in E^*$ with $\|\varphi\| = 1$ such that $\|t^{1-\gamma}h(t)\|_E = |\varphi(t^{1-\gamma}h(t))|$. Then

$$\|t^{1-\gamma}h(t)\|_E = \left| \varphi \left(\frac{\phi}{\Gamma(\gamma)} + \frac{t^{1-\gamma}}{\Gamma(\alpha)}\int_0^t (t - s)^{\alpha-1}v(s)ds \right) \right|.$$

Thus

$$\|t^{1-\gamma}h(t)\|_E \le \frac{t^{1-\gamma}}{\Gamma(\alpha)} \int_0^t (t-s)^{\alpha-1}|\varphi(v(s))|ds$$

$$\le \frac{p^*T^{1-\gamma}}{\Gamma(\alpha)} \int_0^t (t-s)^{\alpha-1}ds$$

$$\le \frac{p^*T^{1-\gamma+\alpha}}{\Gamma(1+\alpha)}$$

$$\le R.$$

Next, let $t_1, t_2 \in I$ such that $t_1 < t_2$ and let $h \in N(u)$, with

$$t_2^{1-\gamma}h(t_2) - t_1^{1-\gamma}h(t_1) \ne 0.$$

Then there exists $\varphi \in E^*$ such that

$$\|t_2^{1-\gamma}h(t_2) - t_1^{1-\gamma}h(t_1)\|_E = |\varphi(t_2^{1-\gamma}h(t_2) - t_1^{1-\gamma}h(t_1))|,$$

and $\|\varphi\| = 1$. Then, we have

$$\|t_2^{1-\gamma}h(t_2) - t_1^{1-\gamma}h(t_1)\|_E$$
$$= |\varphi(t_2^{1-\gamma}h(t_2) - t_1^{1-\gamma}h(t_1))|$$
$$\le \left|\varphi\left(t_2^{1-\gamma}\int_0^{t_2}(t_2-s)^{\alpha-1}\frac{v(s)}{\Gamma(\alpha)}ds - t_1^{1-\gamma}\int_0^{t_1}(t_1-s)^{\alpha-1}\frac{v(s)}{\Gamma(\alpha)}ds\right)\right|.$$

Thus

$$\|t_2^{1-\gamma}h(t_2) - t_1^{1-\gamma}h(t_1)\|_E$$
$$\le t_2^{1-\gamma}\int_{t_1}^{t_2}(t_2-s)^{\alpha-1}\frac{|\varphi(v(s))|}{\Gamma(\alpha)}ds$$
$$+ \int_0^{t_1}|t_2^{1-\gamma}(t_2-s)^{\alpha-1} - t_1^{1-\gamma}(t_1-s)^{\alpha-1}|\frac{|\varphi(v(s))|}{\Gamma(\alpha)}ds$$
$$\le t_2^{1-\gamma}\int_{t_1}^{t_2}(t_2-s)^{\alpha-1}\frac{p(s)}{\Gamma(\alpha)}ds$$
$$+ \int_0^{t_1}|t_2^{1-\gamma}(t_2-s)^{\alpha-1} - t_1^{1-\gamma}(t_1-s)^{\alpha-1}|\frac{p(s)}{\Gamma(\alpha)}ds.$$

Hence, we get

$$\|t_2^{1-\gamma}h(t_2) - t_1^{1-\gamma}h(t_1)\|_E \leq \frac{p^* T^{1-\gamma+\alpha}}{\Gamma(1+\alpha)}(t_2 - t_1)^\alpha$$

$$+ \frac{p^*}{\Gamma(\alpha)} \int_0^{t_1} |t_2^{1-\gamma}(t_2 - s)^{\alpha-1} - t_1^{1-\gamma}(t_1 - s)^{\alpha-1}| ds.$$

This implies that $h \in Q$. Hence $N(Q) \subset Q$.

Step 3. *N has weakly-sequentially closed graph.*

Let (u_n, w_n) be a sequence in $Q \times Q$, with $u_n(t) \to u(t)$ in (E, ω) for each $t \in I$, $w_n(t) \to w(t)$ in (E, ω) for each $t \in I$, and $w_n \in N(u_n)$ for $n \in \{1, 2, \ldots\}$.

We show that $w \in \Omega(u)$. Since $w_n \in \Omega(u_n)$, there exists $v_n \in S_{F \circ u_n}$ such that

$$w_n(t) = \frac{\phi}{\Gamma(\gamma)}t^{\gamma-1} + \int_0^t (t - s)^{\alpha-1}\frac{v_n(s)}{\Gamma(\alpha)}ds.$$

We show that there exists $v \in S_{F \circ u}$ such that, for each $t \in I$,

$$w(t) = \frac{\phi}{\Gamma(\gamma)}t^{\gamma-1} + \int_0^t (t - s)^{\alpha-1}\frac{v(s)}{\Gamma(\alpha)}ds.$$

Since $F(\cdot, \cdot)$ has compact values, there exists a subsequence v_{n_m} such that v_{n_m} is Pettis integrable,

$$v_{n_m}(t) \in F(t, u_n(t)) \text{ a.e. } t \in I,$$

$$v_{n_m}(\cdot) \to v(\cdot) \text{ in } (E, \omega) \text{ as } m \to \infty.$$

As $F(t, \cdot)$ has weakly sequentially closed graph, $v(t) \in F(t, u(t))$. Then by the Lebesgue dominated convergence theorem for the Pettis integral, we obtain

$$|\varphi(w_n(t))| \to |\varphi(\frac{\phi}{\Gamma(\gamma)}t^{\gamma-1} + \int_0^t (t - s)^{\alpha-1}\frac{v(s)}{\Gamma(\alpha)}ds)|,$$

i.e., $w_n(t) \to (Nu)(t)$ in (E, ω). Since this holds, for each $t \in I$, we get $w \in N(u)$.

Step 4. *The implication (1.9) holds.*

Let V be a subset of Q, such that $\overline{V} = \overline{\text{conv}}(\Omega(V) \cup \{0\})$. Obviously $V(t) \subset \overline{\text{conv}}(\Omega(V(t)) \cup \{0\})$ for each $t \in I$. Further, as V is bounded and equicontinuous, the function $t \to v(t) = \beta(V(t))$ is continuous on I. By (8.2.4) and the properties of the measure β, for any $t \in I$ we have

$$t^{1-\gamma}v(t) \le \beta(t^{1-\gamma}(NV)(t) \cup \{0\})$$

$$\le \beta(t^{1-\gamma}(NV)(t))$$

$$\le \beta\{t^{1-\gamma}(Nu)(t) : u \in V\}$$

$$\le \beta\left\{T^{1-\gamma} \int_0^t (t-s)^{\alpha-1}\frac{v(s)}{\Gamma(\alpha)}ds : v(t) \in S_{Fou}, \ u \in V\right\}$$

$$\le \beta\left\{T^{1-\gamma} \int_0^t (t-s)^{\alpha-1}\frac{F(s, V(s))}{\Gamma(\alpha)}ds\right\}$$

$$\le T^{1-\gamma} \int_0^t (t-s)^{\alpha-1}\frac{\beta(V(s))}{\Gamma(\alpha)}ds$$

$$\le T^{1-\gamma} \int_0^t (t-s)^{\alpha-1}\frac{s^{1-\gamma}p(s)v(s)}{\Gamma(\alpha)}ds$$

$$\le \frac{p^*T^{1-\gamma+\alpha}}{\Gamma(1+\alpha)}\|v\|_C.$$

In particular,

$$\|u\|_C \le \frac{p^*T^{1-\gamma+\alpha}}{\Gamma(1+\alpha)}\|v\|_C.$$

By (8.6) it follows that $\|v\|_C = 0$, that is, $v(t) = \beta(V(t)) = 0$ for each $t \in I$, and then V is relatively weakly compact in C. Applying now Theorem 1.7, we conclude that N has a fixed point which is a weak solution of the problem (8.5). $\square$

8.3.3. *Example*

Let

$$E = l^1 = \left\{u = (u_1, u_2, \ldots, u_n, \ldots), \sum_{n=1}^{\infty} |u_n| < \infty\right\}$$

be the Banach space with the norm

$$\|u\|_E = \sum_{n=1}^{\infty} |u_n|.$$

Consider the following problem of Hilfer fractional differential inclusion:

$$
\begin{cases}
(D_0^{\frac{1}{2},\frac{1}{2}} u_n)(t) \in F_n(t, u(t)); & t \in [0,1], \\
(I_0^{\frac{1}{4}} u)(t)|_{t=0} = (1, 0, \dots, 0, \dots),
\end{cases}
\tag{8.8}
$$

where

$$
F_n(t, u(t)) = \frac{ct^2 e^{-4-t}}{1 + \|u(t)\|_E} [u_n(t) - 1, u_n(t)]; \quad t \in [0,1],
$$

with

$$
u = (u_1, u_2, \dots, u_n, \dots) \quad \text{and} \quad c := \frac{e^4}{8} \Gamma\left(\frac{1}{2}\right).
$$

Set

$$
F = (F_1, F_2, \dots, F_n, \dots).
$$

We assume that F is closed and convex valued. Clearly, the function F is continuous.

For each $u \in E$ and $t \in [0,1]$, we have

$$
\|F(t, u(t))\|_{\mathcal{P}} \le ct^2 \frac{1}{e^{t+4}}.
$$

Hence, the hypothesis (8.2.3) is satisfied with $p^* = ce^{-4}$. We shall show that condition (8.6) holds with $T = 1$. Indeed,

$$
\frac{p^* T^{1-\gamma+\alpha}}{\Gamma(1+\alpha)} = \frac{2ce^{-4}}{\Gamma(\frac{1}{2})} = \frac{1}{4} < 1.
$$

Simple computations show that all conditions of Theorem 8.2 are satisfied. It follows that the problem (8.8) has at least one weak solution defined on $[0,1]$.

8.4. Hilfer–Hadamard–Pettis Fractional Differential Equations and Inclusions

8.4.1. *Introduction*

In this section, we discuss the existence of weak solutions for the following problem of Hilfer–Hadamard fractional differential equation:

$$
\begin{cases}
(^H D_1^{\alpha,\beta} u)(t) = f(t, u(t)); & t \in I := [1, T], \\
(^H I_1^{1-\gamma} u)(t)|_{t=1} = \phi,
\end{cases}
\tag{8.9}
$$

where $\alpha \in (0,1)$, $\beta \in [0,1]$, $\gamma = \alpha + \beta - \alpha\beta$, $T > 1$, $\phi \in E$, $f : I \times E \to E$ is a given continuous function, E is a real (or complex) Banach space with norm $\| \cdot \|_E$ and dual E^*, such that E is the dual of a weakly-compactly-generated Banach space X, $^H I_1^{1-\gamma}$ is the left-sided mixed Hadamard integral of order $1 - \gamma$, and $^H D_1^{\alpha,\beta}$ is the Hilfer–Hadamard derivative operator of order α and type β.

Next, we consider the following problem of Hilfer–Hadamard fractional differential inclusion:

$$\begin{cases} (^H D_1^{\alpha,\beta} u)(t) \in F(t, u(t)); & t \in I, \\ (^H I_1^{1-\gamma} u)(t)|_{t=1} = \phi, \end{cases} \tag{8.10}$$

where $F : I \times E \to \mathcal{P}(E)$ is a given multivalued map, and $\mathcal{P}(E)$ is the family of all nonempty subsets of E.

Our goal in this work is to give some existence results for functional Hilfer–Hadamard fractional differential equations and inclusions.

8.4.2. *Existence of weak solutions for Hilfer–Hadamard fractional differential equations*

Let us start in this section by defining what we mean by a weak solution of the problem (8.9).

Definition 8.3. By a weak solution of the problem (8.9) we mean a measurable function $u \in C_{\gamma, \ln}$ that satisfies the condition $(^H I_1^{1-\gamma} u)(1^+) = \phi$, and the equation $(^H D_1^{\alpha,\beta} u)(t) = f(t, u(t))$ on I.

The following hypotheses will be used in the sequel.

(8.3.1) For a.e. $t \in I$, the function $v \to f(t, v)$ is weakly sequentially continuous.

(8.3.2) For each $v \in E$, the function $t \to f(t, v)$ is Pettis integrable a.e. on I.

(8.3.3) There exists $p \in C(I, [0, \infty))$ such that for all $\varphi \in E^*$, we have

$$|\varphi(f(t, u))| \le p(t) \quad \text{for a.e. } t \in I, \text{ and each } u \in E.$$

(8.3.4) For each bounded and measurable set $B \subset E$ and for each $t \in I$, we have

$$\beta(f(t, B)) \le (\ln t)^{1-\gamma} p(t) \beta(B).$$

Set

$$p^* = \sup_{t \in I} p(t).$$

Theorem 8.3. *Assume that the hypotheses (8.3.1)–(8.3.4) hold. If*

$$L := \frac{p^* (\ln T)^{1-\gamma+\alpha}}{\Gamma(1+\alpha)} < 1, \tag{8.11}$$

then the problem (8.9) has at least one weak solution defined on I.

Proof. Consider the operator $N : C_{\gamma, \ln} \to C_{\gamma, \ln}$ defined by

$$(Nu)(t) = \frac{\phi}{\Gamma(\gamma)} (\ln t)^{\gamma-1} + \int_1^t \left(\ln \frac{t}{s}\right)^{\alpha-1} \frac{f(s, u(s))}{s\Gamma(\alpha)} ds. \tag{8.12}$$

First notice that, the hypotheses imply that for each $u \in C_{\gamma, \ln}$, the function $t \mapsto \left(\ln \frac{t}{s}\right)^{\alpha-1} \frac{f(s, u(s))}{s}$, for a.e. $t \in I$, is Pettis integrable. Thus, the operator N is well defined. Let $R > 0$ be such that

$$R > \frac{p^* (\ln T)^{1-\gamma+\alpha}}{\Gamma(1+\alpha)},$$

and consider the set

$$Q = \left\{ u \in C_\gamma : \|u\|_C \le R \text{ and } \|(\ln t_2)^{1-\gamma} u(t_2) - (\ln t_1)^{1-\gamma} u(t_1)\|_E \right.$$

$$\le \frac{p^* (\ln T)^{1-\gamma+\alpha}}{\Gamma(1+\alpha)} \left(\ln \frac{t_2}{t_1}\right)^\alpha$$

$$\left. + \frac{p^*}{\Gamma(\alpha)} \int_1^{t_1} \left| (\ln t_2)^{1-\gamma} \left(\ln \frac{t_2}{s}\right)^{\alpha-1} - (\ln t_1)^{1-\gamma} \left(\ln \frac{t_1}{s}\right)^{\alpha-1} \right| ds \right\}.$$

Clearly, the subset Q is closed, convex and equicontinuous. We shall show that the operator N satisfies all the assumptions of Theorem 1.7. The proof will be given in several steps.

Step 1. *N maps Q into itself.*

Let $u \in Q$, $t \in I$ and assume that $(Nu)(t) \ne 0$. Then there exists $\varphi \in E^*$ such that $\|(\ln t)^{1-\gamma}(Nu)(t)\|_E = |\varphi((\ln t)^{1-\gamma}(Nu)(t))|$. Thus

$$\|(\ln t)^{1-\gamma}(Nu)(t)\|_E$$

$$= \left| \varphi \left(\frac{\phi}{\Gamma(\gamma)} + \frac{(\ln t)^{1-\gamma}}{\Gamma(\alpha)} \int_1^t \left(\ln \frac{t}{s}\right)^{\alpha-1} f(s, u(s)) \frac{ds}{s} \right) \right|.$$

Then

$$\|(\ln t)^{1-\gamma}(Nu)(t)\|_E \leq \frac{(\ln t)^{1-\gamma}}{\Gamma(\alpha)} \int_1^t \left(\ln \frac{t}{s}\right)^{\alpha-1} |\varphi(f(s,u(s)))| \frac{ds}{s}$$

$$\leq \frac{p^*(\ln T)^{1-\gamma}}{\Gamma(\alpha)} \int_0^t \left(\ln \frac{t}{s}\right)^{\alpha-1} \frac{ds}{s}$$

$$\leq \frac{p^*(\ln T)^{1-\gamma+\alpha}}{\Gamma(1+\alpha)}$$

$$\leq R.$$

Next, let $t_1, t_2 \in I$ such that $t_1 < t_2$ and let $u \in Q$, with

$$(\ln t_2)^{1-\gamma}(Nu)(t_2) - (\ln t_1)^{1-\gamma}(Nu)(t_1) \neq 0.$$

Then there exists $\varphi \in E^*$ such that

$$\|(\ln t_2)^{1-\gamma}(Nu)(t_2) - (\ln t_1)^{1-\gamma}(Nu)(t_1)\|_E$$
$$= |\varphi((\ln t_2)^{1-\gamma}(Nu)(t_2) - (\ln t_1)^{1-\gamma}(Nu)(t_1))|,$$

and $\|\varphi\| = 1$. Then

$$\|(\ln t_2)^{1-\gamma}(Nu)(t_2) - (\ln t_1)^{1-\gamma}(Nu)(t_1)\|_E$$
$$= |\varphi((\ln t_2)^{1-\gamma}(Nu)(t_2) - (\ln t_1)^{1-\gamma}(Nu)(t_1))|$$
$$\leq \left|\varphi\left((\ln t_2)^{1-\gamma} \int_1^{t_2} \left(\ln \frac{t_2}{s}\right)^{\alpha-1} \frac{f(s,u(s))}{s\Gamma(\alpha)} ds \right.\right.$$
$$\left.\left. - (\ln t_1)^{1-\gamma} \int_1^{t_1} \left(\ln \frac{t_1}{s}\right)^{\alpha-1} \frac{f(s,u(s))}{s\Gamma(\alpha)} ds\right)\right|$$
$$\leq (\ln t_2)^{1-\gamma} \int_{t_1}^{t_2} \left(\ln \frac{t_2}{s}\right)^{\alpha-1} \frac{|\varphi(f(s,u(s)))|}{s\Gamma(\alpha)} ds$$
$$+ \int_1^{t_1} \left|(\ln t_2)^{1-\gamma} \left(\ln \frac{t_2}{s}\right)^{\alpha-1} - (\ln t_1)^{1-\gamma} \left(\ln \frac{t_1}{s}\right)^{\alpha-1}\right|$$
$$\times \frac{|\varphi(f(s,u(s)))|}{s\Gamma(\alpha)} ds$$

$$\leq (\ln t_2)^{1-\gamma} \int_{t_1}^{t_2} \left(\ln \frac{t_2}{s} \right)^{\alpha-1} \frac{p(s)}{s\Gamma(\alpha)} ds$$

$$+ \int_{1}^{t_1} \left| (\ln t_2)^{1-\gamma} \left(\ln \frac{t_2}{s} \right)^{\alpha-1} - (\ln t_1)^{1-\gamma} \left(\ln \frac{t_1}{s} \right)^{\alpha-1} \right| \frac{p(s)}{s\Gamma(\alpha)} ds.$$

Thus, we get

$$\| (\ln t_2)^{1-\gamma}(Nu)(t_2) - (\ln t_1)^{1-\gamma}(Nu)(t_1) \|_E$$

$$\leq \frac{p^*(\ln T)^{1-\gamma+\alpha}}{\Gamma(1+\alpha)} \left(\ln \frac{t_2}{t_1} \right)^{\alpha}$$

$$+ \frac{p^*}{\Gamma(\alpha)} \int_{1}^{t_1} \left| (\ln t_2)^{1-\gamma} \left(\ln \frac{t_2}{s} \right)^{\alpha-1} - (\ln t_1)^{1-\gamma} \left(\ln \frac{t_1}{s} \right)^{\alpha-1} \right| ds.$$

Hence $N(Q) \subset Q$.

Step 2. *N is weakly-sequentially continuous.*

Let (u_n) be a sequence in Q and let $(u_n(t)) \to u(t)$ in (E, ω) for each $t \in I$. Fix $t \in I$, since f satisfies the assumption (8.3.1), we have $f(t, u_n(t))$ converges weakly to $f(t, u(t))$. Hence the Lebesgue dominated convergence theorem for Pettis integral implies $(Nu_n)(t)$ converges weakly to $(Nu)(t)$ in (E, ω), for each $t \in I$. Thus, $N(u_n) \to N(u)$. Hence, $N : Q \to Q$ is weakly-sequentially continuous.

Step 3. *The implication (1.9) holds.*

Let V be a subset of Q such that $\overline{V} = \overline{\text{conv}}(N(V) \cup \{0\})$. Obviously

$$V(t) \subset \overline{\text{conv}}(NV)(t)) \cup \{0\}), \ \forall t \in I.$$

Further, as V is bounded and equicontinuous, by [172, Lemma 3] the function $t \to v(t) = \beta(V(t))$ is continuous on I. From (8.3.3), (8.3.4), Lemma 2.3 and the properties of the measure β, for any $t \in I$, we have

$$(\ln t)^{1-\gamma} v(t) \leq \beta((\ln t)^{1-\gamma}(NV)(t) \cup \{0\})$$

$$\leq \beta((\ln t)^{1-\gamma}(NV)(t))$$

$$\leq \frac{(\ln T)^{1-\gamma}}{\Gamma(\alpha)} \int_{1}^{t} \left(\ln \frac{t}{s} \right)^{\alpha-1} p(s)\beta(V(s)) ds$$

$$\leq \frac{(\ln T)^{1-\gamma}}{\Gamma(\alpha)} \int_1^t \left(\ln \frac{t}{s}\right)^{\alpha-1} (\ln s)^{1-\gamma} p(s)v(s)ds$$

$$\leq \frac{p^*(\ln T)^{1-\gamma+\alpha}}{\Gamma(1+\alpha)} \|v\|_C.$$

Thus

$$\|v\|_C \leq L\|v\|_C.$$

From (8.11), we get $\|v\|_C = 0$, that is $v(t) = \beta(V(t)) = 0$, for each $t \in I$. and then by [290, Theorem 2], V is weakly-relatively compact in $C_{\gamma,\ln}$. Applying now Theorem 1.7, we conclude that N has a fixed point which is a weak solution of the problem (8.9). □

8.4.3. Existence of weak solutions for Hilfer–Hadamard fractional differential inclusions

Let us start in this section by defining what we mean by a weak solution of the problem (8.10).

Definition 8.4. By a weak solution of the problem (8.10) we mean a measurable function $u \in C_{\gamma,\ln}$ that satisfies the condition $(^H I_1^{1-\gamma} u)(1^+) = \phi$, and the equation $(^H D_1^{\alpha,\beta} u)(t) = h(t)$ on I, where $h \in S_{Fou}$.

Lemma 8.2. *Let $F : I \times E \to E$ be such that $S_{Fou} \subset C_{\gamma,\ln}(I)$ for any $u \in C_{\gamma,\ln}(I)$. Then problem (8.10) is equivalent to the problem of the solutions of the integral equation*

$$u(t) = \frac{\phi}{\Gamma(\gamma)}(\ln t)^{\gamma-1} + (^H I_1^\alpha v)(t),$$

where $v \in S_{Fou}$.

For our purpose we will need the following fixed-point theorem:

Theorem 8.4 ([302]). *Let E be a Banach space with Q a nonempty, bounded, closed, convex and equicontinuous subset of a metrizable locally convex vector space C such that $0 \in Q$. Suppose $T : Q \to \mathcal{P}_{cl,cv}(Q)$ has weakly sequentially closed graph. If the implication*

$$\overline{V} = \overline{\text{conv}}(\{0\} \cup T(V)) \Rightarrow V \text{ is relatively-weakly compact,} \qquad (8.13)$$

holds for every subset $V \subset Q$, then the operator T has a fixed point.

The following hypotheses will be used in the sequel.

(8.4.1) $F : I \times E \to \mathcal{P}_{cp,cl,cv}(E)$ has weakly-sequentially closed graph.

(8.4.2) For each continuous $u : I \to E$, there exists a measurable function $v \in S_{Fou}$ a.e. on I and v is Pettis integrable on I.

(8.4.3) There exists $q \in C(I, [0, \infty))$ such that for all $\varphi \in E^*$, we have

$$\|F(t, u)\|_{\mathcal{P}} = \sup_{v \in S_{Fou}} |\varphi(v)| \le q(t), \quad \text{for a.e. } t \in I, \text{ and each } u \in E.$$

(8.4.4) For each bounded and measurable set $B \subset E$ and for each $t \in I$, we have

$$\beta(F(t, B)) \le (\ln t)^{1-\gamma} q(t) \beta(B).$$

Set

$$q^* = \sup_{t \in I} q(t).$$

Theorem 8.5. *Assume that the hypotheses (8.4.1)–(8.4.4) hold. If*

$$L' := \frac{q^*(\ln T)^{1-\gamma+\alpha}}{\Gamma(1+\alpha)} < 1, \tag{8.14}$$

then the problem (8.10) has at least one weak solution defined on I.

Proof. Consider the multivalued map $N' : C_{\gamma,\ln} \to \mathcal{P}_{cl}(C_{\gamma,\ln})$ defined by

$$(N'u)(t) = \left\{ h \in C_{\gamma,\ln} : h(t) = \frac{\phi}{\Gamma(\gamma)}(\ln t)^{\gamma-1} \right.$$

$$\left. + \int_1^t \left(\ln \frac{t}{s}\right)^{\alpha-1} \frac{v(s)}{s\Gamma(\alpha)} ds; \ v \in S_{Fou} \right\}. \tag{8.15}$$

First notice that, the hypotheses imply that for each $u \in C_{\gamma,\ln}$, there exists a Pettis integrable function $v \in S_{Fou}$, and for each $s \in [1, t]$, the function

$$t \mapsto \left(\ln \frac{t}{s}\right)^{\alpha-1} v(s) \quad \text{for a.e. } t \in I,$$

is Pettis integrable. Thus, the multifunction N' is well defined. Let $R' > 0$ be such that

$$R' > \frac{q^*(\ln T)^{1-\gamma+\alpha}}{\Gamma(1+\alpha)},$$

and consider the set

$$Q' = \left\{ u \in C_{\gamma,\ln} : \|u\|_C \leq R' \text{ and } \|(\ln t_2)^{1-\gamma}u(t_2) - (\ln t_1)^{1-\gamma}u(t_1)\|_E \right.$$

$$\leq \frac{q^*(\ln T)^{1-\gamma+\alpha}}{\Gamma(1+\alpha)} \left(\ln \frac{t_2}{t_1}\right)^{\alpha}$$

$$\left. + \frac{q^*}{\Gamma(\alpha)} \int_1^{t_1} \left| (\ln t_2)^{1-\gamma} \left(\ln \frac{t_2}{s}\right)^{\alpha-1} - (\ln t_1)^{1-\gamma} \left(\ln \frac{t_1}{s}\right)^{\alpha-1} \right| ds \right\}.$$

Clearly, the subset Q' is closed, convex and equicontinuous. We shall show that the operator N' satisfies all the assumptions of Theorem 8.4. The proof will be given in several steps.

Step 1. $N'(u)$ *is convex for each* $u \in Q'$.

For that, let $h_1, h_2 \in N'(u)$. Then there exist $v_1, v_2 \in S_{F \circ u}$ such that, for each $t \in I$, and for any $i = 1, 2$, we have

$$h_i(t) = \frac{\phi}{\Gamma(\gamma)}(\ln t)^{\gamma-1} + \int_1^t \left(\ln \frac{t}{s}\right)^{\alpha-1} \frac{v_i(s)}{s\Gamma(\alpha)} ds.$$

Let $0 \leq \lambda \leq 1$. Then, for each $t \in I$, we have

$$[\lambda h_1 + (1-\lambda)h_2](t) = \frac{\phi}{\Gamma(\gamma)}(\ln t)^{\gamma-1}$$

$$+ \int_1^t \left(\ln \frac{t}{s}\right)^{\alpha-1} \frac{\lambda v_1(s) + (1-\lambda)v_2(s)}{s\Gamma(\alpha)} ds.$$

Since $S_{F \circ u}$ is convex (because F has convex values), it follows that

$$\lambda h_1 + (1-\lambda)h_2 \in N'(u).$$

Step 2. N' *maps* Q' *into itself.*

Take $h \in N'(Q')$. Then there exists $u \in Q'$ with $h \in N'(u)$, and there exists a Pettis integrable $v : I \to E$ with $v(t) \in F(t, u(t))$; for a.e. $t \in I$. Assume that $h(t) \neq 0$, then there exists $\varphi \in E^*$ with $\|\varphi\| = 1$ such that $\|(\ln t)^{1-\gamma}h(t)\|_E = |\varphi((\ln t)^{1-\gamma}h(t))|$. Then

$$\|(\ln t)^{1-\gamma}h(t)\|_E = \left| \varphi \left(\frac{\phi}{\Gamma(\gamma)} + \frac{(/lnt)^{1-\gamma}}{\Gamma(\alpha)} \int_1^t \left(\ln \frac{t}{s}\right)^{\alpha-1} v(s)\frac{ds}{s} \right) \right|.$$

Thus

$$\|(\ln t)^{1-\gamma}h(t)\|_E \leq \frac{(\ln t)^{1-\gamma}}{\Gamma(\alpha)} \int_1^t \left(\ln \frac{t}{s}\right)^{\alpha-1} |\varphi(v(s))| \frac{ds}{s}$$

$$\leq \frac{q^*(\ln T)^{1-\gamma}}{\Gamma(\alpha)} \int_1^t \left(\ln \frac{t}{s}\right)^{\alpha-1} \frac{ds}{s}$$

$$\leq \frac{q^*(\ln T)^{1-\gamma+\alpha}}{\Gamma(1+\alpha)}$$

$$\leq R'.$$

Next, let $t_1, t_2 \in I$ such that $t_1 < t_2$ and let $h \in N'(u)$, with

$$(\ln t_2)^{1-\gamma}h(t_2) - (\ln t_1)^{1-\gamma}h(t_1) \neq 0.$$

Then there exists $\varphi \in E^*$ such that

$$\|(\ln t_2)^{1-\gamma}h(t_2) - (\ln t_1)^{1-\gamma}h(t_1)\|_E = |\varphi((\ln t_2)^{1-\gamma}h(t_2) - (\ln t_1)^{1-\gamma}h(t_1))|,$$

and $\|\varphi\| = 1$. Then, we have

$$\|(\ln t_2)^{1-\gamma}h(t_2) - (\ln t_1)^{1-\gamma}h(t_1)\|_E$$

$$= |\varphi((\ln t_2)^{1-\gamma}h(t_2) - (\ln t_1)^{1-\gamma}h(t_1))|$$

$$\leq \left| \varphi \left((\ln t_2)^{1-\gamma} \int_1^{t_2} \left(\ln \frac{t_2}{s}\right)^{\alpha-1} \frac{v(s)}{s\Gamma(\alpha)} ds - (\ln t_1)^{1-\gamma} \right.\right.$$

$$\left.\left. \times \int_1^{t_1} \left(\ln \frac{t_1}{s}\right)^{\alpha-1} \frac{v(s)}{s\Gamma(\alpha)} ds \right) \right|$$

$$\leq (\ln t_2)^{1-\gamma} \int_{t_1}^{t_2} \left(\ln \frac{t_2}{s}\right)^{\alpha-1} \frac{|\varphi(v(s))|}{s\Gamma(\alpha)} ds$$

$$+ \int_1^{t_1} \left| (\ln t_2)^{1-\gamma}(t_2 - s)^{\alpha-1} - (\ln t_1)^{1-\gamma} \left(\ln \frac{t_1}{s}\right)^{\alpha-1} \right| \frac{|\varphi(v(s))|}{s\Gamma(\alpha)} ds$$

$$\leq (\ln t_2)^{1-\gamma} \int_{t_1}^{t_2} \left(\ln \frac{t_2}{s}\right)^{\alpha-1} \frac{q(s)}{s\Gamma(\alpha)} ds$$

$$+ \int_1^{t_1} \left| (\ln t_2)^{1-\gamma} \left(\ln \frac{t_2}{s}\right)^{\alpha-1} - (\ln t_1)^{1-\gamma} \left(\ln \frac{t_1}{s}\right)^{\alpha-1} \right| \frac{q(s)}{s\Gamma(\alpha)} ds.$$

Hence, we get

$$\|(\ln t_2)^{1-\gamma} h(t_2) - (\ln t_1)^{1-\gamma} h(t_1)\|_E$$

$$\leq \frac{q^*(\ln T)^{1-\gamma+\alpha}}{\Gamma(1+\alpha)} \left(\ln \frac{t_2}{t_1}\right)^{\alpha}$$

$$+ \frac{q^*}{\Gamma(\alpha)} \int_1^{t_1} \left| (\ln t_2)^{1-\gamma} \left(\ln \frac{t_2}{s}\right)^{\alpha-1} - (\ln t_1)^{1-\gamma} \left(\ln \frac{t_1}{s}\right)^{\alpha-1} \right| ds.$$

This implies that $h \in Q'$. Hence $N'(Q') \subset Q'$.

Step 3. N' has weakly-sequentially closed graph.

Let (u_n, w_n) be a sequence in $Q' \times Q'$, with $u_n(t) \to u(t)$ in (E, ω) for each $t \in I$, $w_n(t) \to w(t)$ in (E, ω) for each $(t \in I$, and $w_n \in N'(u_n)$ for $n \in \{1, 2, \ldots\}$.

We show that $w \in \Omega(u)$. Since $w_n \in \Omega(u_n)$, there exists $v_n \in S_{Fou_n}$ such that

$$w_n(t) = \frac{\phi}{\Gamma(\gamma)} (\ln t)^{\gamma-1} + \int_1^t \left(\ln \frac{t}{s}\right)^{\alpha-1} \frac{v_n(s)}{s\Gamma(\alpha)} ds.$$

We show that there exists $v \in S_{Fou}$ such that, for each $t \in I$,

$$w(t) = \frac{\phi}{\Gamma(\gamma)} (\ln t)^{\gamma-1} + \int_1^t \left(\ln \frac{t}{s}\right)^{\alpha-1} \frac{v(s)}{s\Gamma(\alpha)} ds.$$

Since $F(\cdot, \cdot)$ has compact values, there exists a subsequence v_{n_m} such that v_{n_m} is Pettis integrable,

$$v_{n_m}(t) \in F(t, u_n(t)) \text{ a.e. } t \in I,$$

$$v_{n_m}(\cdot) \to v(\cdot) \text{ in } (E, \omega) \text{ as } m \to \infty.$$

As $F(t, \cdot)$ has weakly-sequentially closed graph, $v(t) \in F(t, u(t))$. Then by the Lebesgue dominated convergence theorem for the Pettis integral, we obtain

$$|\varphi(w_n(t))| \to \left| \varphi \left(\frac{\phi}{\Gamma(\gamma)} (\ln t)^{\gamma-1} + \int_1^t \left(\ln \frac{t}{s}\right)^{\alpha-1} \frac{v(s)}{s\Gamma(\alpha)} ds \right) \right|,$$

i.e., $w_n(t) \to (Nu)(t)$ in (E, ω). Since this holds, for each $t \in I$, then we get $w \in N(u)$.

Step 4. *The implication (8.13) holds.*

Let V be a subset of Q', such that $\overline{V} = \overline{\text{conv}}(\Omega(V) \cup \{0\})$. Obviously $V(t) \subset \overline{\text{conv}}(\Omega(V(t)) \cup \{0\})$ for each $t \in I$. Further, as V is bounded and equicontinuous, the function $t \to v(t) = \beta(V(t))$ is continuous on I. By (8.4.4) and the properties of the measure β, for any $t \in I$ we have

$$(\ln t)^{1-\gamma} v(t) \leq \beta((\ln t)^{1-\gamma}(NV)(t) \cup \{0\})$$

$$\leq \beta((\ln t)^{1-\gamma}(NV)(t))$$

$$\leq \beta\{(\ln t)^{1-\gamma}(Nu)(t) : u \in V\}$$

$$\leq \beta \left\{ (\ln T)^{1-\gamma} \int_1^t \left(\ln \frac{t}{s} \right)^{\alpha-1} \frac{v(s)}{s\Gamma(\alpha)} ds : v(t) \in S_{Fou}, \ u \in V \right\}$$

$$\leq \beta \left\{ (\ln T)^{1-\gamma} \int_1^t \left(\ln \frac{t}{s} \right)^{\alpha-1} \frac{F(s, V(s))}{s\Gamma(\alpha)} ds \right\}$$

$$\leq (\ln T)^{1-\gamma} \int_1^t \left(\ln \frac{t}{s} \right)^{\alpha-1} \frac{\beta(V(s))}{s\Gamma(\alpha)} ds$$

$$\leq (\ln T)^{1-\gamma} \int_1^t \left(\ln \frac{t}{s} \right)^{\alpha-1} \frac{(\ln s)^{1-\gamma} q(s) v(s)}{s\Gamma(\alpha)} ds$$

$$\leq \frac{q^*(\ln T)^{1-\gamma+\alpha}}{\Gamma(1+\alpha)} \|v\|_C.$$

In particular,

$$\|v\|_C \leq L' \|v\|_C.$$

By (8.14) it follows that $\|v\|_C = 0$, that is, $v(t) = \beta(V(t)) = 0$ for each $t \in I$, and then V is relatively-weakly compact in C. Applying now Theorem 8.4, we conclude that N' has a fixed point which is a weak solution of the problem (8.10). □

8.4.4. *Examples*

Let

$$E = l^1 = \left\{ u = (u_1, u_2, \ldots, u_n, \ldots), \sum_{n=1}^{\infty} |u_n| < \infty \right\}$$

be the Banach space with the norm

$$\|u\|_E = \sum_{n=1}^{\infty} |u_n|.$$

Example 1. Consider the following problem of Hilfer–Hadamard fractional differential equation:

$$\begin{cases} ({}^H D_1^{\frac{1}{2},\frac{1}{2}} u_n)(t) = f_n(t, u(t)); & t \in [1, e], \\ ({}^H I_1^{\frac{1}{4}} u)(t)|_{t=1} = (2^{-1}, 2^{-2}, \ldots, 2^{-n}, \ldots), \end{cases} \tag{8.16}$$

where

$$f_n(t, u(t)) = \frac{ct^2}{1 + \|u(t)\|_E} \frac{u_n(t)}{e^{t+4}}; \quad t \in [1, e],$$

with

$$u = (u_1, u_2, \ldots, u_n, \ldots) \quad \text{and} \quad c := \frac{e^3}{8} \Gamma\left(\frac{1}{2}\right).$$

Set

$$f = (f_1, f_2, \ldots, f_n, \ldots).$$

Clearly, the function f is continuous.

For each $u \in E$ and $t \in [1, e]$, we have

$$\|f(t, u(t))\|_E \le ct^2 \frac{1}{e^{t+4}}.$$

Hence, the hypothesis (8.3.3) is satisfied with $p^* = ce^{-3}$. We shall show that condition (8.2) holds with $T = e$. Indeed,

$$\frac{p^*(\ln T)^{1-\gamma+\alpha}}{\Gamma(1+\alpha)} = \frac{2ce^{-3}}{\Gamma(\frac{1}{2})} = \frac{1}{4} < 1.$$

Simple computations show that all conditions of Theorem 8.3 are satisfied. It follows that the problem (8.16) has at least one weak solution defined on $[1, e]$.

Example 2. Consider now the following problem of Hilfer–Hadamard fractional differential inclusion

$$\begin{cases} ({}^H D_1^{\frac{1}{2},\frac{1}{2}} u_n)(t) \in F_n(t, u(t)); & t \in [1, e], \\ ({}^H I_1^{\frac{1}{4}} u)(t)|_{t=1} = (1, 0, \ldots, 0, \ldots), \end{cases} \tag{8.17}$$

where

$$F_n(t, u(t)) = \frac{ct^2 e^{-4-t}}{1 + \|u(t)\|_E} [u_n(t) - 1, u_n(t)]; \quad t \in [1, e],$$

with

$$u = (u_1, u_2, \ldots, u_n, \ldots) \quad \text{and} \quad c := \frac{e^3}{8} \Gamma\left(\frac{1}{2}\right).$$

Set

$$F = (F_1, F_2, \ldots, F_n, \ldots).$$

We assume that F is closed and convex valued. Clearly, the function F is continuous.

For each $u \in E$ and $t \in [1, e]$, we have

$$\|F(t, u(t))\|_{\mathcal{P}} \le ct^2 \frac{1}{e^{t+4}}.$$

Hence, the hypothesis (8.4.3) is satisfied with $q^* = ce^{-3}$. We shall show that condition (2.23) holds with $T = e$. Indeed,

$$\frac{q^*(\ln T)^{1-\gamma+\alpha}}{\Gamma(1+\alpha)} = \frac{2ce^{-3}}{\Gamma(\frac{1}{2})} = \frac{1}{4} < 1.$$

Simple computations show that all conditions of Theorem 8.5 are satisfied. It follows that the problem (8.17) has at least one weak solution defined on $[1, e]$.

8.5. Hilfer–Pettis Fractional Differential Equations with Maxima

8.5.1. *Introduction*

In this section, by applying Mönch's fixed-point theorem associated with the technique of measure of weak noncompactness, we present some results concerning the existence of weak solutions for some functional Hilfer differential equations with maxima.

Differential equations with maximum arise naturally when solving practical problems. For example, many problems in the control theory correspond to the maximal deviation of the regulated quantity. The existence and uniqueness of solutions of differential equations with maxima is considered in [218,221,305], and the references therein. Recently, considerable attention has been given to the existence of solutions of initial and boundary value problems for fractional differential equations with Hilfer fractional derivative; see [238,239]. In this section, we discuss the existence of weak solutions for the following problem of Hilfer fractional differential equation:

$$\begin{cases} (D_0^{\alpha,\beta} u)(t) = f(t, \max_{0 \le \tau \le t} \|u(\tau)\|); & t \in I := [0, T], \\ (I_0^{1-\gamma} u)(t)|_{t=0} = \phi, \end{cases} \tag{8.18}$$

where $\alpha \in (0,1)$, $\beta \in [0,1]$, $\gamma = \alpha+\beta-\alpha\beta$, $T > 0$, $\phi \in E$, $f : I \times [0,\infty) \to E$ is a given continuous function, E is a real (or complex) Banach space with norm $\|\cdot\|$ and dual E^*, such that E is the dual of a weakly-compactly-generated Banach space X, $I_0^{1-\gamma}$ is the left-sided mixed Riemann–Liouville integral of order $1 - \gamma$, and $D_0^{\alpha,\beta}$ is the generalized Riemann–Liouville derivative operator of order α and type β, introduced by Hilfer in [238].

8.5.2. *Existence of weak solutions*

Definition 8.5. By a weak solution of the problem (8.18) we mean a measurable function $u \in C_\gamma$ that satisfies the condition $(I_0^{1-\gamma}u)(0^+) = \phi$, and the equation $(D_0^{\alpha,\beta}u)(t) = f(t,u(t))$ on I.

The following hypotheses will be used in the sequel.

(8.5.1) For a.e. $t \in I$, the function $v \to f(t,v)$ is weakly-sequentially continuous.

(8.5.2) For each $v \in [0,\infty)$, the function $t \to f(t,v)$ is Pettis integrable a.e. on I.

(8.5.3) There exists $p \in C(I, [0,\infty))$ such that for all $\varphi \in E^*$, we have

$$|\varphi(f(t,u))| \le \frac{p(t)u}{1 + u + \|\varphi\|}; \text{ for a.e. } t \in I, \text{ and each } u \in [0,\infty),$$

(8.5.4) For each bounded and measurable set $B \subset E$ and for each $t \in I$, we have

$$\beta(f(t,\|B\|)) \le t^{1-r}p(t)\beta(B),$$

where $\|B\| = \left\{ \max_{0\le\tau\le t} \|u(\tau)\|; \ u(t) \in B, \ t \in I \right\}.$

Set

$$p^* = \sup_{t\in I} p(t).$$

Theorem 8.6. *Assume that the hypotheses (8.5.1)–(8.5.4) hold. If*

$$L := \frac{p^* T^{1-\gamma+\alpha}}{\Gamma(1 + \alpha)} < 1, \tag{8.19}$$

then the problem (8.18) has at least one weak solution defined on I.

Proof. Consider the operator $N : C_\gamma \to C_\gamma$ defined by

$$(Nu)(t) = \frac{\phi}{\Gamma(\gamma)}t^{\gamma-1} + \frac{1}{\Gamma(\alpha)}\int_0^t (t-s)^{\alpha-1}f(s, \max_{0 \le \tau \le s}\|u(\tau)\|)ds. \quad (8.20)$$

First notice that, the hypotheses imply that for each $u \in C_\gamma$, the function $t \mapsto (t-s)^{\alpha-1}f(s, u(s))$ for a.e. $t \in I$ is Pettis integrable. Thus, the operator N is well defined. Let $R > 0$ be such that

$$R > \frac{p^* T^{1-\gamma+\alpha}}{\Gamma(1+\alpha)},$$

and consider the set

$$Q = \left\{ u \in C_\gamma : \|u\|_C \le R \text{ and } \|t_2^{1-\gamma}u(t_2) - t_1^{1-\gamma}u(t_1)\| \right.$$

$$\le \frac{p^* T^{1-\gamma+\alpha}}{\Gamma(1+\alpha)}(t_2 - t_1)^\alpha + \frac{p^*}{\Gamma(\alpha)}$$

$$\left. \times \int_0^{t_1} |t_2^{1-\gamma}(t_2-s)^{\alpha-1} - t_1^{1-\gamma}(t_1-s)^{\alpha-1}|ds \right\}.$$

Clearly, the subset Q is closed, convex and equicontinuous. We shall show that the operator N satisfies all the assumptions of Theorem 1.9. The proof will be given in several steps.

Step 1. *N maps Q into itself.*

Let $u \in Q$, $t \in I$ and assume that $(Nu)(t) \ne 0$. Then there exists $\varphi \in E^*$ such that $\|t^{1-\gamma}(Nu)(t)\| = |\varphi(t^{1-\gamma}(Nu)(t))|$. Thus

$$\|t^{1-\gamma}(Nu)(t)\| = \left| \varphi\left(\frac{\phi}{\Gamma(\gamma)} + \frac{t^{1-\gamma}}{\Gamma(\alpha)}\int_0^t (t-s)^{\alpha-1}f(s, \max_{0 \le \tau \le s}\|u(\tau)\|)ds \right) \right|.$$

Then

$$\|t^{1-\gamma}(Nu)(t)\| \le \frac{t^{1-\gamma}}{\Gamma(\alpha)}\int_0^t (t-s)^{\alpha-1}|\varphi(f(s, \max_{0 \le \tau \le s}\|u(\tau)\|))|ds$$

$$\le \frac{p^* T^{1-\gamma}}{\Gamma(\alpha)}\int_0^t (t-s)^{\alpha-1}ds$$

$$\le \frac{p^* T^{1-\gamma+\alpha}}{\Gamma(1+\alpha)}$$

$$\le R.$$

Next, let $t_1, t_2 \in I$ such that $t_1 < t_2$ and let $u \in Q$, with

$$t_2^{1-\gamma}(Nu)(t_2) - t_1^{1-\gamma}(Nu)(t_1) \neq 0.$$

Then there exists $\varphi \in E^*$ such that

$$\|t_2^{1-\gamma}(Nu)(t_2) - t_1^{1-\gamma}(Nu)(t_1)\| = |\varphi(t_2^{1-\gamma}(Nu)(t_2) - t_1^{1-\gamma}(Nu)(t_1))|,$$

and $\|\varphi\| = 1$. Then

$$\|t_2^{1-\gamma}(Nu)(t_2) - t_1^{1-\gamma}(Nu)(t_1)\| = |\varphi(t_2^{1-\gamma}(Nu)(t_2) - t_1^{1-\gamma}(Nu)(t_1))|$$

$$\leq \left| \varphi \left(t_2^{1-\gamma} \int_0^{t_2} (t_2 - s)^{\alpha-1} \frac{f(s, u(s))}{\Gamma(\alpha)} ds \right. \right.$$

$$\left. \left. -t_1^{1-\gamma} \int_0^{t_1} (t_1 - s)^{\alpha-1} \frac{f(s, \max_{0 \leq \tau \leq s} \|u(\tau)\|)}{\Gamma(\alpha)} ds \right) \right|.$$

Then

$$\|t_2^{1-\gamma}(Nu)(t_2) - t_1^{1-\gamma}(Nu)(t_1)\|$$

$$\leq t_2^{1-\gamma} \int_{t_1}^{t_2} (t_2 - s)^{\alpha-1} \frac{|\varphi(f(s, \max_{0 \leq \tau \leq s} \|u(\tau)\|))|}{\Gamma(\alpha)} ds$$

$$+ \int_0^{t_1} |t_2^{1-\gamma}(t_2 - s)^{\alpha-1} - t_1^{1-\gamma}(t_1 - s)^{\alpha-1}| \frac{|\varphi(f(s, \max_{0 \leq \tau \leq s} \|u(\tau)\|))|}{\Gamma(\alpha)} ds$$

$$\leq t_2^{1-\gamma} \int_{t_1}^{t_2} (t_2 - s)^{\alpha-1} \frac{p(s)}{\Gamma(\alpha)} ds$$

$$+ \int_0^{t_1} |t_2^{1-\gamma}(t_2 - s)^{\alpha-1} - t_1^{1-\gamma}(t_1 - s)^{\alpha-1}| \frac{p(s)}{\Gamma(\alpha)} ds.$$

Thus, we get

$$\|t_2^{1-\gamma}(Nu)(t_2) - t_1^{1-\gamma}(Nu)(t_1)\|$$

$$\leq \frac{p^* T^{1-\gamma+\alpha}}{\Gamma(1+\alpha)} (t_2 - t_1)^{\alpha} + \frac{p^*}{\Gamma(\alpha)}$$

$$\times \int_0^{t_1} |t_2^{1-\gamma}(t_2 - s)^{\alpha-1} - t_1^{1-\gamma}(t_1 - s)^{\alpha-1}| ds.$$

Hence $N(Q) \subset Q$.

Step 2. N *is weakly-sequentially continuous.*

Let (u_n) be a sequence in Q and let $(u_n(t)) \to u(t)$ in (E, ω) for each $t \in I$. Fix $t \in I$, since f satisfies the assumption (8.5.1), we have $f(t, \max_{0 \le \tau \le t} \|u_n(\tau)\|)$ converges weakly to $f(t, \max_{0 \le \tau \le t} \|u(\tau)\|)$. Hence the Lebesgue dominated convergence theorem for Pettis integral implies $(Nu_n)(t)$ converges weakly to $(Nu)(t)$ in (E, ω), for each $t \in I$. Thus, $N(u_n) \to N(u)$. Hence, $N : Q \to Q$ is weakly-sequentially continuous.

Step 3. *The implication (1.9) holds.*

Let V be a subset of Q such that $\overline{V} = \overline{\mathrm{conv}}(N(V) \cup \{0\})$. Obviously

$$V(t) \subset \overline{\mathrm{conv}}(NV)(t)) \cup \{0\}), \quad \forall t \in I.$$

Further, as V is bounded and equicontinuous, by [172, Lemma 3] the function $t \to v(t) = \beta(V(t))$ is continuous on I. From (8.5.3), (8.5.4), Lemma 2.3 and the properties of the measure β, for any $t \in I$, we have

$$
\begin{aligned}
t^{1-\gamma} v(t) &\le \beta(t^{1-\gamma}(NV)(t) \cup \{0\}) \\
&\le \beta(t^{1-\gamma}(NV)(t)) \\
&\le \frac{T^{1-\gamma}}{\Gamma(\alpha)} \int_0^t |t-s|^{\alpha-1} p(s) \beta(V(s)) ds \\
&\le \frac{T^{1-\gamma}}{\Gamma(\alpha)} \int_0^t |t-s|^{\alpha-1} s^{1-\gamma} p(s) v(s) ds \\
&\le \frac{p^* T^{1-\gamma+\alpha}}{\Gamma(1+\alpha)} \|v\|_C.
\end{aligned}
$$

Thus

$$\|v\|_C \le L \|v\|_C.$$

From (8.19), we get $\|v\|_C = 0$, that is $v(t) = \beta(V(t)) = 0$, for each $t \in I$, and then by [290, Theorem 2], V is weakly-relatively compact in C. Applying now Theorem 1.9, we conclude that N has a fixed point which is a weak solution of the problem (8.18). $\qquad\square$

8.5.3. *Example*

Let

$$E = l^1 = \left\{ u = (u_1, u_2, \ldots, u_n, \ldots), \sum_{n=1}^{\infty} |u_n| < \infty \right\}$$

be the Banach space with the norm

$$\|u\| = \sum_{n=1}^{\infty} |u_n|.$$

Consider the following problem of Hilfer fractional differential equation:

$$\begin{cases} (D_0^{\frac{1}{2}, \frac{1}{2}} u_n)(t) = f_n(t, \max_{0 \leq \tau \leq t} \|u(\tau)\|); & t \in [0, 1], \\ (I_0^{\frac{1}{4}} u)(t)|_{t=0} = (2^{-1}, 2^{-2}, \ldots, 2^{-n}, \ldots), \end{cases} \tag{8.21}$$

where

$$f_n(t, u(t)) = \frac{ct^2}{1 + \|u(t)\|} \frac{u_n(t)}{e^{t+4}}, \quad t \in [0, 1],$$

with

$$c := \frac{e^4}{8} \Gamma\left(\frac{1}{2}\right), \quad u = (u_1, u_2, \ldots, u_n, \ldots),$$

and

$$\max_{0 \leq \tau \leq t} \|u(\tau)\| = \left(\max_{0 \leq \tau \leq t} \|u_1(\tau)\|, \max_{0 \leq \tau \leq t} \|u_2(\tau)\|, \ldots, \max_{0 \leq \tau \leq t} \|u_n(\tau)\|, \ldots\right).$$

Set

$$f = (f_1, f_2, \ldots, f_n, \ldots).$$

Clearly, the function f is continuous.

For each $u \in E$ and $t \in [0, 1]$, we have

$$\left\| f(t, \max_{0 \leq \tau \leq t} \|u(\tau)\|) \right\| \leq ct^2 \frac{1}{e^{t+4}}.$$

Hence, the hypothesis (8.5.3) is satisfied with $p^* = ce^{-4}$. We shall show that condition (8.19) holds with $T = 1$. Indeed,

$$\frac{p^* T^{1-\gamma+\alpha}}{\Gamma(1+\alpha)} = \frac{2ce^{-4}}{\Gamma(\frac{1}{2})} = \frac{1}{4} < 1.$$

Simple computations show that all the conditions of Theorem 8.6 are satisfied. It follows that the problem (8.21) has at least one weak solution defined on $[0, 1]$.

8.6. Notes and Remarks

The results of Chapter 8 are taken from [1,39–44,57,60].

Chapter 9

Implicit Hilfer–Pettis Fractional Differential Equations

9.1. Introduction

In this chapter, we present some existence of weak solutions for some implicit fractional differential equations of Hilfer type, by applying Mönch's fixed-point theorem associated with the technique of measure of weak non-compactness.

9.2. Implicit Hilfer–Pettis Fractional Differential Equations

9.2.1. Introduction

In this section, we discuss the existence of weak solutions for the following problem of implicit Hilfer fractional differential equation:

$$\begin{cases} (D_0^{\alpha,\beta}u)(t) = f(t, u(t), (D_0^{\alpha,\beta}u)(t)); \quad t \in I := [0, T], \\ (I_0^{1-\gamma}u)(t)|_{t=0} = \phi, \end{cases} \tag{9.1}$$

where $\alpha \in (0, 1)$, $\beta \in [0, 1]$, $\gamma = \alpha + \beta - \alpha\beta$, $T > 0$, $\phi \in E$, $f : I \times E \times E \to E$ is a given continuous function, E is a real (or complex) Banach space with norm $\|\cdot\|_E$ and dual E^*, such that E is the dual of a weakly-compactly-generated Banach space X, $I_0^{1-\gamma}$ is the left-sided mixed Riemann–Liouville integral of order $1 - \gamma$, and $D_0^{\alpha,\beta}$ is the generalized Riemann–Liouville derivative operator of order α and type β, introduced by Hilfer in [238].

9.2.2. Existence of weak solutions

Definition 9.1. By a weak solution of the problem (9.1) we mean a measurable function $u \in C_\gamma$ that satisfies the condition $(I_0^{1-\gamma}u)(0^+) = \phi$, and the equation $(D_0^{\alpha,\beta}u)(t) = f(t, u(t), (D_0^{\alpha,\beta}u)(t))$ on I.

The following hypotheses will be used in the sequel.

(9.1.1) For a.e. $t \in I$, the functions $v \to f(t, v, w)$ and $w \to f(t, v, w)$ are weakly-sequentially continuous.

(9.1.2) For each $v, w \in E$, the function $t \to f(t, v, w)$ is Pettis integrable a.e. on I.

(9.1.3) There exists $p \in C(I, [0, \infty))$ such that for all $\varphi \in E^*$, we have

$$|\varphi(f(t, u, v))| \leq p(t)\|\varphi\| \quad \text{for a.e. } t \in I, \text{ and each } u, v \in E.$$

(9.1.4) For each bounded and measurable set $B \subset E$ and for each $t \in I$, we have

$$\beta(f(t, B, D_0^{\alpha, \beta} B)) \leq t^{1-r} p(t)\beta(B),$$

where $D_0^{\alpha, \beta} B = \{D_0^{\alpha, \beta} w : w \in B\}$.

Set

$$p^* = \sup_{t \in I} p(t).$$

Theorem 9.1. *Assume that the hypotheses (9.1.1)–(9.1.4) hold. If*

$$L := \frac{p^* T^{1-\gamma+\alpha}}{\Gamma(1+\alpha)} < 1, \tag{9.2}$$

then the problem (9.1) has at least one weak solution defined on I.

Proof. Consider the operator $N : C_\gamma \to C_\gamma$ defined by:

$$(Nu)(t) = \frac{\phi}{\Gamma(\gamma)} t^{\gamma-1} + (I_0^\alpha g)(t), \tag{9.3}$$

where $g \in C_\gamma$ such that

$$g(t) = f\left(t, \frac{\phi}{\Gamma(\gamma)} t^{\gamma-1} + (I_0^\alpha g)(t), g(t)\right).$$

First notice that, the hypotheses imply that for each $u, g \in C_\gamma$, the function

$$t \mapsto (t-s)^{\alpha-1} g(s)$$

is Pettis integrable over I, and

$$t \mapsto f\left(t, \frac{\phi}{\Gamma(\gamma)} t^{\gamma-1} + (I_0^\alpha g)(t), g(t)\right) \quad \text{for a.e. } t \in I,$$

is Pettis integrable. Thus, the operator N is well defined. Let $R > 0$ be such that

$$R > \frac{p^* T^{1-\gamma+\alpha}}{\Gamma(1+\alpha)},$$

and consider the set

$$Q = \left\{ u \in C_\gamma : \|u\|_C \leq R \text{ and } \|t_2^{1-\gamma} u(t_2) - t_1^{1-\gamma} u(t_1)\|_E \right.$$

$$\leq \frac{p^* T^{1-\gamma+\alpha}}{\Gamma(1+\alpha)} (t_2 - t_1)^\alpha + \frac{p^*}{\Gamma(\alpha)}$$

$$\left. \times \int_0^{t_1} |t_2^{1-\gamma}(t_2 - s)^{\alpha-1} - t_1^{1-\gamma}(t_1 - s)^{\alpha-1}| ds \right\}.$$

Clearly, the subset Q is closed, convex and equicontinuous. We shall show that the operator N satisfies all the assumptions of Theorem 1.7. The proof will be given in several steps.

Step 1. *N maps Q into itself.*

Let $u \in Q$, $t \in I$ and assume that $(Nu)(t) \neq 0$. Then there exists $\varphi \in E^*$ such that $\|t^{1-\gamma}(Nu)(t)\|_E = |\varphi(t^{1-\gamma}(Nu)(t))|$. Thus

$$\|t^{1-\gamma}(Nu)(t)\|_E = \left| \varphi \left(\frac{\phi}{\Gamma(\gamma)} + \frac{t^{1-\gamma}}{\Gamma(\alpha)} \int_0^t (t-s)^{\alpha-1} g(s) ds \right) \right|,$$

where $g \in C_\gamma$ such that

$$g(t) = f\left(t, \frac{\phi}{\Gamma(\gamma)} t^{\gamma-1} + (I_0^\alpha g)(t), g(t) \right).$$

Then

$$\|t^{1-\gamma}(Nu)(t)\|_E \leq \frac{t^{1-\gamma}}{\Gamma(\alpha)} \int_0^t (t-s)^{\alpha-1} |\varphi(g(s))| ds$$

$$\leq \frac{p^* T^{1-\gamma}}{\Gamma(\alpha)} \int_0^t (t-s)^{\alpha-1} ds$$

$$\leq \frac{p^* T^{1-\gamma+\alpha}}{\Gamma(1+\alpha)}$$

$$\leq R.$$

Next, let $t_1, t_2 \in I$ such that $t_1 < t_2$ and let $u \in Q$, with

$$t_2^{1-\gamma}(Nu)(t_2) - t_1^{1-\gamma}(Nu)(t_1) \neq 0.$$

Then there exists $\varphi \in E^*$ such that

$$\|t_2^{1-\gamma}(Nu)(t_2) - t_1^{1-\gamma}(Nu)(t_1)\|_E = |\varphi(t_2^{1-\gamma}(Nu)(t_2) - t_1^{1-\gamma}(Nu)(t_1))|,$$

and $\|\varphi\| = 1$. Then

$$\|t_2^{1-\gamma}(Nu)(t_2) - t_1^{1-\gamma}(Nu)(t_1)\|_E$$
$$= |\varphi(t_2^{1-\gamma}(Nu)(t_2) - t_1^{1-\gamma}(Nu)(t_1))|$$
$$\leq \left|\varphi\left(t_2^{1-\gamma}\int_0^{t_2}(t_2-s)^{\alpha-1}\frac{g(s)}{\Gamma(\alpha)}ds - t_1^{1-\gamma}\int_0^{t_1}(t_1-s)^{\alpha-1}\frac{g(s)}{\Gamma(\alpha)}ds\right)\right|,$$

where $g \in C_\gamma$ such that

$$g(t) = f\left(t, \frac{\phi}{\Gamma(\gamma)}t^{\gamma-1} + (I_0^\alpha g)(t), g(t)\right).$$

Then

$$\|t_2^{1-\gamma}(Nu)(t_2) - t_1^{1-\gamma}(Nu)(t_1)\|_E$$
$$\leq t_2^{1-\gamma}\int_{t_1}^{t_2}(t_2-s)^{\alpha-1}\frac{|\varphi(g(s))|}{\Gamma(\alpha)}ds$$
$$+ \int_0^{t_1}|t_2^{1-\gamma}(t_2-s)^{\alpha-1} - t_1^{1-\gamma}(t_1-s)^{\alpha-1}|\frac{|\varphi(g(s))|}{\Gamma(\alpha)}ds$$
$$\leq t_2^{1-\gamma}\int_{t_1}^{t_2}(t_2-s)^{\alpha-1}\frac{p(s)}{\Gamma(\alpha)}ds$$
$$+ \int_0^{t_1}|t_2^{1-\gamma}(t_2-s)^{\alpha-1} - t_1^{1-\gamma}(t_1-s)^{\alpha-1}|\frac{p(s)}{\Gamma(\alpha)}ds.$$

Thus, we get

$$\|t_2^{1-\gamma}(Nu)(t_2) - t_1^{1-\gamma}(Nu)(t_1)\|_E$$
$$\leq \frac{p^*T^{1-\gamma+\alpha}}{\Gamma(1+\alpha)}(t_2-t_1)^\alpha$$
$$+ \frac{p^*}{\Gamma(\alpha)}\int_0^{t_1}|t_2^{1-\gamma}(t_2-s)^{\alpha-1} - t_1^{1-\gamma}(t_1-s)^{\alpha-1}|ds.$$

Hence $N(Q) \subset Q$.

Step 2. *N is weakly-sequentially continuous.*

Let (u_n) be a sequence in Q and let $(u_n(t)) \to u(t)$ in (E, ω) for each $t \in I$. Fix $t \in I$, since f satisfies the assumption (9.1.1), we have $f(t, u_n(t), (D_0^{\alpha,\beta}u_n)(t))$ converges weakly to $f(t, u(t), (D_0^{\alpha,\beta}u)(t))$. Hence the Lebesgue dominated convergence theorem for Pettis integral implies $(Nu_n)(t)$ converges weakly to $(Nu)(t)$ in (E,ω), for each $t \in I$. Thus, $N(u_n) \to N(u)$. Hence, $N : Q \to Q$ is weakly-sequentially continuous.

Step 3. *The implication (1.9) holds.*

Let V be a subset of Q such that $\overline{V} = \overline{\mathrm{conv}}(N(V) \cup \{0\})$. Obviously

$$V(t) \subset \overline{\mathrm{conv}}(NV)(t)) \cup \{0\}), \ \forall t \in I.$$

Further, as V is bounded and equicontinuous, by [172, Lemma 3] the function $t \to v(t) = \beta(V(t))$ is continuous on I. From (9.1.3), (9.1.4), Lemma 1.7 and the properties of the measure β, for any $t \in I$, we have

$$t^{1-\gamma}v(t) \le \beta(t^{1-\gamma}(NV)(t) \cup \{0\})$$
$$\le \beta(t^{1-\gamma}(NV)(t))$$
$$\le \frac{T^{1-\gamma}}{\Gamma(\alpha)}\int_0^t |t-s|^{\alpha-1}p(s)\beta(V(s))ds$$
$$\le \frac{T^{1-\gamma}}{\Gamma(\alpha)}\int_0^t |t-s|^{\alpha-1}s^{1-\gamma}p(s)v(s)ds$$
$$\le \frac{p^*T^{1-\gamma+\alpha}}{\Gamma(1+\alpha)}\|v\|_C.$$

Thus

$$\|v\|_C \le L\|v\|_C.$$

From (9.2), we get $\|v\|_C = 0$, that is $v(t) = \beta(V(t)) = 0$, for each $t \in I$, and then by [290, Theorem 2], V is weakly-relatively compact in C. Applying now Theorem 1.7, we conclude that N has a fixed point which is a weak solution of the problem (9.1). □

9.2.3. *Example*

Let

$$E = l^1 = \left\{ u = (u_1, u_2, \ldots, u_n, \ldots), \sum_{n=1}^{\infty} |u_n| < \infty \right\}$$

be the Banach space with the norm

$$\|u\|_E = \sum_{n=1}^{\infty} |u_n|.$$

Consider the following problem of Hilfer fractional differential equation:

$$\begin{cases} (D_0^{\frac{1}{2},\frac{1}{2}} u_n)(t) = f_n(t, u(t), (D_0^{\frac{1}{2},\frac{1}{2}} u_n)(t)), & t \in [1, e], \\ (I_0^{\frac{1}{4}} u)(t)|_{t=0} = (0, 0, \dots, 0, \dots), \end{cases} \quad (9.4)$$

where

$$f_n(t, u(t), v(t)) = \frac{ct^2}{1 + \|u(t)\|_E + \|v(t)\|_E} \frac{u_n(t)}{e^{t+4}}, \quad t \in [0, 1],$$

with

$$u = (u_1, u_2, \dots, u_n, \dots) \quad \text{and} \quad c := \frac{e^4}{8} \Gamma\left(\frac{1}{2}\right).$$

Set

$$f = (f_1, f_2, \dots, f_n, \dots).$$

Clearly, the function f is continuous.

For each $u, v \in E$ and $t \in [0, 1]$, we have

$$\|f(t, u(t), v(t))\|_E \le ct^2 \frac{1}{e^{t+4}}.$$

Hence, the hypothesis (9.1.3) is satisfied with $p^* = ce^{-4}$. We shall show that condition (9.2) holds with $T = 1$. Indeed,

$$\frac{p^* T^{1-\gamma+\alpha}}{\Gamma(1+\alpha)} = \frac{2ce^{-4}}{\Gamma(\frac{1}{2})} = \frac{1}{4} < 1.$$

Simple computations show that all the conditions of Theorem 9.1 are satisfied. It follows that the problem (9.4) has at least one weak solution defined on $[0, 1]$.

9.3. Implicit Hilfer–Hadamard–Pettis Fractional Differential Equations

9.3.1. *Introduction*

In this section, we discuss the existence of weak solutions for the following problem of implicit Hilfer–Hadamard fractional differential equation:

$$\begin{cases} (^H D_1^{\alpha,\beta} u)(t) = f(t, u(t), (^H D_1^{\alpha,\beta} u)(t)); & t \in I := [1, T], \\ (^H I_1^{1-\gamma} u)(t)|_{t=1} = \phi, \end{cases} \quad (9.5)$$

where $\alpha \in (0,1)$, $\beta \in [0,1]$, $\gamma = \alpha + \beta - \alpha\beta$, $T > 0$, $\phi \in E$, $f : I \times E \times E \to E$ is a given continuous function, E is a real (or complex) Banach space with norm $\|\cdot\|_E$ and dual E^*, such that E is the dual of a weakly-compactly-generated Banach space X, $^H I_1^{1-\gamma}$ is the left-sided mixed Hadamard integral of order $1 - \gamma$, and $^H D_1^{\alpha,\beta}$ is the Hilfer–Hadamard fractional derivative of order α and type β.

9.3.2. *Existence of weak solutions*

Definition 9.2. By a weak solution of the problem (9.5) we mean a measurable function $u \in C_{\gamma,\ln}$ that satisfies the condition $(^H I_1^{1-\gamma} u)(1^+) = \phi$, and the equation $(^H D_1^{\alpha,\beta} u)(t) = f(t, u(t), (^H D_1^{\alpha,\beta} u)(t))$ on I.

The following hypotheses will be used in the sequel.

(9.2.1) For a.e. $t \in I$, the functions $v \to f(t,v,w)$ and $w \to f(t,v,w)$ are weakly-sequentially continuous.
(9.2.2) For each $v, w \in E$, the function $t \to f(t,v,w)$ is Pettis integrable a.e. on I.
(9.2.3) There exists $p \in C(I, [0, \infty))$ such that for all $\varphi \in E^*$, we have

$$|\varphi(f(t,u,v))| \leq p(t)\|u\|_E \quad \text{for a.e. } t \in I, \text{ and each } u, v \in E.$$

(9.2.4) For each bounded and measurable set $B \subset E$ and for each $t \in I$, we have

$$\beta(f(t, B, ^H D_1^{\alpha,\beta} B)) \leq (\ln t)^{1-\gamma} p(t)\beta(B),$$

where $^H D_1^{\alpha,\beta} B = \{^H D_1^{\alpha,\beta} w : w \in B\}$.

Set

$$p^* = \sup_{t \in I} p(t).$$

Theorem 9.2. *Assume that the hypotheses (9.2.1)–(9.2.4) hold. If*

$$L := \frac{p^* (\ln T)^{1-\gamma+\alpha}}{\Gamma(1+\alpha)} < 1, \tag{9.6}$$

then the problem (9.5) has at least one weak solution defined on I.

Proof. Consider the operator $N : C_{\gamma,\ln} \to C_{\gamma,\ln}$ defined by

$$(Nu)(t) = \frac{\phi}{\Gamma(\gamma)}(\ln t)^{\gamma-1} + ({}^H I_1^\alpha g)(t), \qquad (9.7)$$

where $g \in C_{\gamma,\ln}$ such that

$$g(t) = f\left(t, \frac{\phi}{\Gamma(\gamma)}(\ln t)^{\gamma-1} + ({}^H I_1^\alpha g)(t), g(t)\right).$$

First notice that, the hypotheses imply that for each $u, g \in C_{\gamma,\ln}$, the function

$$t \mapsto \left(\ln \frac{t}{s}\right)^{\alpha-1} g(s)$$

is Pettis integrable over I, and

$$t \mapsto f\left(t, \frac{\phi}{\Gamma(\gamma)}(\ln t)^{\gamma-1} + ({}^H I_1^\alpha g)(t), g(t)\right) \quad \text{for a.e. } t \in I,$$

is Pettis integrable. Thus, the operator N is well defined. Let $R > 0$ be such that

$$R > \frac{p^*(\ln T)^{1-\gamma+\alpha}}{\Gamma(1+\alpha)},$$

and consider the set

$$Q = \left\{ u \in C_{\gamma,\ln} : \|u\|_C \le R \text{ and } \|(\ln t_2)^{1-\gamma}u(t_2) - (\ln t_1)^{1-\gamma}u(t_1)\|_E \right.$$

$$\le \frac{p^*(\ln T)^{1-\gamma+\alpha}}{\Gamma(1+\alpha)}\left(\ln \frac{t_2}{t_1}\right)^\alpha$$

$$\left. + \frac{p^*}{\Gamma(\alpha)}\int_1^{t_1}\left|(\ln t_2)^{1-\gamma}\left(\ln \frac{t_2}{s}\right)^{\alpha-1} - (\ln t_1)^{1-\gamma}\left(\ln \frac{t_1}{s}\right)^{\alpha-1}\right|ds \right\}.$$

Clearly, the subset Q is closed, convex and equicontinuous. We shall show that the operator N satisfies all the assumptions of Theorem 1.7. The proof will be given in several steps.

Step 1. N *maps Q into itself.*

Let $u \in Q$, $t \in I$ and assume that $(Nu)(t) \ne 0$. Then there exists $\varphi \in E^*$ such that $\|(\ln t)^{1-\gamma}(Nu)(t)\|_E = |\varphi((\ln t)^{1-\gamma}(Nu)(t))|$. Thus

$$\|(\ln t)^{1-\gamma}(Nu)(t)\|_E = \left|\varphi\left(\frac{\phi}{\Gamma(\gamma)} + \frac{(\ln t)^{1-\gamma}}{\Gamma(\alpha)}\int_1^t \left(\ln \frac{t}{s}\right)^{\alpha-1} g(s)\frac{ds}{s}\right)\right|,$$

where $g \in C_{\gamma, \ln}$ such that

$$g(t) = f\left(t, \frac{\phi}{\Gamma(\gamma)}(\ln t)^{\gamma-1} + ({}^H I_1^\alpha g)(t), g(t)\right).$$

Then

$$\|(\ln t)^{1-\gamma}(Nu)(t)\|_E \leq \frac{(\ln t)^{1-\gamma}}{\Gamma(\alpha)} \int_1^t \left(\ln \frac{t}{s}\right)^{\alpha-1} |\varphi(g(s))| \frac{ds}{s}$$

$$\leq \frac{p^*(\ln T)^{1-\gamma}}{\Gamma(\alpha)} \int_1^t \left(\ln \frac{t}{s}\right)^{\alpha-1} \frac{ds}{s}$$

$$\leq \frac{p^*(\ln T)^{1-\gamma+\alpha}}{\Gamma(1+\alpha)}$$

$$\leq R.$$

Next, let $t_1, t_2 \in I$ such that $t_1 < t_2$ and let $u \in Q$, with

$$(\ln t_2)^{1-\gamma}(Nu)(t_2) - (\ln t_1)^{1-\gamma}(Nu)(t_1) \neq 0.$$

Then there exists $\varphi \in E^*$ such that

$$\|(\ln t_2)^{1-\gamma}(Nu)(t_2) - (\ln t_1)^{1-\gamma}(Nu)(t_1)\|_E$$
$$= |\varphi((\ln t_2)^{1-\gamma}(Nu)(t_2) - (\ln t_1)^{1-\gamma}(Nu)(t_1))|,$$

and $\|\varphi\| = 1$. Then

$$\|(\ln t_2)^{1-\gamma}(Nu)(t_2) - (\ln t_1)^{1-\gamma}(Nu)(t_1)\|_E$$
$$= |\varphi((\ln t_2)^{1-\gamma}(Nu)(t_2) - (\ln t_1)^{1-\gamma}(Nu)(t_1))|$$
$$\leq \left|\varphi\left((\ln t_2)^{1-\gamma}\int_1^{t_2}\left(\ln \frac{t_2}{s}\right)^{\alpha-1}\frac{g(s)}{s\Gamma(\alpha)}ds\right.\right.$$
$$\left.\left. -(\ln t_1)^{1-\gamma}\int_1^{t_1}\left(\ln \frac{t_1}{s}\right)^{\alpha-1}\frac{g(s)}{s\Gamma(\alpha)}ds\right)\right|,$$

where $g \in C_{\gamma, \ln}$ such that

$$g(t) = f\left(t, \frac{\phi}{\Gamma(\gamma)}(\ln t)^{\gamma-1} + ({}^H I_1^\alpha g)(t), g(t)\right).$$

Then

$$\|(\ln t_2)^{1-\gamma}(Nu)(t_2) - (\ln t_1)^{1-\gamma}(Nu)(t_1)\|_E$$

$$\leq (\ln t_2)^{1-\gamma} \int_{t_1}^{t_2} \left(\ln \frac{t_2}{s}\right)^{\alpha-1} \frac{|\varphi(g(s))|}{s\Gamma(\alpha)} ds$$

$$+ \int_1^{t_1} \left|(\ln t_2)^{1-\gamma}\left(\ln \frac{t_2}{s}\right)^{\alpha-1} - (\ln t_1)^{1-\gamma}\left(\ln \frac{t_1}{s}\right)^{\alpha-1}\right| \frac{|\varphi(g(s))|}{s\Gamma(\alpha)} ds$$

$$\leq (\ln t_2)^{1-\gamma} \int_{t_1}^{t_2} \left(\ln \frac{t_2}{s}\right)^{\alpha-1} \frac{p(s)}{s\Gamma(\alpha)} ds$$

$$+ \int_1^{t_1} \left|(\ln t_2)^{1-\gamma}\left(\ln \frac{t_2}{s}\right)^{\alpha-1} - (\ln t_1)^{1-\gamma}\left(\ln \frac{t_1}{s}\right)^{\alpha-1}\right| \frac{p(s)}{s\Gamma(\alpha)} ds.$$

Thus, we get

$$\|(\ln t_2)^{1-\gamma}(Nu)(t_2) - (\ln t_1)^{1-\gamma}(Nu)(t_1)\|_E$$

$$\leq \frac{p^*(\ln T)^{1-\gamma+\alpha}}{\Gamma(1+\alpha)} \left(\ln \frac{t_2}{t_1}\right)^{\alpha}$$

$$+ \frac{p^*}{\Gamma(\alpha)} \int_1^{t_1} \left|(\ln t_2)^{1-\gamma}\left(\ln \frac{t_2}{s}\right)^{\alpha-1} - (\ln t_1)^{1-\gamma}\left(\ln \frac{t_1}{s}\right)^{\alpha-1}\right| ds.$$

Hence $N(Q) \subset Q$.

Step 2. *N is weakly-sequentially continuous.*

Let (u_n) be a sequence in Q and let $(u_n(t)) \to u(t)$ in (E, ω) for each $t \in I$. Fix $t \in I$, since f satisfies the assumption (9.2.1), we have $f(t, u_n(t), (^H D_1^{\alpha,\beta}, u_n)(t))$ converges weakly to $f(t, u(t), (D_0^{\alpha,\beta}u)(t))$. Hence the Lebesgue dominated convergence theorem for Pettis integral implies $(Nu_n)(t)$ converges weakly to $(Nu)(t)$ in (E, ω), for each $t \in I$. Thus, $N(u_n) \to N(u)$. Hence, $N : Q \to Q$ is weakly-sequentially continuous.

Step 3. *The implication (1.9) holds.*

Let V be a subset of Q such that $\overline{V} = \overline{\text{conv}}(N(V) \cup \{0\})$. Obviously

$$V(t) \subset \overline{\text{conv}}(NV)(t)) \cup \{0\}), \quad \forall t \in I.$$

Further, as V is bounded and equicontinuous, by [172, Lemma 3] the function $t \to v(t) = \beta(V(t))$ is continuous on I. From (9.2.3), (9.2.4),

Lemma 1.7 and the properties of the measure β, for any $t \in I$, we have

$$(\ln t)^{1-\gamma} v(t) \leq \beta((\ln t)^{1-\gamma}(NV)(t) \cup \{0\})$$

$$\leq \beta((\ln t)^{1-\gamma}(NV)(t))$$

$$\leq \frac{(\ln T)^{1-\gamma}}{\Gamma(\alpha)} \int_1^t \left(\ln \frac{t}{s}\right)^{\alpha-1} p(s)\beta(V(s))\frac{ds}{s}$$

$$\leq \frac{(\ln T)^{1-\gamma}}{\Gamma(\alpha)} \int_1^t \left(\ln \frac{t}{s}\right)^{\alpha-1} (\ln s)^{1-\gamma} p(s)v(s)\frac{ds}{s}$$

$$\leq \frac{p^*(\ln T)^{1-\gamma+\alpha}}{\Gamma(1+\alpha)} \|v\|_C.$$

Thus

$$\|v\|_C \leq L\|v\|_C.$$

From (9.6), we get $\|v\|_C = 0$, that is $v(t) = \beta(V(t)) = 0$, for each $t \in I$, and then by [290, Theorem 2], V is weakly-relatively compact in C. Applying now Theorem 1.7, we conclude that N has a fixed point which is a weak solution of the problem (9.5). □

9.3.3. *Example*

Let

$$E = l^1 = \left\{ u = (u_1, u_2, \ldots, u_n, \ldots), \sum_{n=1}^{\infty} |u_n| < \infty \right\}$$

be the Banach space with the norm

$$\|u\|_E = \sum_{n=1}^{\infty} |u_n|.$$

Consider the following problem of Hilfer–Hadamard fractional differential equation

$$\begin{cases} ({}^H D_1^{\frac{1}{2},\frac{1}{2}} u_n)(t) = f_n(t, u(t), ({}^H D_1^{\frac{1}{2},\frac{1}{2}} u_n)(t)); & t \in [1, e], \\ ({}^H I_1^{\frac{1}{4}} u)(t)|_{t=1} = (0, 0, \ldots, 0, \ldots), \end{cases} \tag{9.8}$$

where

$$f_n(t, u(t), v(t)) = \frac{ct^2}{1 + \|u(t)\|_E + \|v(t)\|_E} \frac{u_n(t)}{e^{t+4}}, \quad t \in [1, e],$$

with

$$u = (u_1, u_2, \ldots, u_n, \ldots), \quad \text{and} \quad c := \frac{e^3}{8}\sqrt{\pi}.$$

Set

$$f = (f_1, f_2, \ldots, f_n, \ldots).$$

Clearly, the function f is continuous.

For each $u, v \in E$ and $t \in [1, e]$, we have

$$\|f(t, u(t), v(t))\|_E \leq ct^2 \frac{1}{e^{t+4}}.$$

Hence, the hypothesis (9.2.3) is satisfied with $p^* = ce^{-3}$. We shall show that condition (9.6) holds with $T = e$. Indeed,

$$\frac{p^*(\ln T)^{1-\gamma+\alpha}}{\Gamma(1+\alpha)} = \frac{2ce^{-3}}{\sqrt{\pi}} = \frac{1}{4} < 1.$$

Simple computations show that all the conditions of Theorem 9.2 are satisfied. It follows that the problem (9.8) has at least one weak solution defined on $[1, e]$.

9.4. Implicit Hilfer–Pettis Fractional Differential Equations with Not Instantaneous Impulses

9.4.1. *Introduction*

In this section, we discuss the existence of weak solutions for the following problem of implicit Hilfer fractional differential equation with not instantaneous impulses

$$\begin{cases} (D_{s_k}^{\alpha,\beta}u)(t) = f(t, u(t), D_{s_k}^{\alpha,\beta}u(t)); & \text{if } t \in I_k, \ k = 0, \ldots, m, \\ u(t) = g_k(t, u(t_k^-)); & \text{if } t \in J_k, \ k = 1, \ldots, m, \\ (I_1^{1-\gamma}u)(t)|_{t=0} = \phi_0, \end{cases} \quad (9.9)$$

where $I_0 := [0, t_1]$, $J_k := (t_k, s_k]$, $I_k := (s_k, t_{k+1}]$; $k = 1, \ldots, m$, $\alpha \in (0, 1)$, $\beta \in [0, 1]$, $\gamma = \alpha + \beta - \alpha\beta$, $T > 0$, $\phi_0 \in E$, $f : I_k \times E \times E \to E$, $g_k : J_k \times E \to E$ are given continuous functions such that $(I_{s_k}^{1-\gamma}g_k)(t, u(t_k^-))|_{t=s_k} = \phi_k \in E$; $k = 1, \ldots, m$, E is a real (or complex) Banach space with norm $\|\cdot\|_E$ and dual E^*, such that E is the dual of a weakly-compactly-generated Banach space X, $I_{s_k}^{1-\gamma}$ is the left-sided mixed Riemann–Liouville integral of order $1 - \gamma \in (0, 1]$, and $D_{s_k}^{\alpha,\beta}$ is the generalized Riemann–Liouville derivative operator of order α and type β, introduced by Hilfer in [238], $0 = s_0 < t_1 \le s_1 < t_2 \le s_2 < \cdots \le s_{m-1} < t_m \le s_m < t_{m+1} = T$.

9.4.2. *Existence of weak solutions*

Let $\mathcal{C}$ be the Banach space of all continuous functions v from $I := [0, T]$ into E with the supremum (uniform) norm

$$\|v\|_\infty := \sup_{t \in I} \|v(t)\|_E.$$

Denote by

$$\mathcal{PC} = \{u : I \to E : u \in \mathcal{C}(I_0 \cup_{k=1}^m (t_k, t_{k+1})), \ u(t_k^-) = u(t_k)\},$$

the Banach space equipped with the standard supremum norm.

As usual, $AC(I)$ denotes the space of absolutely continuous functions from I into E. We denote by $AC^1(I)$ the space defined by

$$AC^1(I) := \left\{ w : I \to E : \frac{d}{dt} w(t) \in AC(I) \right\}.$$

By $C_\gamma(I)$, $C_\gamma^1(I)$, $\mathcal{PC}_\gamma(I)$, and $\mathcal{PC}_\gamma^1(I)$, we denote the weighted spaces of continuous functions defined by

$$C_\gamma(I) = \{w : (0, T] \to E : t^{1-\gamma} w(t) \in \mathcal{C}\},$$

with the norm

$$\|w\|_{C_\gamma} := \sup_{t \in I} \|t^{1-\gamma} w(t)\|_E,$$

and

$$C_\gamma^1(I) = \left\{ w \in C : \frac{dw}{dt} \in C_\gamma \right\},$$

with the norm

$$\|w\|_{C_\gamma^1} := \|w\|_\infty + \|w'\|_{C_\gamma}.$$

$$\mathcal{PC}_\gamma(I) = \{w : (0, T] \to E : t^{1-\gamma} w(t) \in \mathcal{PC}\},$$

with the norm

$$\|w\|_{\mathcal{PC}_\gamma} := \sup_{t \in I} \|t^{1-\gamma} w(t)\|_E,$$

and

$$\mathcal{PC}_\gamma^1(I) = \left\{ w \in \mathcal{PC} : \frac{dw}{dt} \in \mathcal{PC}_\gamma \right\},$$

with the norm

$$\|w\|_{\mathcal{PC}^1_\gamma} := \|w\|_\infty + \|w'\|_{\mathcal{PC}_\gamma}.$$

In what follows, we denote $\|w\|_{\mathcal{PC}_\gamma}$ by $\|w\|_{\mathcal{PC}}$.

Definition 9.3. By a weak solution of the problem (9.9) we mean a measurable function $u \in \mathcal{PC}_\gamma$ that satisfies the condition $(I_0^{1-\gamma}u)(t)|_{t=0} = \phi_0$, and the equations $(D_{s_k}^{\alpha,\beta}u)(t) = f(t, u(t), (D_{s_k}^{\alpha,\beta}u)(t))$ on I_k; $k = 0, \ldots, m$, and $u(t) = g_k(t, u(t_k^-))$ on J_k; $k = 1, \ldots, m$.

The following hypotheses will be used in the sequel.

(9.3.1) The function $f : I_k \times E \times E \to E$; $k = 0, \ldots, m$ is weakly Carathéodory.
(9.3.2) There exist $p_k \in C(I_k, [0, \infty))$; $k = 0, \ldots, m$, such that for all $\varphi \in E^*$, we have

$$|\varphi(f(t, u, v))| \le p_k(t)\|\varphi\| \quad \text{for a.e. } t \in I_k, \text{ and each } u, v \in E.$$

and for each bounded and measurable set $B \subset E$, we have

$$\beta(f(t, B, D_{s_k}^{\alpha,\beta}B)) \le t^{1-\gamma}p_k(t)\beta(B) \quad \text{for each } t \in I_k; \ k = 0, \ldots, m,$$

where $D_{s_k}^{\alpha,\beta}B = \{D_{s_k}^{\alpha,\beta}w : w \in B\}$.
(9.3.3) There exist $q_k \in C(J_k, [0, \infty))$; $k = 1, \ldots, m$, such that for all $\varphi \in E^*$, we have

$$|\varphi(g_k(t, u_{t_k^-}))| \le q_k(t)\|\varphi\| \quad \text{for a.e. } t \in J_k; \ k = 1, \ldots, m.$$

Set

$$p^* = \max_{k=0,\ldots,m} \sup_{t \in I_k} p_k(t), \quad q^* = \max_{k=1,\ldots,m} \sup_{t \in J_k} q_k(t).$$

Theorem 9.3. *Assume that the hypotheses (9.3.1)–(9.6.3) hold. If*

$$L := \frac{p^* T^{1-\gamma+\alpha}}{\Gamma(1+\alpha)} < 1, \tag{9.10}$$

then the problem (9.9) has at least one weak solution defined on I.

Proof. Transform the problem (9.9) into a fixed-point equation. Consider the operator $N : \mathcal{PC}_\gamma \to \mathcal{PC}_\gamma$ defined by

$$(Nu)(t) = \begin{cases} \dfrac{\phi_k}{\Gamma(\gamma)} t^{\gamma-1} + \displaystyle\int_{s_k}^t (t-s)^{\alpha-1} \dfrac{h(s)}{\Gamma(\alpha)} \, ds; & \text{if } t \in I_k, \ k = 0, \ldots, m, \\[2mm] g_k(t, u(t_k^-)); & \text{if } t \in J_k, \ k = 1, \ldots, m, \end{cases} \tag{9.11}$$

where $h \in C_\gamma(I_k, E)$; $k = 0, \ldots, m$, with

$$h(t) = f\left(t, \frac{\phi_k}{\Gamma(\gamma)} t^{\gamma-1} + (I_{s_k}^\alpha h)(t), h(t)\right).$$

First notice that, the hypotheses imply that $(t - s^{\alpha-1} \frac{h(s)}{s}$ for all $t \in I_k$ is Pettis integrable, and for each $u \in \mathcal{PC}_\gamma$, the function

$$t \mapsto f\left(t, \frac{\phi_k}{\Gamma(\gamma)} t^{\gamma-1} + \int_{s_k}^t (t-s)^{\alpha-1} \frac{h(s)}{\Gamma(\alpha)} \, ds, h(t)\right)$$

is Pettis integrable over I_k. Thus, the operator N is well defined.

Let $R > 0$ be such that

$$R \geq \max\left\{\frac{p^* T^{1-\gamma+\alpha}}{\Gamma(1+\alpha)}, q^* T^{1-\gamma}\right\},$$

and consider the set

$$Q = \Bigg\{ u \in \mathcal{PC}_\gamma : \|u\|_{\mathcal{PC}} \leq R \text{ and } \|t_2^{1-\gamma} u(t_2) - t_1^{1-\gamma} u(t_1)\|_E$$

$$\leq \frac{p^*}{\Gamma(1+\alpha)} T^{1-\gamma} |t_2 - t_1|^\alpha + \frac{p^*}{\Gamma(\alpha)}$$

$$\times \int_1^{t_1} |t_2^{1-\gamma}(t_2 - s)^{\alpha-1} - t_1^{1-\gamma}(t_1 - s)^{\alpha-1}| \, ds$$

$$\text{and } \|t_2^{1-\gamma} u(t_2) - t_1^{1-\gamma} u(t_1)\|_E$$

$$\leq \|t_2^{1-\gamma} g_k(t_2, u(t_k^-)) - t_1^{1-\gamma} g_k(t_1, u(t_k^-))\|_E; \text{ on } J_k, \ k = 1, \ldots, m \Bigg\}.$$

Clearly, the subset Q is closed, convex and equicontinuous. We shall show that the operator N satisfies all the assumptions of Theorem 1.7. The proof will be given in several steps.

Step 1. *N maps Q into itself.*

Let $u \in Q$, $t \in I$ and assume that $(Nu)(t) \neq 0$. Then there exists $\varphi \in E^*$ such that $\|t^{1-\gamma}(Nu)(t)\|_E = |\varphi(t^{1-\gamma}(Nu)(t))|$. Thus, for each $t \in I_k$, $k = 0, \ldots, m$, we have

$$\|t^{1-\gamma}(Nu)(t)\|_E = \left|\varphi\left(\frac{\phi_k}{\Gamma(\gamma)} + \frac{t^{1-\gamma}}{\Gamma(\alpha)}\int_0^t (t-s)^{\alpha-1}h(s)ds\right)\right|,$$

where $h \in C_\gamma(I_k)$; $k = 0, \ldots, m$, with

$$h(t) = f\left(t, \frac{\phi_k}{\Gamma(\gamma)}t^{\gamma-1} + (I_0^\alpha h)(t), h(t)\right).$$

Then

$$\|t^{1-\gamma}(Nu)(t)\|_E \leq \frac{t^{1-\gamma}}{\Gamma(\alpha)}\int_0^t (t-s)^{\alpha-1}|\varphi(h(s))|ds$$

$$\leq \frac{p^* T^{1-\gamma}}{\Gamma(\alpha)}\int_0^t (t-s)^{\alpha-1}ds$$

$$\leq \frac{p^* T^{1-\gamma+\alpha}}{\Gamma(1+\alpha)}$$

$$\leq R.$$

Also, for each $t \in J_k$; $k = 1, \ldots, m$, it is clear that

$$\|t^{1-\gamma}(Nu)(t)\|_E \leq q^* T^{1-\gamma} \leq R.$$

Hence,

$$\|N(u)\|_{\mathcal{PC}} \leq R.$$

Next, let $t_1, t_2 \in I_k$; $k = 0, \ldots, m$, such that $t_1 < t_2$ and let $u \in Q$, with

$$(\ln t_2)^{1-r}(Nu)(t_2) - (\ln t_1)^{1-r}(Nu)(t_1) \neq 0.$$

Then there exists $\varphi \in E^*$ such that

$$\|t_2^{1-\gamma}(Nu)(t_2) - t_1^{1-\gamma}(Nu)(t_1)\|_E = |\varphi(t_2^{1-\gamma}(Nu)(t_2) - t_1^{1-\gamma}(Nu)(t_1))|,$$

and $\|\varphi\| = 1$. Then

$$\|t_2^{1-\gamma}(Nu)(t_2) - t_1^{1-\gamma}(Nu)(t_1)\|_E$$

$$= \varphi(t_2^{1-\gamma}(Nu)(t_2) - t_1^{1-\gamma}(Nu)(t_1)),$$

$$\leq \left|\varphi\left(t_2^{1-\gamma}\int_0^{t_2}(t_2-s)^{\alpha-1}\frac{h(s)}{\Gamma(\alpha)}ds - t_1^{1-\gamma}\int_0^{t_1}(t_1-)^{\alpha-1}\frac{h(s)}{\Gamma(\alpha)}ds\right)\right|,$$

where $h \in C_\gamma(I_k)$ with

$$h(t) = f\left(t, \frac{\phi_k}{\Gamma(\gamma)}t^{\gamma-1} + (I_0^\alpha h)(t), h(t)\right).$$

Then

$$\|t_2^{1-\gamma}(Nu)(t_2) - t_1^{1-\gamma}(Nu)(t_1)\|_E$$

$$\leq t_2^{1-\gamma}\int_{t_1}^{t_2}|t_2 - s|^{\alpha-1}\frac{|\varphi(h(s))|}{\Gamma(\alpha)}ds$$

$$+ \int_1^{t_1}|t_2^{1-\gamma}(t_2 - s)^{\alpha-1} - t_1^{1-\gamma}(t_1 - s)^{\alpha-1}|\frac{|\varphi(h(s))|}{\Gamma(\alpha)}ds$$

$$\leq t_2^{1-\gamma}\int_{t_1}^{t_2}|t_2 - s|^{\alpha-1}\frac{p(s)}{\Gamma(\alpha)}ds$$

$$+ \int_1^{t_1}|t_2^{1-\gamma}(t_2 - s)^{\alpha-1} - t_1^{1-\gamma}(t_1 - s)^{\alpha-1}|\frac{p(s)}{\Gamma(\alpha)}ds.$$

Thus, we get

$$\|t_2^{1-\gamma}(Nu)(t_2) - t_1^{1-\gamma}(Nu)(t_1)\|_E$$

$$\leq \frac{p^*}{\Gamma(1+\alpha)}T^{1-\gamma}|t_2 - t_1|^\alpha$$

$$+ \frac{p^*}{\Gamma(\alpha)}\int_1^{t_1}|t_2^{1-\gamma}(t_2 - s)^{\alpha-1} - t_1^{1-\gamma}(t_1 - s)^{\alpha-1}|ds.$$

Also, for $t_1, t_2 \in J_k$; $k = 1, \ldots, m$, such that $t_1 < t_2$ and let $u \in Q$, with

$$t_2^{1-\gamma}(Nu)(t_2) - t_1^{1-\gamma}(Nu)(t_1) \neq 0,$$

then there exists $\varphi \in E^*$ such that

$$\|t_2^{1-\gamma}(Nu)(t_2) - t_1^{1-\gamma}(Nu)(t_1)\|_E \leq \|t_2^{1-\gamma}g_k(t_2, u(t_k^-)) - t_1^{1-\gamma}g_k(t_1, u(t_k^-))\|_E.$$

Hence $N(Q) \subset Q$.

Step 2. *N is weakly-sequentially continuous.*

Let (u_n) be a sequence in Q and let $(u_n(t)) \rightarrow u(t)$ in (E, ω) for each $t \in I$. Fix $t \in I$, since f satisfies the assumption (9.3.1), we have $f(t, u_n(t), D_{s_k}^{\alpha,\beta}u_n(t))$ converges weakly to $f(t, u(t), D_{s_k}^{\alpha,\beta}u(t))$ on I_k; $k = 0, \ldots, m$. Hence the Lebesgue dominated convergence theorem for Pettis integral implies $(Nu_n)(t)$ converges weakly to $(Nu)(t)$ in (E, ω), for each

$t \in I$. Thus, $N(u_n) \rightharpoonup N(u)$. Hence, $N : Q \to Q$ is weakly-sequentially continuous.

Step 3. *The implication (1.9) holds.*

Let V be a subset of Q such that $\overline{V} = \overline{\operatorname{conv}}(N(V) \cup \{0\})$. Obviously

$$V(t) \subset \overline{\operatorname{conv}}(NV)(t)) \cup \{0\}), \ \forall t \in I.$$

Further, as V is bounded and equicontinuous, by [172, Lemma 3] the function $t \to v(t) = \beta(V(t))$ is continuous on I. From (9.3.2), Lemma 2.3 and the properties of the measure β, for any $t \in I_k$; $k = 0, \ldots, m$, we have

$$
\begin{aligned}
t^{1-\gamma} v(t) &\leq \beta(t^{1-\gamma}(NV)(t) \cup \{0\}) \\
&\leq \beta(t^{1-\gamma}(NV)(t)) \\
&\leq \frac{T^{1-\gamma}}{\Gamma(\alpha)} \int_0^t |t-s|^{\alpha-1} p(s) \beta(V(s)) ds \\
&\leq \frac{T^{1-\gamma}}{\Gamma(\alpha)} \int_0^t |t-s|^{\alpha-1} s^{1-\gamma} p(s) v(s) ds \\
&\leq \frac{p^* T^{1-\gamma+\alpha}}{\Gamma(1+\alpha)} \|v\|_{\mathcal{PC}}.
\end{aligned}
$$

Thus

$$\|v\|_{\mathcal{PC}} \leq L \|v\|_{\mathcal{PC}}.$$

From (9.10), we get $\|v\|_{\mathcal{PC}} = 0$, that is $v(t) = \beta(V(t)) = 0$, for each $t \in I$, and then by [290, Theorem 2], V is relatively-weakly compact in $\mathcal{PC}_\gamma$. Applying now Theorem 1.7, we conclude that N has a fixed point which is a solution of the problem (9.9). $\qquad\square$

9.4.3. *Example*

Let

$$E = l^1 = \left\{ u = (u_1, u_2, \ldots, u_n, \ldots), \sum_{n=1}^\infty |u_n| < \infty \right\}$$

be the Banach space with the norm

$$\|u\|_E = \sum_{n=1}^\infty |u_n|.$$

As an application of our results we consider the following problem of implicit Hilfer fractional differential equation:

$$\begin{cases} (D_0^{\frac{1}{2},\frac{1}{2}} u_n)(t) = f_n(t, u(t), (D_0^{\frac{1}{2},\frac{1}{2}} u)(t)); & t \in [0,1], \\ u(t) = g(t, e^-); & t \in (1,2], \\ (D_2^{\frac{1}{2},\frac{1}{2}} u_n)(t) = f_n(t, u(t), (D_2^{\frac{1}{2},\frac{1}{2}} u)(t)); & t \in (2,3], \\ (I_0^{\frac{1}{4}} u)(t)|_{t=0} = 0, \end{cases} \qquad (9.12)$$

where

$$f_n(t, u(t), (D_k^{\frac{1}{2},\frac{1}{2}} u)(t)) = \frac{ct^2 (e^{-7} + e^{-t-5})}{1 + \|u\|_E + \|D_k^{\frac{1}{2},\frac{1}{2}} u\|_E} u_n(t);$$

$$t \in [0,1] \cup (2,3], \quad k \in \{0,2\},$$

$$g(t, e^-) = \frac{e^{-2t}}{1+e}; \ t \in (1,2],$$

with

$$u = (u_1, u_2, \ldots, u_n, \ldots) \quad \text{and} \quad c := \frac{e^5}{4 \times 3^{\frac{13}{4}}} \Gamma\left(\frac{1}{2}\right).$$

Set $f = (f_1, f_2, \ldots, f_n, \ldots)$. Clearly, the function f is continuous. For each $u \in E$ and $t \in [0,1] \cup (2,3]$, we have

$$\|f(t, u(t), (D_k^{\frac{1}{2},\frac{1}{2}})(t))\|_E \leq ct^2 \left(e^{-7} + \frac{1}{e^{t+5}}\right); \quad k \in \{0,2\}.$$

Hence, the hypothesis (9.6.2) is satisfied with $p^* = 18ce^{-5}$.

We shall show that condition (9.10) holds with $T = 3$. Indeed,

$$\frac{p^* T^{\frac{5}{4}}}{\Gamma(\frac{3}{2})} = \frac{36c3^{\frac{5}{4}} e^{-5}}{\Gamma(\frac{1}{2})} = \frac{1}{2} < 1.$$

Simple computations show that all the conditions of Theorem 9.3 are satisfied. It follows that the problem (9.12) has at least one solution on $[0,3]$.

9.5. Implicit Hilfer–Pettis Fractional Differential Equations with Retarded and Advanced Arguments

9.5.1. *Introduction*

In this section, we discuss the existence of weak solutions for the following boundary value problem for implicit Pettis–Hadamard fractional differential equation

$$\begin{cases} u(t) = \phi(t); & t \in [1-\alpha, 1], \\ ({}^H D_1^r u)(t) = f(t, u_t, ({}^H D_1^r u)(t)); & t \in I := [1, e], \\ u(t) = \psi(t); & t \in [e, e+\beta], \end{cases} \quad (9.13)$$

where $\alpha, \beta > 0$, $r \in (1, 2]$, $f : I \times C[-\alpha, \beta] \times E \to E$ is a given continuous function, $\phi \in C[1-\alpha, 1]$ with $\phi(1) = 0$, $\psi \in C[e, e+\beta]$ with $\psi(e) = 0$, E is a real (or complex) Banach space with norm $\| \cdot \|_E$ and dual E^*, such that E is the dual of a weakly-compactly-generated Banach space X, $C[-\alpha, \beta]$ is the space of continuous functions from $[-\alpha, \beta]$ to E.

9.5.2. *Existence of weak solutions*

Definition 9.4. A function $u \in C^2([1-\alpha, e+\beta], E)$, is said to be a weak solution of the problem (9.13) if u satisfies the equation $({}^H D_1^r u)(t) = f(t, u_t, ({}^H D_1^r u)(t))$ on I, and the conditions $u(t) = \phi(t)$, $\phi(1) = 0$ on $[1-\alpha, 1]$, and $u(t) = \psi(t)$, $\psi(e) = 0$ on $[e, e+\beta]$.

The following hypotheses will be used in the sequel.

(9.4.1) For a.e. $t \in I$, the functions $v \to f(t, v, \cdot)$ and $w \to f(t, \cdot, w)$ are weakly-sequentially continuous.

(9.4.2) For a.e. $v \in C[1-\alpha, e+\beta]$, and $w \in E$, the function $t \to f(t, v, w)$ is Pettis integrable a.e. on I.

(9.4.3) There exists $p \in C(I, [0, \infty))$ such that for all $\varphi \in E^*$, we have

$$|\varphi(f(t, u, v))| \le p(t)\|\varphi\|$$

for a.e. $t \in I$, and each $u \in C[1-\alpha, e+\beta]$, and $v \in E$.

(9.4.4) For each bounded and measurable set $B \subset C[1-\alpha, e+\beta]$ and for each $t \in I$, we have

$$\mu(f(t, B, {}^H D_1^r B)) \le p(t)\mu(B),$$

where ${}^H D_1^r B = \{{}^H D_1^r w : w \in B \cap C(I)\}$.

Set

$$p^* = \sup_{t \in I} p(t).$$

Theorem 9.4. *Assume that the hypotheses (9.4.1)–(9.4.4) hold. If*

$$L := \frac{2p^*}{\Gamma(1+r)} < 1, \tag{9.14}$$

then the problem (9.13) has at least one solution defined on I.

Proof. Transform the problem (9.13) into a fixed-point equation. Consider the operator $N : C[1-\alpha, e+\beta] \to C[1-\alpha, e+\beta]$ defined by

$$(Nu)(t) = \begin{cases} \phi(t); & t \in [1-\alpha, 1], \\ -\int_1^e G(t,s)\frac{g(s)}{s}ds; & t \in I, \\ \psi(t); & t \in [e, e+\beta], \end{cases} \tag{9.15}$$

where $g \in C(I)$ with

$$g(t) = f\left(t, {}^H I_1^r g_t, g(t)\right).$$

First notice that, the functions ϕ and ψ are continuous, and the hypotheses imply that for all $t \in I$, the functions $t \mapsto G(\cdot, t)$, and $t \mapsto g(t)$ are Pettis integrables, over I. Thus, the operator N is well defined.

In the sequel we denote $\|w\|_{C[1-\alpha,e+\beta]}$ by $\|w\|_C$.

Let $R > 0$ be such that

$$R > \max\left\{\frac{2p^*}{\Gamma(1+r)}, \|\phi\|_{C[1-\alpha,1]}, \|\psi\|_{C[e,e+\beta]}\right\},$$

and consider the set

$$Q = \left\{u \in C : \|u\|_C \le R \text{ and } \|u(t_2) - u(t_1)\|_E \le p^* \right.$$
$$\left. \times \int_1^e |G(t_2,s) - G(t_1,s)|\frac{ds}{s}\right\}.$$

Clearly, the subset Q is closed, convex and equicontinuous. We shall show that the operator N satisfies all the assumptions of Theorem 1.7. The proof will be given in several steps.

Step 1. *N maps Q into itself.*

Let $u \in Q$; $t \in I$ and assume that $(Nu)(t) \neq 0$. Then there exists $\varphi \in E^*$ such that for each $t \in I$, we have $\|(Nu)(t)\|_E = |\varphi(|(Nu)(t))|$. Thus

$$\|(Nu)(t)\|_E = \left| \varphi \left(\int_1^e G(t,s)g(s)\frac{ds}{s} \right) \right|,$$

where $g \in C(I)$, with

$$g(t) = f\left(t, {}^H I_1^r g_t, g(t)\right).$$

If $t \in [1 - \alpha, 1]$, then

$$\|N(u)(t)\|_E \leq \|\phi\|_{[1-\alpha,1]} \leq R,$$

also, if $t \in [e, e + \beta]$, then

$$\|N(u)(t)\|_E \leq \|\psi\|_{[e,e+\beta]} \leq R.$$

For each $t \in I$, we have

$$\int_1^e |G(t,s)| \frac{ds}{s} \leq \frac{1}{\Gamma(r)} \left[\int_1^t \left(\log \frac{t}{s}\right)^{r-1} \frac{ds}{s} + (\log t)^{r-1} \int_1^e \left(\log \frac{e}{s}\right)^{r-1} \frac{ds}{s} \right]$$

$$\leq \frac{2}{\Gamma(r)} \int_1^e \left(\log \frac{e}{s}\right)^{r-1} \frac{ds}{s} = \frac{2}{\Gamma(1+r)}. \qquad (9.16)$$

Thus, for each $t \in I$, we have

$$\|(Nu)(t)\|_E \leq \int_1^e |G(t,s)| \frac{|\varphi(g(s))|}{s} ds$$

$$\leq \frac{2p^*}{\Gamma(1+r)}$$

$$\leq R.$$

Hence, if $u \in Q$; $t \in [\tilde{1} - \alpha, e + \beta]$, we have

$$\|(Nu)(t)\|_E \leq R.$$

Next, let $t_1, t_2 \in I$ such that $t_1 < t_2$ and let $u \in Q$, with

$$(Nu)(t_2) - (Nu)(t_1) \neq 0.$$

Then there exists $\varphi \in E^*$ such that

$$\|(Nu)(t_2) - (Nu)(t_1)\|_E = |\varphi((Nu)(t_2) - (Nu)(t_1))|$$

and $\|\varphi\| = 1$. Thus

$$\|(Nu)(t_2) - (Nu)(t_1)\|_E = |\varphi((Nu)(t_2) - (Nu)(t_1))|$$

$$\leq \left| \varphi \left(\int_1^e (G(t_2, s) - G(t_1, s))g(s)\frac{ds}{s} \right) \right|,$$

where $g \in C(I)$, with

$$g(t) = f\left(t, {}^H I_1^\tau g_t, g(t)\right).$$

Thus, we get

$$\|(Nu)(t_2) - (Nu)(t_1)\|_E \leq \int_1^e |G(t_2, s) - G(t_1, s)| |g(s)| \frac{ds}{s}$$

$$\leq p^* \int_1^e |G(t_2, s) - G(t_1, s)| \frac{ds}{s}.$$

Hence $N(Q) \subset Q$.

Step 2. *N is weakly-sequentially continuous.*

Let (u_n) be a sequence in Q and let $(u_n(t)) \to u(t)$ in (E, ω) for each $t \in [1 - \alpha, e + \beta]$. Fix $t \in [1 - \alpha, e + \beta]$, since f satisfies the assumption (9.4.1), we have $f(t, u_{nt}, {}^H D_1 u_n(t))$ converges weakly to $f(t, u_t, {}^H D_1 u(t))$. Hence the Lebesgue dominated convergence theorem for Pettis integral implies $(Nu_n)(t)$ converges weakly to $(Nu)(t)$ in (E, ω), for each $t \in [1 - \alpha, e + \beta]$. Thus, $N(u_n) \to N(u)$. Hence, $N : Q \to Q$ is weakly-sequentially continuous.

Step 3. *The implication (1.9) holds.*

Let V be a subset of Q such that $\overline{V} = \overline{\text{conv}}(N(V) \cup \{0\})$. Obviously

$$V(t) \subset \overline{\text{conv}}(NV)(t)) \cup \{0\}), \quad \forall t \in [1 - \alpha, e + \beta].$$

Further, as V is bounded and equicontinuous, by [172, Lemma 3] the function $t \to v(t) = \mu(V(t))$ is continuous on $[1-\alpha, e+\beta]$. From (9.4.3), (9.4.4), Lemma 1.7 and the properties of the measure μ, for any $t \in [1 - \alpha, e + \beta]$, we have

$$v(t) \leq \mu((NV)(t) \cup \{0\})$$

$$\leq \mu((NV)(t))$$

$$\leq \int_1^e |G(t, s)| \frac{p(s)\mu(V(s))}{s} ds$$

$$\le \frac{2}{\Gamma(1+r)} \int_1^e \frac{p(s)v(s)}{s} ds$$

$$\le \frac{2p^*}{\Gamma(1+r)} \|v\|_C.$$

Thus

$$\|v\|_C \le L\|v\|_C.$$

From (9.14), we get $\|v\|_C = 0$, that is $v(t) = \mu(V(t)) = 0$, for each $t \in [1-\alpha, e+\beta]$, and then by [290, Theorem 2], V is relatively-weakly compact in $C[1-\alpha, e+\beta]$. From Theorem 1.7, we conclude that N has a fixed point which is a solution of the problem (9.13). $\qquad\square$

9.5.3. *Example*

Let

$$E = l^1 = \left\{ u = (u_1, u_2, \ldots, u_n, \ldots), \sum_{n=1}^{\infty} |u_n| < \infty \right\}$$

be the Banach space with the norm

$$\|u\|_E = \sum_{n=1}^{\infty} |u_n|.$$

As an application of our results we consider the following problem of implicit Hadamard fractional differential equation:

$$\begin{cases} u(t) = 1 - e^{t-1}; & t \in [-2, 1], \\ (^H D_1^{\frac{3}{2}} u_n)(t) = f_n(t, u_t, (^H D_1^{\frac{3}{2}} u)(t)); & t \in [1, e], \\ u(t) = (\ln t) - 1; & t \in [e, 2e], \end{cases} \qquad (9.17)$$

where

$$f_n(t, u_t, (^H D_1^{\frac{3}{2}} u)(t)) = \frac{ct^2}{1 + \|u\|_{C[-3,e]} + \|^H D_1^{\frac{3}{2}} u\|_E}$$

$$\times \left(e^{-7} + \frac{1}{e^{t+5}} \right) u_n(t); \quad t \in [1, e],$$

with

$$u = (u_1, u_2, \ldots, u_n, \ldots) \quad \text{and} \quad c := \frac{e^4}{12} \Gamma\left(\frac{1}{2}\right).$$

Set $f = (f_1, f_2, \ldots, f_n, \ldots)$. Clearly, the function f is continuous.

For each $u \in E$ and $t \in [1, e]$, we have

$$\|f(t, u(t), ({}^H D_1^{\frac{3}{2}})(t))\|_E \leq \frac{ct^2}{1 + \|u\|_{C[-3,e]} + \|{}^H D_1^{\frac{3}{2}} u\|_E} \left(e^{-7} + \frac{1}{e^{t+5}} \right).$$

Hence, the hypothesis (9.4.3) is satisfied with $p^* = ce^{-4}$,

We shall show that condition (9.14) holds. Indeed,

$$\frac{2p^*}{\Gamma(1+r)} = \frac{2c}{e^4 \Gamma(\frac{5}{2})} = \frac{1}{2} < 1.$$

Simple computations show that all the conditions of Theorem 9.4 are satisfied. It follows that the problem (9.17) has at least one solution on $[-2, 2e]$.

9.6. Notes and Remarks

The results of Chapter 9 are taken from [46,52,61,65].

Bibliography

[1] S. Abbas, R.P. Agarwal, M. Benchohra, and N. Benkhettou, Hilfer–Hadamard fractional differential equations and inclusions under weak topologies, *Progress in Fract. Differ. Appl.* **4** (4) (2018), 247–261.

[2] S. Abbas, E. Alaidarous, W. Albarakati, and M. Benchohra, Upper and lower solutions method for partial Hadamard fractional integral equations and inclusions, *Discus. Math. Diff. Incl., Contr. Optim.* **35** (2015), 105–122.

[3] S. Abbas, A. Arara, and M. Benchohra, Existence and global stability results for Volterra type fractional Hadamard–Stieltjes partial integral equations, (to appear in Filomat).

[4] S. Abbas, W. Albarakati, and M. Benchohra, Existence and attractivity results for Volterra type nonlinear multi-delay Hadamard–Stieltjes fractional integral equations, *PanAmer. Math. J.* **16** (1) (2016), 1–17.

[5] S. Abbas, W.A. Albarakati, M. Benchohra, M.A. Darwish, and E.M. Hilal, New existence and stability results for partial fractional differential inclusions with multiple delay, *Ann. Polon. Math.* **114** (2015), 81–100.

[6] S. Abbas, W.A. Albarakati, M. Benchohra, and E. M. Hilal, Global existence and stability results for partial fractional random differential equations, *J. Appl. Anal.* **21** (2)(2015), 79–87.

[7] S. Abbas, W. Albarakati, M. Benchohra, and J. Henderson, Ulam–Hyers–Rassias stability for multi-delay fractional Hadamard integral equations in Fréchet spaces with random effects (Submitted).

[8] S. Abbas, W.A. Albarakati, M. Benchohra, and J. Henderson, Existence and Ulam stabilities for Hadamard fractional integral equations with random effects, *Electron. J. Differential Equations* **2016** (5) (2016), 1–12.

[9] S. Abbas, W. Albarakati, M. Benchohra, and G.M. N'Guérékata, Existence and Ulam stabilities for Hadamard fractional integral equations in Fréchet spaces, *J. Fract. Calc. Appl.* **7** (2) (2016), 1–12.

[10] S. Abbas, E. Alaidarous, M. Benchohra, and J.J. Nieto, Existence and stability of solutions for Hadamard-Stieltjes fractional integral equations, *Discrete Dyn. Nature Soc.*, **2015** (2015), Article ID 317094, 6 pp.

[11] S. Abbas, W. Albarakati, M. Benchohra, and J.J. Nieto, Existence and global stability results for Volterra type fractional Hadamard partial integral equations, *Commun. Math. Anal.* **21** (1) (2018), 42–53.

[12] S. Abbas, W. Albarakati, M. Benchohra, and A. Petruşel, Existence and Ulam stability results for Hadamard partial fractional integral inclusions via Picard operators, *Stud. Univ. Babes-Bolyai Math.* **61** (4) (2016), 409–420.

[13] S. Abbas, W. Albarakati, M. Benchohra, and S. Sivasundaram, Dynamics and stability of Fredholm type fractional order Hadamard integral equations, *J. Nonlinear Stud.* **22** (4) (2015), 673–686.

[14] S. Abbas, W.A. Albarakati, M. Benchohra, and S. Sivasundaram, On the solutions of Pettis partial Hadamard-Stieltjes fractional integral equations, *Nonlinear Stud.* **23** (2) (2016), 333–344.

[15] S. Abbas, W. Albarakati, M. Benchohra, and J.J. Trujillo, Ulam stabilities for partial Hadamard fractional integral equations, *Arab. J. Math.* **5** (1) (2016), 1–7.

[16] S. Abbas, W. Albarakati, M. Benchohra, and Y. Zhou, Weak solutions for partial pettis Hadamard fractional integral equations with random effects, *J. Int. Equ. Appl.* **29** (4) (2017) 473–491.

[17] S. Abbas, W. Albarakati, M. Benchohra, and Y. Zhou, Weak solutions for partial random Hadamard fractional integral equations with multiple delay, *Integral Equations and Operator Theory* **29** (4) (2017), 473–791.

[18] S. Abbas and M. Benchohra, *Advanced Functional Evolution Equations and Inclusions*, Developments in Mathematics, Springer, Cham, 2015.

[19] S. Abbas and M. Benchohra, Some stability concepts for Darboux problem for partial fractional differential equations on unbounded domain, *Fixed Point Theory* **16** (1) (2015), 3–14.

[20] S. Abbas and M. Benchohra, Uniqueness and Ulam stabilities results for partial fractional differential equations with not instantaneous impulses, *Appl. Math. Comput.* **257** (2015), 190–198.

[21] S. Abbas and M. Benchohra, On the generalized Ulam–Hyers–Rassias stability for Darboux problem for partial fractional implicit differential equations, *Appl. Math. E-Notes* **14** (2014), 20–28.

[22] S. Abbas and M. Benchohra, Upper and lower solutions method for Darboux problem for fractional order implicit impulsive partial hyperbolic differential equations, *Acta Univ. Palacki. Olomuc.* **51** (2) (2012), 5–18.

[23] S. Abbas and M. Benchohra, A global uniqueness result for fractional order implicit differential equations, *Comment. Math. Univ. Carolin.* **53** (4) (2012), 605–614.

[24] S. Abbas and M. Benchohra, Darboux problem for implicit impulsive partial hyperbolic differential equations, *Electron. J. Differential Equations* **2011** (2011), 15 pp.

[25] S. Abbas, M. Benchohra, and N. Hamidi, Successive approximations for the Darboux problem for implicit partial differential equations, *PanAmer. Math. J.* **28** (3) (2018), 1–10.

[26] S. Abbas and M. Benchohra, Some stability concepts for Darboux problem for partial fractional differential equations on unbounded domain, *Fixed Point Theory* **16** (1) (2015), 3–14.

[27] S. Abbas and M. Benchohra, Uniqueness and Ulam stabilities results for partial fractional differential equations with not instantaneous impulses, *Appl. Math. Comput.* **257** (2015), 190–198.

[28] S. Abbas and M. Benchohra, Nonlinear fractional order Riemann–Liouville Volterra–Stieltjes partial integral equations on unbounded domains, *Commun. Math. Anal.* **14** (1) (2013), 104–117.

[29] S. Abbas and M. Benchohra, Existence and stability of nonlinear fractional order Riemann–Liouville– Volterra–Stieltjes multi-delay integral equations, *J. Integral Equations Appl.* **25** (2) (2013), 143–158.

[30] S. Abbas and M. Benchohra, Global stability results for nonlinear partial fractional order Riemann-Liouville– Volterra–Stieltjes functional integral equations, *Math. Sci. Res. J.* **16** (4) (2012), 82–92.

[31] S. Abbas and M. Benchohra, Ulam–Hyers stability for the Darboux problem for partial fractional differential and integro-differential equations via Picard operators, *Results. Math.* **65** (1-2) (2014), 67–79.

[32] S. Abbas and M. Benchohra, Ulam stabilities for the Darboux problem for partial fractional differential inclusions, *Demonstr. Math.* **XLVII** (4) (2014), 826–838.

[33] A. Salim, S. Abbas, M. Benchohra and E. Karapinar, Global stability results for Voltera–Hadamard random partial fractional integral equations, *Rendiconti del Circolo Matematico di Palermo Series 2*, (to appear).

[34] S. Abbas and M. Benchohra, On the generalized Ulam–Hyers–Rassias stability for Darboux problem for partial fractional implicit differential equations, *Appl. Math. E-Notes* **14** (2014), 20–28.

[35] S. Abbas and M. Benchohra, Existence and Ulam stabilities for Hilfer fractional differential equations in Banach spaces (Submitted).

[36] S. Abbas and M. Benchohra, Weak solutions for partial Pettis Hadamard fractional integral inclusions (Submitted).

[37] S. Abbas and M. Benchohra, Weak solutions for Fredholm type pettis Hadamard partial fractional integral equations (Submitted).

[38] S. Abbas and M. Benchohra, Weak solutions for implicit pettis Hadamard fractional differential equations with finite delay (Submitted).

[39] S. Abbas and M. Benchohra, Successive approximations for implicit Hadamard fractional differential equations (Submitted).

[40] S. Abbas and M. Benchohra, Weak solutions for implicit Hadamard fractional differential equations with not instantaneous impulses (Submitted).

[41] S. Abbas and M. Benchohra, Weak solutions for fractional Hilfer differential equations (Submitted).

[42] S. Abbas and M. Benchohra, Weak solutions for functional differential inclusions involving the Hilfer fractional derivative (Submitted).

[43] S. Abbas and M. Benchohra, Weak solutions for Hilfer fractional differential equations with Maxima (Submitted).

[44] S. Abbas, M. Benchohra, F. Berhoun, and J.J. Nieto, Weak solutions for impulsive implicit Hadamard fractional differential equations, *Adv. Dyn. Syst. Appl.* **13** (1) (2018), 1–18.

[45] S. Abbas, M. Benchohra, and M. Bohner, Weak solutions for implicit differential equations of Hilfer–Hadamard fractional derivative, *Adv. Dyn. Syst. Appl.* **12** (1) (2017), 1–16.

[46] S. Abbas, M. Benchohra, and M.A. Darwish, New stability results for partial fractional differential inclusions with not instantaneous impulses, *Frac. Calc. Appl. Anal.* **18** (1) (2015), 172–191.

[47] S. Abbas, M. Benchohra, and M.A. Darwish, Fractional differential inclusions of Hilfer and Hadamard types in Banach spaces, *Discuss. Math. Differ. Incl. Control Optim.* **37** (2) (2017), 187–204.

[48] S. Abbas, M. Benchohra, and M.A. Darwish, Asymptotic stability for implicit differential equations involving Hilfer fractional derivative, *Panamer. Math. J.* **27** (3) (2017), 40–52.

[49] S. Abbas, M. Benchohra, M.A. Darwish, and J.J. Nieto, Weak solutions for partial pettis Hadamard fractional integral equations, *Ital. J. Pure Appl. Math.*, (to appear).

[50] S. Abbas, M. Benchohra, and P. Eloe, Hilfer and Hilfer-Hadamard fractional differential equations in Banach spaces (Submitted).

[51] S. Abbas, M. Benchohra, and J.R. Graef, Weak solutions to implicit differential equations involving the Hilfer fractional derivative, *Nonlinear Dyn. Syst. Theory* **18** (1) (2018), 1–11.

[52] S. Abbas, M. Benchohra, J.R. Graef, and J. Henderson, *Implicit Fractional Differential and Integral Equations: Existence and Stability*, De Gruyter, Berlin, 2018.

[53] S. Abbas, M. Benchohra, and N. Hamidi, Hilfer and Hilfer-Hadamard fractional differential equations with random effects, *Libertas Math.* **37** (1) (2018), 45–64.

[54] S. Abbas, M. Benchohra, N. Hamidi, and J. Henderson, Caputo-Hadamard fractional differential equations in Banach spaces, *Frac. Calc. Appl. Anal.* **21** (4) (2018), 1027–1045.

[55] S. Abbas, M. Benchohra, and A. Hammoudi, Upper, lower solutions method and extremal solutions for impulsive discontinuous partial fractional differential inclusions, *Panamer. Math. J.* **24** (1) (2014), 31–52.

[56] S. Abbas and M. Benchohra, and J. Henderson, Partial Hadamard fractional integral equations, *Adv. Dyn. Syst. Appl.* **10** (2) (2015), 97–107.

[57] S. Abbas, M. Benchohra, and J. Henderson, Asymptotic behavior of solutions of nonlinear fractional order Riemann-Liouville Volterra-Stieltjes quadratic integral equations, *Int. Electron. J. Pure Appl. Math.* **4** (3) (2012), 195–209.

[58] S. Abbas, M. Benchohra, and J. Henderson, Ulam stability for partial fractional integral inclusions via Picard operators, *J. Frac. Calc. Appl.* **5** (2) (2014), 133–144.

[59] S. Abbas, M. Benchohra, and J. Henderson, Weak solutions for implicit fractional differential equations of Hadamard type, *Adv. Dyn. Syst. Appl.* **11** (1) (2016), 1–13.

[60] S. Abbas, M. Benchohra, and J. Henderson, Weak solutions for implicit Hilfer fractional differential equations with not instantaneous impulses, *Commun. Math. Anal.* **20** (2) (2017), 48–61.

[61] S. Abbas, M. Benchohra, J. Henderson, and J.E. Lazreg, Measure of noncompactness and impulsive Hadamard fractional implicit differential equations in Banach spaces, *Math. Eng. Science Aerospace* **8** (3) (2017), 1–19.

[62] S. Abbas, M. Benchohra, J. Henderson, and J.E. Lazreg, Weak solution for a coupled system of partial Pettis Hadamard fractional integral equations, *Adv. Theory Nonlinear Anal. Appl.* **1** (2) (2017), 136–146.

[63] S. Abbas, M. Benchohra, J.E. Lazreg, A. Alsaedi, and Y.Zhou, Existence and Ulam stability for fractional differential equations of Hilfer-Hadamard type, *Adv. Difference Equ.* **2017** (2017), 180. doi 10.1186/s13662-017-1231-1.

[64] S. Abbas, M. Benchohra, J.E. Lazreg, and G.M. N'Guérékata, Weak solution for implicit Pettis Hadamard fractional differential equations with retarded and avanced arguments. *Nonlinear Stud.* **24** (2017), 355–365.

[65] S. Abbas, M. Benchohra, J.E. Lazreg, and G.M. N'Guérékata, Hilfer and Hadamard functional random fractional differential inclusions. *Cubo* **19** (1) (2017), 17–38.

[66] S. Abbas, M. Benchohra, J.E. Lazreg, and J.J. Nieto, On a Coupled system of Hilfer and Hilfer-Hadamard fractional differential equations in Banach spaces, *Nonlinear Funct. Anal.* **2018** (2018), Article ID 12, 12 pp.

[67] S. Abbas, M. Benchohra, J.E. Lazreg, and Y. Zhou, A Survey on Hadamard and Hilfer fractional differential equations: Analysis and Stability, *Chaos, Solitons Fractals* **102** (2017) 47–71.

[68] S. Abbas, M. Benchohra, and J.J. Nieto, Functional implicit hyperbolic fractional order differential equations with delay, *Afr. Diaspora J. Math.* **15** (1) (2013), 74–96.

[69] S. Abbas, M. Benchohra, and J.J. Nieto, Global attractivity of solutions for nonlinear fractional order Riemann-Liouville Volterra-Stieltjes partial integral equations, *Electron. J. Qual. Theory Differ. Equ.* **81** (2012), 1–15.

[70] S. Abbas, M. Benchohra, and G.M. N'Guérékata, *Advanced Fractional Differential and Integral Equations*, Nova Science Publishers, New York, 2015.

[71] S. Abbas, M. Benchohra, and G.M. N'Guérékata, *Topics in Fractional Differential Equations*, Developments in Mathematics, Vol. 27, Springer, New York, 2012.

[72] S. Abbas, M. Benchohra, and A. Petruşel, Ulam stabilities for the Darboux problem for partial fractional differential inclusions via picard operators, *Electron. J. Qual. Theory Differ. Equ.* **2014** (51) (2014), 13 pp.

[73] S. Abbas, M. Benchohra, and A. Petruşel, Ulam stability for Hilfer type fractional differential inclusions via the weakly Picard operator theory, *Frac. Calc. Appl. Anal.* **20** (2) (2017), 384–398.

[74] S. Abbas, M. Benchohra, M. Rivero, and J.J. Trujillo, Existence and stability results for nonlinear fractional order Riemann–Liouville Volterra–Stieltjes quadratic integral equations, *Appl. Math. Comput.* **247** (2014), 319–328.

[75] S. Abbas, M. Benchohra, and S. Sivasundaram, Ulam stability for partial fractional differential inclusions with multiple delay and impulses via picard operators, *J. Nonlinear Stud.* **20** (4) (2013), 623–641.

[76] S. Abbas, M. Benchohra, and S. Sivasundaram, Dynamics and Ulam stability for Hilfer type fractional differential equations, *Nonlinear Stud.* **23** (4) (2016), 627–637.

[77] S. Abbas, M. Benchohra, and B.A. Slimani, Existence and Ulam stabilities for partial fractional random differential inclusions with nonconvex right hand side, *PanAmer. Math. J.* **25** (1) (2015), 95–110.

[78] S. Abbas, M. Benchohra, and B.A. Slimani, Partial hyperbolic implicit differential equations with variable times impulses, *Stud. Univ. Babeş-Bolyai Math.* **60** (1) (2015), 61–73.

[79] S. Abbas, R.P. Agarwal, M. Benchohra, and B.A. Slimani, Global stability for implicit Hilfer–Hadamard fractional differential equations (Submitted).

[80] S. Abbas, M. Benchohra, and J.J. Trujillo, Upper and lower solutions method for partial fractional differential inclusions with not instantaneous impulses, *Prog. Frac. Diff. Appl.* **1** (1) (2015), 11–22.

[81] S. Abbas, M. Benchohra, and A.N. Vityuk, On fractional order derivatives and Darboux problem for implicit differential equations, *Frac. Calc. Appl. Anal.* **15** (2012), 168–182.

[82] S. Abbas, M. Benchohra, and Y. Zhou, Hilfer and Hadamard random fractional differential equations in Fréchet spaces (Submitted).

[83] S. Abbas, M. Benchohra, and Y. Zhou, Coupled Hilfer fractional differential systems with random effects, *Adv. Difference Equ.* **2018** (369) (2018), 12 pp.

[84] R.P. Agarwal and B. Ahmad, Existence theory for anti-periodic boundary value problems of fractional differential equations and inclusions, *Comput. Math. Appl.* **62** (2011), 1200–1214.

[85] R.P. Agarwal, M. Benchohra, and B.A. Slimani, Existence results for differential equations with fractional order and impulses, *Mem. Differ. Equ. Math. Phys.* **44** (1) (2008), 1–21.

[86] R.P. Agarwal, M. Meehan, and D. O'Regan, *Fixed Point Theory and Applications*, Cambridge University Press, Cambridge, 2001.

[87] B. Ahmad, A. Alsaedi, S.K. Ntouyas, J. Tariboon, *Hadamard-Type Fractional Differential Equations, Inclusions and Inequalities*. Springer, Cham, 2017.

[88] B. Ahmad and J.R. Graef, Coupled systems of nonlinear fractional differential equations with nonlocal boundary conditions, *Panamer. Math. J.* **19** (2009), 29–39.

[89] B. Ahmad, J. Henderson, and R. Luca, *Boundary Value Problems for Fractional Differential Equations and Systems*, World Scientific, USA, 2021.

[90] B. Ahmad and J.J. Nieto, Existence of solutions for nonlocal boundary value problems of higher-order nonlinear fractional differential equations, *Abstrac. Appl. Anal.* **2009** (2009), Article ID 494720, 9 pp.

[91] B. Ahmad and J.J. Nieto, Existence of solutions for impulsive anti-periodic boundary value problems of fractional order, *Taiwaness J. Math.* **15** (3) (2011), 981–993.

[92] B. Ahmad and J.J. Nieto, Anti-periodic fractional boundary value problems with nonlinear term depending on lower order derivative, *Fract. Calc. Appl. Anal.* **15** (2012), 451–462.

[93] B. Ahmad and J.J. Nieto, Anti-periodic fractional boundary value problems, *Comput. Math. Appl.* **62** (3) (2011), 1150–1156.

[94] B. Ahmad, J.J. Nieto, A. Alsaedi, and N. Mohamad, On a new Class of anti-periodic fractional boundary value problems, *Abst. Appl. Anal.* **2013** (2013), 7 pp.

[95] B. Ahmad and S. Sivasundaram, Existence results for nonlinear impulsive hybrid boundary value problems involving fractional differential equations, *Nonlinear Anal. Hybrid Syst.* **3** (3) (2009), 251–258.

[96] B. Ahmad and S. Sivasundaram, Theory of fractional differential equations with three-point boundary conditions, *Commun. Appl. Anal.* **12** (2008), 479–484.

[97] K.K. Akhmerov, M.I. Kamenskii, A.S. Potapov, A.E. Rodkina, and B.N. Sadovskii, *Measures of Noncompactness and Condensing Operators*, Birkhäuser Verlag, Basel, 1992.

[98] W. Albarakati, M. Benchohra, J.E. Lazreg, and J.J. Nieto, Anti-periodic boundary value problem for nonlinear implicit fractional differential equations with impulses, *Analele Univ. Oradea Fasc. Mat.* **XXV** (1) (2018), 13–24.

[99] S. Aljoudi, B. Ahmad, J.J. Nieto, and A. Alsaedi, A coupled system of Hadamard type sequential fractional differential equations with coupled strip conditions, *Chaos, Solitons Fractals*, **91** (2016), 39–46.

[100] S. Aljoudi, B. Ahmad, J.J. Nieto, and A. Alsaedi, On coupled Hadamard type sequential fractional differential equations with variable coefficients and nonlocal integral boundary conditions. *Filomat* **31** (2017), 6041–6049.

[101] R. Almeida, S. Pooseh, and D. F. M. Torres, *Computational Methods in the Fractional Calculus of Variations*, World Scientific Publishing, USA, 2015.

[102] R. Almeida, D. Tavares, and D.F.M. Torres, *The Variable-Order Fractional Calculus of Variations*, Springer International Publishing, 2019.

[103] S. Almezel, Q.H. Ansari, and M.A. Khamsi, *Topics in Fixed Point Theory*, Springer-Verlag, New York, 2014.

[104] J.C. Alvàrez, Measure of noncompactness and fixed points of nonexpansive condensing mappings in locally convex spaces, *Rev. Real. Acad. Cienc. Exact. Fis. Natur. Madrid* **79** (1985), 53–66.

[105] C. Alsina and R. Ger, On some inequalities and stability results related to the exponential function, *J. Inequal. Appl.* **2**, (1998), 373–380.

[106] G.A. Anastassiou, *Generalized Fractional Calculus: New Advancements and Applications*, Springer International Publishing, Switzerland, 2021.

[107] G.A. Anastassiou, *Advances on Fractional Inequalities*, Springer, New York, 2011.

[108] G.A. Anastassiou and I.K. Argyros, *Functional Numerical Methods: Applications to Abstract Fractional Calculus*, Springer, 2018.

[109] J. Andres and L. Górniewicz, *Topological Fixed Point Principles for Boundary Value Problems,* Kluwer Academic Publishers, Dordrecht, 2003.

[110] T. Aoki, On the stability of the linear transformation in Banach spaces. *J. Math. Soc. Japan* **2** (1950), 64–66.

[111] J. Appell, Implicit functions, nonlinear integral equations, and the measure of noncompactness of the superposition operator. *J. Math. Anal. Appl.* **83**, (1981), 251–263.

[112] J. Appell, J. Banaś, and N. Merentes, *Bounded Variation and Around*, De Gruyter Studies in Nonlinear Analysis and Applications, Vol. 17, Walter de Gruyter GmbH, Berlin, Germany, 2014.

[113] J.P. Aubin and A. Cellina, *Differential Inclusions*, Springer-Verlag, Berlin, 1984.

[114] A. Atangana, *Fractional Operators with Constant and Variable Order with Application to Geo-hydrology.* Academic Press, London, 2018.

[115] J.-P. Aubin, H. Frankowska, *Set-Valued Analysis,* Birkhäuser, Basel, 1990.

[116] C. Avramescu, Some remarks on a fixed point theorem of Krasnoselskii, *Electron. J. Qual. Theory Differ. Equ.* **5** (2003), 1–15.

[117] A. Babakhani and V. Daftardar-Gejji, Existence of positive solutions for multi-term non-autonomous fractional differential equations with polynomial coefficients. *Electron. J. Differential Equations* **2006** (129) (2006), 12 pp.

[118] A. Babakhani and V. Daftardar-Gejji, Existence of positive solutions for N-term non-autonomous fractional differential equations, *Positivity* **9** (2) (2005), 193–206.

[119] Z. Bai, On positive solutions of a nonlocal fractional boundary value problem, *Nonlinear Anal.* **72** (2) (2010), 916–924, 2010.

[120] D.D. Bainov and S.G. Hristova, Integral inequalities of Gronwall type for piecewise continuous functions, *J. Appl. Math. Stoc. Anal.* **10** (1997), 89–94.

[121] D.D. Bainov and P.S. Simeonov, *Systems with Impulsive Effect,* Horwood, Chichester, 1989.

[122] D.D. Bainov and P.S. Simeonov, *Impulsive Differential Equations: Periodic Solutions and Applications,* Pitman Monographs and Surveys in Pure and Applied Mathematics, Vol. 66, Longman Scientific & Technical and John Wiley & Sons, Inc., New York, 1993.

[123] D. Baleanu, K. Diethelm, E. Scalas, and J.J. Trujillo, *Fractional Calculus Models and Numerical Methods*, World Scientific Publishing, New York, 2012.

[124] D. Baleanu, Z.B. Güvenç, and J.A.T. Machado, *New Trends in Nanotechnology and Fractional Calculus Applications*, Springer, New York, 2010.

[125] D. Baleanu and A.M. Lopes, *Handbook of Fractional Calculus with Applications. Volume 7: Applications in Engineering, Life and Social Sciences, Part A*, De Gruyter, Berlin, 2019.

[126] D. Baleanu and A.M. Lopes, *Handbook of Fractional Calculus with Applications. Volume 8: Applications in Engineering, Life and Social Sciences, Part B*, De Gruyter, Berlin, 2019.

[127] D. Baleanu, J.A.T. Machado, and A.C.-J. Luo, *Fractional Dynamics and Control*, Springer, 2012.

[128] J. Banaś and K. Goebel, *Measures of Noncompactness in Banach Spaces*, Marcel Dekker, New York, 1980.

[129] J. Banaś and M. Mursaleen, *Sequence Spaces and Measures of Noncompactness with Applications to Differential and Integral Equations*, Springer, New Delhi, 2014.

[130] J. Banaś and L. Olszowy, Measures of noncompactness related to monotonicity, *Comment. Math. (Prace Mat.)* **41** (2001), 13–23.

[131] J. Banaś and T. Zając, A new approach to the theory of functional integral equations of fractional order. *J. Math. Anal. Appl.* **375** (2011), 375–387.

[132] M. Belmekki and M. Benchohra, Existence results for fractional order semilinear functional differential equations, *Proc. A. Razmadze Math. Inst.* **146** (2008), 9–20.

[133] M. Benchohra and F. Berhoun, Impulsive fractional differential equations with variable times, *Comput. Math. Appl.* **59** (2010), 1245–1252.

[134] M. Benchohra and F. Berhoun, Impulsive fractional differential equations with state dependent delay, *Commun. Appl. Anal.* **14** (2) (2010), 213–224.

[135] M. Benchohra, F. Berhoun, N. Hamidi, and J.J. Nieto, Fractional differential inclusions with anti-periodic boundary conditions, *Nonlinear Anal. Forum* **19** (2014), 27–35.

[136] M. Benchohra, F. Berhoun, and G.M. N'Guérékata, Bounded solutions for fractional order differential equations on the half-line, *Bull. Math. Anal. Appl.* **146** (4) (2012), 62–71.

[137] M. Benchohra and S. Bouriah, Existence and stability results for nonlinear implicit fractional differential equations with impulses, *Mem. Differ. Equ. Math. Phys.* **69** (2016), 15–31.

[138] M. Benchohra, S. Bouriah, and M. Darwish, Nonlinear boundary value problem for implicit differential equations of fractional order in Banach spaces, *Fixed Point Theory* **18** (2) (2017), 457–470.

[139] M. Benchohra, S. Bouriah, and J.Henderson, Existence and stability results for nonlinear implicit neutral fractional differential equations with finite delay and impulses, *Comm. Appl. Nonlinear Anal.* **22** (1) (2015), 46–67.

[140] M. Benchohra, S. Bouriah, J.E. Lazreg, and J. J. Nieto, Nonlinear implicit Hadamard fractional differential equations with delay in Banach spaces, *Acta Univ. Palack. Olomuc. Fac. Rerum Natur. Math.* **55** (2016), 15–26.

[141] M. Benchohra, S. Bouriah, J.J. Nieto, Existence of periodic solutions for nonlinear implicit Hadamard's fractional differential equations. *Revista Real Academia de Ciencias Exactas, Fisicas Nat. Serie A: Mat.* **112** (2018), 25–35.

[142] M. Benchohra, A. Cabada, and D. Seba, An existence result for non-linear fractional differential equations on Banach spaces, *Bound. Value Probl.* **2009** (2009), Article ID 628916, 11 pp.

[143] M. Benchohra, J.R. Graef, and S. Hamani, Existence results for boundary value problems with nonlinear fractional differential equations, *Appl. Anal.* **87** (7) (2008), 851–863.

[144] M. Benchohra, J. Graef, and F.-Z. Mostefai, Weak solutions for boundary-value problems with nonlinear fractional differential inclusions, *Nonlinear Dyn. Syst. Theory* **11** (3) (2011), 227–237.

[145] M. Benchohra and S. Hamani, Boundary value problems for differential inclusions with fractional order, *Discus. Mathem. Diff. Incl., Contr. Optim.* **28** (2008), 147–164.

[146] M. Benchohra, S. Hamani, and S.K. Ntouyas, Boundary value problems for differential equations with fractional order, *Surv. Math. Appl.* **3** (2008), 1–12.

[147] M. Benchohra, N. Hamidi, and J. Henderson, Fractional differential equations with anti-periodic boundary conditions *Numer. Funct. Anal. Optim.* **34** (4) (2013), 404–414.

[148] M. Benchohra, S. Hamani, J.J. Nieto, and B.A. Slimani, Existence of solutions to differential inclusions with fractional order and impulses, *Electron. J. Differential Equations* **2010** (80) (2010), 1–18.

[149] M. Benchohra and B. Hedia, Positive solutions for boundary value problems with fractional order, *Inter. J. Adv. Math. Sci.* **1** (1) (2013), 12–22.

[150] M. Benchohra, J. Henderson, and F-Z. Mostefai, Weak solutions for hyperbolic partial fractional differential inclusions in Banach spaces, *Comput. Math. Appl.* **64** (2012), 3101–3107.

[151] M. Benchohra, J. Henderson, and S.K. Ntouyas, *Impulsive Differential Equations and Inclusions*, Vol. 2, Hindawi Publishing Corporation, New York, 2006.

[152] M. Benchohra, J. Henderson, S.K. Ntouyas, and A. Ouahab, Existence results for fractional order functional differential equations with infinite delay, *J. Math. Anal. Appl.* **338** (2) (2008), 1340–1350.

[153] M. Benchohra, J. Henderson, and D. Seba, Measure of noncompactness and fractional differential equations in Banach spaces, *Commun. Appl. Anal.* **12** (4) (2008), 419–428.

[154] M. Benchohra and J.E. Lazreg, Nonlinear fractional implicit differential equations. *Commun. Appl. Anal.* **17** (2013), 471–482.

[155] M. Benchohra and J.E. Lazreg, Existence and uniqueness results for nonlinear implicit fractional differential equations with boundary conditions, *Rom. J. Math. Comput. Sci.* **4** (1) (2014), 60–72.

[156] M. Benchohra and J.E. Lazreg, Existence results for nonlinear implicit fractional differential equations, *Surv. Math. Appl.* **9** (2014), 79–92.

[157] M. Benchohra and J.E. Lazreg, Existence results for nonlinear implicit fractional differential equations with impulses, *Commun. Appl. Anal.* **19** (2015), 413–426.

[158] M. Benchohra and J.E. Lazreg, On stability for nonlinear implicit fractional differential equations, *Matematiche (Catania)* **70** (2) (2015), 49–61.

[159] M. Benchohra and J.E. Lazreg, Existence and Ulam stability for nonlinear implicit fractional differential equations with Hadamard derivatives, *Stud. Univ. Babe-s-Bolyai Math.* **62** (1) (2017), 27–38.

[160] M. Benchohra, J.E. Lazreg, and G.M. N'Guérékata, Nonlinear implicit Hadamard's fractional differential equations on Banach space with Retarded and Advanced arguments, *Internat. J. Evol. Equat.* **10** (2018), 283–295.

[161] M. Benchohra and D. Seba, Impulsive fractional differential equations in Banach Spaces, *Electron. J. Qual. Theory Differ. Equ. Spec. Ed.* **I** (8) (2009), 1–14.

[162] M. Benchohra and B.A. Slimani, Existence and uniqueness of solutions to impulsive fractional differential equations, *Electron. J. Differential Equations* **10** (2009), 1–11.

[163] M. Benchohra and M.S. Souid, Integrable solutions for implicit fractional order differential equations, *Transylv. J. Math. Mech.* **6** (2) (2014), 101–107.

[164] M. Benchohra and M.S. Souid, L^1-Solutions for implicit fractional order differential equations with nonlocal condition, *Filomat* **30** (6) (2016), 1485–1492.

[165] M. Benchohra and M.S. Souid, Integrable solutions For implicit fractional order functional differential equations with infinite delay, *Arch. Math. (Brno) Tomus* **51** (2015), 13–22.

[166] A. Bica, V.A. Caus, and S. Muresan, Application of a trapezoid inequality to neutral Fredholm integro-differential equations in Banach spaces, *J. Inequal. Pure Appl. Math.* **7** (2006), Art. 173.

[167] H.F. Bohnenblust and S. Karlin, *On a Theorem of Ville. Contribution to the Theory of Games*, Annals of Mathematics Studies, Vol. 24. Priceton University Press, Princeton, 1950, pp. 155–160.

[168] D. Bothe, Multivalued perturbation of m-accretive differential inclusions, *Isr. J. Math.* **108** (1998), 109–138.

[169] M. F. Bota-Boriceanu and A. Petruşel, Ulam-Hyers stability for operatorial equations and inclusions, *Analele Univ. I. Cuza Iasi* **57** (2011), 65–74.

[170] J. Brzdek, D. Popa, I. Rasa, and B. Xu, *Ulam Stability of Operators,* Mathematical Analysis and Its Applications. Academic Press, London, 2018.

[171] D. Bugajewski and S. Szufla, Kneser's theorem for weak solutions of the Darboux problem in a Banach space, *Nonlinear Anal.* **20** (2) (1993), 169–173.

[172] T.A. Burton and C. Kirk, A fixed point theorem of Krasnoselskii–Schaefer type. *Math. Nachr.* **189** (1989), 23–31.

[173] T.A. Burton and T. Furumochi, A note on stability by Schauder's theorem, *Funkcial. Ekvac.* **44** (2001), 73–82.

[174] P. L. Butzer, A.A. Kilbas, and J.J. Trujillo. Fractional calculus in the mellin setting and Hadamard-type fractional integrals, *J. Math. Anal. Appl.* **269** (2002), 1–27.

[175] P.L. Butzer, A.A. Kilbas, and J.J. Trujillo. Mellin transform analysis and integration by parts for Hadamard-type fractional integrals, *J. Math. Anal. Appl.* **270** (2002), 1–15.

[176] L. Byszewski, Theorems about existence and uniqueness of solutions of a semilinear evolution nonlocal Cauchy problem, *J. Math. Anal. Appl.* **162** (1991), 494–505.

[177] L. Byszewski, Existence and uniqueness of mild and classical solutions of semilinear functional-differential evolution nonlocal Cauchy problem. *Selected Problems of Mathematics*, 50th Anniv. Cracow Univ. Technol. Anniv. Issue, 6, Cracow Univ. Technol., Krakow, 1995, pp. 25–33.

[178] L. Byszewski and V. Lakshmikantham, Theorem about the existence and uniqueness of a solution of a nonlocal abstract Cauchy problem in a Banach space, *Appl. Anal.* **40** (1991), 11–19.

[179] C. Castaing and M. Valadier, *Convex Analysis and Measurable Multifunctions*, Lecture Notes in Mathematics, Vol. 580, Springer-Verlag, Berlin, 1977.

[180] E. Capelas de Oliveira, *Solved Exercises in Fractional Calculus*, Springer International Publishing, Switzerland, 2019.

[181] K. Cao and Y. Chen, *Fractional Order Crowd Dynamics: Cyber-Human System Modeling and Control*, De Gruyter, Berlin, 2018.

[182] C. Cattani, H.M. Srivastava, and X.J. Yang, *Fractional Dynamics*, De Gruyter, 2016.

[183] A. Cernea, Arcwise connectedness of the solution set of a nonclosed nonconvex integral inclusion, *Miskolc Math. Notes* **9** (1) (2008), 33–39.

[184] S. Chakraverty, B. Tapaswini, and D. Behera, *Fuzzy arbitrary order system. Fuzzy Fractional Differential Equations and Applications*, John Wiley & Sons, Inc., Hoboken, NJ, 2016.

[185] Y. Chang, A. Anguraj and P. Karthikeyan, Existence results for initial value problems with integral condition for impulsive fractional differential equations. *J. Fract. Calc. Appl.* **2** (7) (2012), 1–10.

[186] H. Y. Chen, Successive approximations for solutions of functional integral equations, *J. Math. Anal. Appl.* **80** (1981), 19–30.

[187] A. Chen and Y. Chen, Existence of Solutions to Anti-periodic boundary value problem for nonlinear fractional differential equations with impulses. *Adv. Difference. Equat.* **2011** (2011), 17 pp.

[188] J. Chen, F. Liu, I. Turner, and V. Anh, The fundamental and numerical solutions of the Riesz space-fractional reaction-dispersion equation, *ANZIAM J.* **50** (1) (2008), 45–57

[189] Y. J. Cho, Th. M. Rassias, and R. Saadati, *Stability of Functional Equations in Random Normed Spaces*, Vol. 52 Science-Business Media, Springer, 2013.

[190] C. Corduneanu, *Integral Equations and Applications*, Cambridge University Press, 1991.

[191] H. Covitz and S. B. Nadler Jr, Multivalued contraction mappings in generalized metric spaces, *Israel J. Math.* **8** (1970), 5–11.

[192] R. F. Curtain and A. J. Pritchard, *Functional Analysis in Modern Applied Mathematics* Academic Press. 1977.

[193] T. Człapiński, Global convergence of successive approximations of the Darboux problem for partial functional differential equations with infinite delay, *Opuscula Math.* **34** (2) (2014), 327–338.

[194] V. Daftardar-Gejji, *Fractional Calculus and Fractional Differential Equations*, Birkhäuser, Basel, 2019.

[195] G. Darbo, Punti uniti in transformazioni a condominio non compatto, *Rend Sem. Mat. Univ. Padova* **24** (1955), 84–92.

[196] K. Deimling, *Nonlinear Functional Analysis*, Springer-Verlag, 1985.

[197] K. Deimling, *Multivalued Differential Equations*, Walter De Gruyter, Berlin, 1992.

[198] K. Deng, Exponential decay of solutions of semilinear parabolic equations with nonlocal initial conditions, *J. Math. Anal. Appl.* **179** (1993), 630–637.

[199] F.S. De Blasi, On the property of the unit sphere in a Banach space, *Bull. Math. Soc. Sci. Math. R.S. Roumanie* **21** (1977), 259–262.

[200] H. Dutta, A.O. Akdemir, and A. Atangana, *Fractional Order Analysis: Theory, Methods and Applications*, Wiley, Hoboken, NJ, 2020.

[201] F.S. De Blasi and J. Myjak, Some generic properties of functional differential equations in Banach space, *J. Math. Anal. Appl.* **80** (1981), 19–30.

[202] B.C. Dhage, Multi-valued condensing random operators and functional random integral inclusions, *Opuscula Math.* **31**(1) (2011), 27–48.

[203] B.C. Dhage, S.V. Badgire, and S.K. Ntouyas, Periodic boundary value problems of second order random differential equations, *Electron. J. Qual. Theory Diff. Equ.* **21** (2009), 1–14.

[204] S. Djebali, L. Górniewicz, and A. Ouahab, *Existence and Structure of Solution Sets for Impulsive Differential Inclusions: A Survey*. Lecture Notes in Nonlinear Analysis, Julius Schauder University, Centre for Nonlinear Studies, Nicolaus Copernicus University, Vol. 13, 2012.

[205] K. Diethelm, *The Analysis of Fractional Differential Equations*, Lecture Notes in Mathematics, Springer, 2010.

[206] A. El-Sayed and F. Gaafar, Stability of a nonlinear non-autonomous fractional order systems with different delays and non-local conditions. *Adv. Difference Equ.* **47** (1) (2011), 12 pp.

[207] H.W. Engl, A general stochastic fixed-point theorem for continuous random operators on stochastic domains, *J. Math. Anal. Appl.* **66** (1978), 220–231.

[208] N. Engheta, Fractional curl operator in electromagnetics. *Microwave Opt. Tech. Lett.* **17** (1) (1998), 86–91.

[209] L. Faina, The generic property of global covergence of successive approximations for functional differential equations with infinite delay, *Commun. Appl. Anal.* **3** (1999), 219–234.

[210] M. Faryad and Q.A. Naqvi, Fractional rectangular waveguide, *Progress Electromagnetics Res.* **75** (2007), 383–396.

[211] M. Fečkan, J.R. Wang, and M. Pospíšil, *Fractional-Order Equations and Inclusions*, De Gruyter, 2017.

[212] M. Francesco, *Fractional Calculus: Theory and Applications*, MDPI, 2018.

[213] Z. Gao; A computing method on stability intervals of time-delay for fractional-order retarded systems with commensurate time-delays, *Automatica* **50** (2014), 1611–1616.

[214] P. Gavruta, A generalization of the Hyers–Ulam–Rassias stability of approximately additive mappings, *J. Math. Anal. Appl.* **184** (1994), 431–436.

[215] K. Goebel, *Concise Course on Fixed Point Theorems*, Yokohama Publishers, Japan, 2002.

[216] S.G. Georgiev, *Fractional Dynamic Calculus and Fractional Dynamic Equations on Time Scales,* Springer Nature, 2018.

[217] L. Georgiev and V.G. Angelov, On the existence and uniqueness of solutions for maximum equations, *Glasnik Matematički*, **37** (2) (2002), 275–281.

[218] L. Górniewicz, *Topological Fixed Point Theory of Multivalued Mappings*, Mathematics and its Applications, Vol. 495, Kluwer Academic Publishers, Dordrecht, 1999.

[219] L. Gorniewicz and T. Pruszko, On the set of solutions of the Darboux problem for some hyperbolic equations, *Bull. Acad. Polon. Sci. Math. Astronom. Phys.* **38** (1980), 279–285.

[220] C. Goodrich and A.C. Peterson, *Discrete Fractional Calculus*, Springer International Publishing, New York, 2016.

[221] J.R. Graef, J. Henderson, and A. Ouahab, *Impulsive Differential Inclusions. A Fixed Point Approach*, De Gruyter, Berlin, 2013.

[222] A. Granas and J. Dugundji, *Fixed Point Theory*, Springer-Verlag, New York, 2003.

[223] A. Guezane-Lakoud and R. Khaldi, Solvability of a fractional boundary value problem with fractional integral condition, *Nonlinear Anal.* **75** (2012), 2692–2700.

[224] B. Guo, X. Pu, and F. Huang, *Fractional Partial Differential Equations and Their Numerical Solutions*, World Scientific Publishing, Hackensack, 2015.

[225] D.J. Guo, V. Lakshmikantham, and X. Liu, *Nonlinear Integral Equations in Abstract Spaces,* Kluwer Academic Publishers, Dordrecht, 1996.

[226] Z. Guo and M. Liu, Existence, and uniqueness of solutions for fractional order integrodifferential equations with nonlocal initial conditions. *Panamer. Math. J.* **21** (2011), 51–61.

[227] J. Hadamard, Essai sur l'étude des fonctions données par leur développment de Taylor, *J. Pure Appl. Math.* **4** (8) (1892), 101–186.

[228] J. Hale and J. Kato, Phase space for retarded equations with infinite delay, *Funkcial. Ekvac.* **21** (1978), 11–41.

[229] J. Hale and S.M. Verduyn Lunel, *Introduction to Functional Differential Equations*, Applied Mathematicals Sciences, Vol. 99, Springer-Verlag, New York, 1993.

[230] X. Han, X. Ma, and G. Dai, Solutions to fourth-order random differential equations with periodic boundary conditions, *Electron. J. Differential Equations*, **235** (2012) 1–9.

[231] A. Harrat, J.J. Nieto, and A. Debbouche, Solvability and optimal controls of impulsive Hilfer fractional delay evolution inclusions with Clarke subdifferential, *J. Comput. Appl. Math.* **344** (2018), 725–737.

[232] J. Henderson and R. Luca, *Boundary Value Problems for Systems of Differ-ential, Difference and Fractional Equations — Positive Solutions*, Elsevier, 2016.

[233] J. Henderson and A. Ouahab, Impulsive differential inclusions with frac-tional order, *Comput. Math. Appl.* **59** (2010), 1191–1226.

[234] J. Henderson and A. Ouahab, A Filippov's theorem, some existence results and the compactness of solution sets of impulsive fractional order differen-tial inclusions, *Mediterr. J. Math.* **9** (3) (2012), 453–485.

[235] R. Hermann, *Fractional Calculus: An Introduction For Physicists*, World Scientific Publishing, 2011.

[236] N. Heymans and I. Podlubny, Physical interpretation of initial condi-tions for fractional differential equations with Riemann–Liouville fractional derivatives, *Rheol. Acta* **45** (2006), 765–771.

[237] R. Hilfer, *Applications of Fractional Calculus in Physics*, World Scientific Publishing, Singapore, 2000.

[238] R. Hilfer, Threefold introduction to fractional derivatives, *Anomalous Transport: Foundations and Applications,* 17–73, 2008.

[239] Ch. Horvath, Measure of Non-compactness and multivalued mappings in complete metric topological spaces, *J. Math. Anal. Appl.* **108** (1985), 403–408.

[240] Y. Hino, S. Murakami, and T. Naito, *Functional Differential Equations with Infinite Delay*, Springer-Verlag, Berlin, 1991.

[241] Sh. Hu and N. Papageorgiou, *Handbook of Multivalued Analysis, Volume I: Theory*, Kluwer Academic Publishers, Dordrecht, 1997.

[242] D.H. Hyers, *On the stability of the linear functional equation, Proc. Natl. Acad. Sci. USA* **27** (1941), 222–224.

[243] D.H. Hyers, G. Isac, and Th.M. Rassias, *Stability of Functional Equations in Several Variables*, Birkhäuser, 1998.

[244] D.H. Hyers, On the stability of the linear functional equation, *Proc. Natl. Acad. Sci.* **27** (1941), 222–224.

[245] S. Itoh, Random fixed point theorems with applications to random differen-tial equations in Banach spaces, *J. Math. Anal. Appl.,* **67** (1979), 261–273.

[246] S.-M. Jung, *Hyers–Ulam–Rassias Stability of Functional Equations in Mathematical Analysis*, Hadronic Press, Palm Harbor, 2001.

[247] S.-M. Jung, *Hyers–Ulam–Rassias Stability of Functional Equations in Non-linear Analysis*, Springer, New York, 2011.

[248] S.-M. Jung, A fixed point approach to the stability of a Volterra integral equation, *Fixed Point Theory Appl.* **2007** (2007), Article ID 57064, 9 pp.

[249] Rabha W. Ibrahim, Stability for Univalent Solutions of Complex Fractional Differential Equations, *Proc. Pakistan Acad. Sci.* **49** (2012), 227–232.

[250] B. Jin, *Fractional Differential Equations: An Approach via Fractional Derivatives*, Springer International Publishing, Switzerland, 2021.

[251] K.W. Jun and H.M. Kim, On the stability of an n-dimensional quadratic and additive functional equation, *Math. Inequal. Appl.* **19** (9) (2006), 854–858.

[252] S.M. Jung, On the Hyers–Ulam stability of the functional equations that have the quadratic property, *J. Math. Anal. Appl.* **222** (1998), 126–137.

[253] S.M. Jung, Hyers–Ulam stability of linear differential equations of first order, II, *Appl. Math. Lett.* **19** (2006), 854–858.

[254] S.M. Jung and K.S. Lee, Hyers–Ulam stability of first order linear partial differential equations with constant coefficients, *Math. Inequal. Appl.* **10** (2007), 261–266.

[255] B. Jin, *Fractional Differential Equations: An Approach via Fractional Derivatives*, Springer International Publishing, Switzerland, 2021.

[256] K. Karthikeyan and J.J. Trujillo, Existence and uniqueness results for fractional integrodifferential equations with boundary value conditions, *Commun. Nonlinear Sci. Numer. Simulat.* **17** (2012) 4037–4043.

[257] A. Kochubei and Y. Luchko, *Handbook of Fractional Calculus with Applications. Volume 1: Basic Theory*, De Gruyter, Berlin, 2019.

[258] A. Kochubei, Y. Luchko, *Handbook of Fractional Calculus with Applications. Volume 2: Fractional Differential Equations*, De Gruyter, Berlin, 2019.

[259] A.A. Kilbas, Hadamard-type fractional calculus. *J. Korean Math. Soc.* **38** (6) (2001) 1191–1204.

[260] A.A. Kilbas, B. Bonilla, and J. Trujillo, Nonlinear differential equations of fractional order in a space of integrable functions. *Dokl. Ross. Akad. Nauk* **37** (4) (2000), 445–449.

[261] A.A. Kilbas and S.A. Marzan, Nonlinear differential equations with the Caputo fractional derivative in the space of continuously differentiable functions, *Diff. Equat.* **41** (2005), 84–89.

[262] A.A. Kilbas, H.M. Srivastava, and Juan J. Trujillo, *Theory and Applications of Fractional Differential Equations*, North-Holland Mathematics Studies, Vol. 204. Elsevier Science B.V., Amsterdam, 2006.

[263] W.A. Kirk and B. Sims, *Handbook of Metric Fixed Point Theory*, Springer-Science + Business Media, B.V, Dordrecht, 2001.

[264] G.H. Kim, On the stability of functional equations with square-symmetric operation, *Math. Inequal. Appl.* **17** (4) (2001), 257–266.

[265] M. Kisielewicz, *Differential Inclusions and Optimal Control*, Kluwer Academic Publishers, Dordrecht, The Netherlands, 1991.

[266] W.A. Kirk and B. Sims, *Handbook of Metric Fixed Point Theory*, Springer-Science + Business Media, B.V, Dordrecht, 2001.

[267] D. Kumar, *Fractional Calculus in Medical and Health Science*, CRC Press, Boca Raton, 2021.

[268] K. Kuratowski, Sur les espaces complets, *Fund. Math.* **15** (1930), 301–309.

[269] V. Lakshmikantham, D.D. Bainov, and P.S. Simeonov; *Theory of Impulsive Differntial Equations*, Worlds Scientific Publishing, Singapore, 1989.

[270] V. Lakshmikantham, S. Leela, and J. Vasundhara, *Theory of Fractional Dynamic Systems*, Cambridge University Press, Cambridge, 2009.

[271] V. Lakshmikantham and J. Vasundhara Devi. Theory of fractional differential equations in a Banach space, *Eur. J. Pure Appl. Math.* **1** (2008), 38–45.

[272] V. Lakshmikantham and A.S. Vatsala, Basic theory of fractional differential equations, *Nonlinear Anal.* **69** (2008), 2677–2682.

[273] V. Lakshmikantham and A.S. Vatsala, General uniqueness and monotone iterative technique for fractional differential equations, *Appl. Math. Lett.* **21** (2008) 828–834.

[274] V. Lakshmikantham and A.S. Vatsala, Theory of fractional differential inequalities and applications, *Commun. Appl. Anal.* **11** (3–4) (2007), 395–402.

[275] V.L. Lazăr, Fixed point theory for multivalued φ–contractions, *Fixed Point Theory Appl.* **2011** (50) (2011), 12 pp.

[276] A. Lasota and Z. Opial, An application of the Kakutani–Ky Fan theorem in the theory of ordinary differential equations, *Bull. Acad. Pol. Sci. Ser. Sci. Math. Astronom. Phys.* **13** (1965), 781–786.

[277] Y. Li, Y.Q. Chen, and I. Podlubny, Stability of fractional-order nonlinear dynamic systems: Lyapunov direct method and generalized Mittag-Leffler stability, *Comput. Math. Appl.* **59** (2010), 1810–1821.

[278] Y. Li, Y.Q. Chen, and I. Podlubny, Mittag-Leffler stability of fractional order nonlinear dynamic systems, *Automatica* **45** (2009), 1965–1969.

[279] L. Liu, F. Guo, C. Wu, and Y. Wu, Existence theorems of global solutions for nonlinear Volterra type integral equations in Banach spaces, *J. Math. Anal. Appl.* **309** (2005), 638–649.

[280] Q. Liu, F. Liu, I. Turner, and V. Anh, Numerical simulation for the 3D seep age flow with fractional derivatives in porous media, *IMA J. Appl. Math.* **74** (2) (2009), 201–229.

[281] Lizhen Chen and Zhenbin Fan, On mild solutions to fractional differential equations with nonlocal conditions, *Electron. J. Qual. Theory Differ. Equ.* 53 (2011), 1–13.

[282] A.J. Luo and V. Afraimovich, *Long-Range Interactions, Stochasticity and Fractional Dynamics*, Springer, New York, 2010.

[283] C. Li, F. Zeng, and F. Liu, Spectral approximations to the fractional integral and derivative, *Fract. Calc. Appl. Anal.* **15** (3) (2012), 383–406.

[284] F. Mainardi, *Fractional Calculus and Waves in Linear Viscoelasticity: An Introduction to Mathematical Models*, World Scientific Publishing, 2010.

[285] M. Martelli, A Rothe's type theorem for noncompact acyclic-valued map, *Boll. Un. Math. Ital.,* **11** (1975), 70–76.

[286] C. Milici, G. Draganescu, and J.A.T. Machado, *Introduction to Fractional Differential Equations*, Springer International Publishing, 2019.

[287] V.D. Milman and A.A. Myshkis, On the stability of motion in the presence of impulses, *Sib. Math. J.* **1** (1960), 233–237 (in Russian).

[288] K.S. Miller and B. Ross, *An Introduction to the Fractional Calculus and Differential Equations*, John Wiley & Sons, New York, 1993.

[289] A.R. Mitchell and Ch. Smith. An existence theorem for weak solutions of differential equations in Banach spaces, In: V. Lakshmikantham, (ed.) Nonlinear Equations in Abstract Spaces. Academic Press, New York, 1978, pp. 387–403.

[290] H. Mönch, Boundary value problems for nonlinear ordinary differential
       equations of second order in Banach spaces, *Nonlinear Anal.* **4** (1980),
       985–999.
[291] S.A. Murad and S. Hadid, An existence and uniqueness theorem for frac-
       tional differential equation with integral boundary condition, *J. Frac. Calc.
       Appl.* **3** (6), (2012), 1–9.
[292] S.B. Nadler Jr., Multivalued contraction mappings, *Pacific J. Math.,* **30**
       (1969), 475–488.
[293] Q.A. Naqvi and M. Abbas, Complex and higher order fractional curl oper-
       ator in electromagnetics, *Optics Communi.* **241** (2004), 349–355.
[294] G.M. N'Guérékata, A Cauchy problem for some fractional abstract differ-
       ential equation with non local conditions, *Nonlinear Anal.* **70** (5) (2009),
       1873–1876.
[295] G.M. N'Guérékata, Corrigendum: A Cauchy problem for some fractional
       differential equations, *Commun. Math. Anal.* **7** (2009), 11–11.
[296] A. Nowak, Applications of random fixed point theorem in the theory of gen-
       eralized random differential equations, *Bull. Polish. Acad. Sci.* **34** (1986),
       487–494.
[297] M. Obloza, Hyers stability of the linear differential equation, *Rocznik Nauk-
       Dydakt. Prace Mat.* **13** (1993), 259–270.
[298] K.B. Oldham and J. Spanier, *The Fractional Calculus: Theory and Applica-
       tion of Differentiation and Integration to Arbitrary Order*, Academic Press,
       New York, 1974.
[299] M.D. Ortigueira, *Fractional Calculus for Scientists and Engineers.* Lecture
       Notes in Electrical Engineering, Vol. 84. Springer, Dordrecht, 2011.
[300] M.D. Ortigueira and D. Valério, *Fractional Signals and Systems*, De
       Gruyter, 2020.
[301] D. O'Regan, Fixed point theory for weakly sequentially continuous map-
       ping, *Math. Comput. Model.* **27** (5) (1998), 1–14.
[302] D. O'Regan, Weak solutions of ordinary differential equations in Banach
       spaces, *Appl. Math. Lett.* **12** (1999), 101–105.
[303] D. O'Regan and R. Precup; Fixed point theorems for set-valued maps and
       existence principles for integral inclusions, *J. Math. Anal. Appl.* **245** (2000),
       594–612.
[304] D. Otrocol, Hybrid differential equations with maxima via Picard operators
       theory, *Stud. Univ. Babeş-Bolyai Math.* **61** (4) (2016), 421–428.
[305] D. Otrocol and I.A. Rus, Functional differential equations with "maxima"
       via weakly picard operators theory, *Bull. Math. Soc. Sci. Math. Roumanie,*
       51 (99) (2008), 253–261.
[306] B.G. Pachpatte, On nonlinear integral and discrete inequalities in two inde-
       pendent variables, *Bul. Sti. Tech. Inst. Politehn. Timisoara* **40** (54) (1995),
       29–38.
[307] B.G. Pachpatte, On Volterra-Fredholm integral equation in two variables,
       Demonstr. Math. **XL** (4) (2007), 839–852.
[308] B.G. Pachpatte, On Fredholm type integrodifferential equation, *Tamkang
       J. Math.* **39** (1) (2008), 85–94.

[309] B.G. Pachpatte, On Fredholm type integral equation in two variables *Differ. Equ. Appl.* **1** (2009), 27–39.

[310] A. Petruşel, Multivalued weakly Picard operators and applications, *Sci. Math. Japon.* **59** (2004), 167–202.

[311] N.A. Perestyuk, V.A. Plotnikov, A.M. Samoilenko, and N.V. Skripnik, *Differential Equation with Impulse Effects, Multivalued Right-Hand Sides with Discontinuities*, Walter de Gruyter, Berlin, 2011.

[312] Ivo Petras, *Fractional-Order Nonlinear Systems: Modeling, Analysis and Simulation* Springer, Heidelberg, 2011.

[313] I. Petras, *Handbook of Fractional Calculus with Applications. Volume 6: Applications in Control*, De Gruyter, Berlin, 2019.

[314] T.P. Petru, A. Petruşel, and J.-C. Yao, Ulam-Hyers stability for operatorial equations and inclusions via nonself operators, *Taiwanese J. Math.* **15** (2011), 2169–2193.

[315] I. Podlubny, *Fractional Differential Equations*, Academic Press, San Diego, 1999.

[316] S. Pooseh, R. Almeida, and D. Torres. Expansion formulas in terms of integer-order derivatives for the hadamard fractional integral and derivative. *Numer. Funct. Anal. Optim.* **33** (3) (2012), 301–319.

[317] Y. Povstenko, *Linear Fractional Diffusion-Wave Equation for Scientists and Engineers*, Birkhäuser Mathematics, Springer, New York, 2015.

[318] Y. Povstenko, *Fractional Thermoelasticity*, Solid Mechanics and Its Applications, Vol. 219, Springer, New York, 2015.

[319] M.D. Qassim, K.M. Furati, and N.-e. Tatar, On a differential equation involving Hilfer-Hadamard fractional derivative, *Abst. Appl. Anal.* **2012** (2012), Article ID 391062, 17 pp.

[320] M.D. Qassim and N.-E. Tatar, Well-posedness and stability for a differential problem with Hilfer–Hadamard fractional derivative, *Abst. Appl. Anal.* **2013** (2013), Article ID 605029, 12 pp.

[321] Th. M. Rassias, On the stability of the linear mapping in Banach spaces, *Proc. Amer. Math. Soc.* **72** (1978), 297–300.

[322] J.M. Rassias, *Functional Equations, Difference Inequalities and Ulam Stability Notions (F.U.N.)*, Nova Science Publishers, Inc., New York, 2010.

[323] Th.M. Rassias, *Handbook of Functional Equations, Stability Theory*, Springer, 2014.

[324] Th.M. Rassias and J. Brzdek, *Functional Equations in Mathematical Analysis*, Springer, New York 2012.

[325] S.S. Ray, A new approach for the application of Adomian decomposition method for the solution of fractional space diffusion equation with insulated ends, *Appl. Math. Comput.* **202** (2) (2008), 544–549.

[326] S.S. Ray and A.K. Gupta, *Wavelet Methods for Solving Partial Differential Equations and Fractional Differential Equations*, CRC Press, Taylor & Francis Group, 2018.

[327] B. Ross, *Fractional Calculus and Its Applications, Proceedings of the International Conference*, New Haven, Springer-Verlag, New York, 1974.

[328] I.A. Rus, Ulam stability of ordinary differential equations, *Studia Univ. Babes-Bolyai, Math.* **LIV** (4) (2009), 125–133.

[329] I.A. Rus, Remarks on Ulam stability of the operatorial equations, *Fixed Point Theory* **10** (2009), 305–320.

[330] I.A. Rus, Fixed points, upper and lower fixed points: abstract Gronwall lemmas, *Carpathian J. Math.* **20** (2004), 125–134.

[331] I.A. Rus, Picard operators and applications *Sci. Math. Jpn.* **58** (2003), 191–219.

[332] I.A. Rus, *Generalized Contractions and Applications*, Cluj University Press, Cluj-Napoca, 2001.

[333] I.A. Rus, Weakly Picard operators and applications, *Semin. Fixed Point Theory, Cluj-Napoca* **2** (2001), 41–57.

[334] I.A. Rus, Ulam stabilities of ordinary differential equations in a Banach space, *Carpathian J. Math.* **26** (2010), 103–107.

[335] I.A. Rus, A. Petruşel, and A. Sîtămărian, Data dependence of the fixed points set of some multivalued weakly picard operators, *Nonlinear Anal.* **52** (2003), 1947–1959.

[336] L. Rybinski, On Carathédory type selections, *Fund. Math.* **125** (1985), 187–193.

[337] J. Sabatier,P. Lanusse, P. Melchior, and A. Oustaloup, *Fractional Order Differentiation and Robust Control Design, CRONE, H-infinity and Motion Control*, Intelligent Systems, Control and Automation: Science and Engineering, Vol. 77 **77**, Springer, New York, 2015.

[338] S. Saha Ray and S. Sahoo, *Generalized Fractional Order Differential Equations Arising in Physical Models*, CRC Press, Boca Raton, 2019.

[339] P. Sahoo, T. Barman, and J.P. Davim, *Fractal Analysis in Machining*, Springer, New York, 2011.

[340] S.G. Samko, A.A. Kilbas, and O.I. Marichev, *Fractional Integrals and Derivatives. Theory and Applications*, Gordon and Breach, Yverdon, 1993.

[341] A.M. Samoĭlenko and N.A. Perestyuk, *Impulsive Differential Equations* World Scientific Publishing, Singapore, 1995.

[342] J.S. Shin, Global convergence of successive approximations of solutions for functional differential equations with infinite delay, *Tôhoku Math. J.* **39** (1986), 557–574.

[343] E. Shishkina and S.M. Sitnik, *Transmutations, Singular and Fractional Differential Equations with Applications to Mathematical Physics*, Academic Press, 2020.

[344] I.M. Stamova and G.T. Stamov, *Functional and Impulsive Differential Equations of Fractional Order — Qualitative Analysis and Applications*, CRS Press, 2017.

[345] X. Su and L. Liu, Existence of solution for boundary value problem of nonlinear fractional differential equation, *Appl. Math.* **22** (3) (2007) 291–298.

[346] Z.Z. Sun and G. Gao, *Fractional Differential Equations: Finite Difference Methods*, De Gruyter, 2020.

[347] V.E. Tarasov, *Handbook of Fractional Calculus with Applications. Volume 4: Applications in Physics, Part A*, De Gruyter, Berlin, 2019.

[348] V.E. Tarasov, *Handbook of Fractional Calculus with Applications. Volume 5: Applications in Physics, Part B*, De Gruyter, Berlin, 2019.

[349] V.E. Tarasov, *Fractional Dynamics: Application of Fractional Calculus to Dynamics of particles, Fields and Media*, Springer, Heidelberg, Higher Education Press, Beijing, 2010.

[350] K. Tas, D. Baleanu, and J.A. Tenreiro Machado *Mathematical Methods in Engineering: Applications in Dynamics of Complex Systems*, Springer Nature, Switzerland AG, 2019.

[351] J.M.A. Toledano, T.D. Benavides, and G.L. Acedo, *Measures of Noncompactness in Metric Fixed Point Theory*, Birkhauser, Basel, 1997.

[352] Ž. Tomovski, R. Hilfer, and H.M. Srivastava, Fractional and operational calculus with generalized fractional derivative operators and Mittag-Leffler type functions, *Integral Transf. Spec. Funct.* **21** (11) (2010), 797–814.

[353] V. Uchaikin and R. Sibatov, *Fractional Kinetics in Solids: Anomalous Charge Transport in Semiconductors, Dielectrics and Nanosystems*, World Scientific Publishing, 2013.

[354] S.M. Ulam, *Problems in Modern Mathematics*, Chapter 6, John Wiley & Sons, New York, 1940.

[355] S.M. Ulam, *A Collection of Mathematical Problems*, Interscience Publishers, New York, 1968.

[356] S. Umarov, *Introduction to Fractional and Pseudo-Differential Equations with Singular Symbols*, Developments in Mathematics, Vol. 41, Springer, New York, 2015.

[357] A.N. Vityuk and A.V. Golushkov, Existence of solutions of systems of partial differential equations of fractional order, *Nonlinear Oscil.* **7** (3) (2004), 318–325.

[358] A.N. Vityuk and A.V. Mykhailenko, The Darboux problem for an implicit fractional-order differential equation, *J. Math. Sci.* **175** (4) (2011), 391–401.

[359] V. Vyawahare and P.S.V. Nataraj, *Fractional-Order Modeling of Nuclear Reactor: From Subdiffusive Neutron Transport to Control-Oriented Models. A Systematic Approach*, Springer, Singapore, 2018.

[360] G. Wang, B. Ahmad, L. Zhang, and J.J. Nieto, Comments on the concept of existence of solution for impulsive fractional differential equations, *Electron. Commun. Nonlinear Sci. Numer. Simulat.* **19**, (2014), 401–403.

[361] J. Wang, M. Fečkan, and Y. Zhou, Ulam's type stability of impulsive ordinary differential equations, *J. Math. Anal. Appl.* **395** (2012), 258–264.

[362] G. Wang, B. Ahmad, and L. Zhang, Impulsive anti-periodic boundary value problem for nonlinear differential equations of fractional order. *Nonlinear Anal.* **74** (2011), 792–804.

[363] G. Wang, B. Ahmad, and L. Zhang, Some existence results for impulsive nonlinear fractional differential equations with mixed boundary conditions, *Comput. Math. Appl.* **62** (2011), 1389–1397.

[364] F. Wang and Z. Liu, Anti-periodic fractional boundary value problems for nonlinear differential equations of fractional order, *Adv. Difference Equat.* (2012), 12 pp.

[365] J. Wang, L. Lv, and Y. Zhou, Ulam stability and data dependence for fractional differential equations with Caputo derivative, *Electron. J. Qual. Theory Differ. Equat.* **63** (2011), 1–10.

[366] J. Wang and Y. Zhang, Existence and stability of solutions to nonlinear impulsive differential equations in β-normed spaces, *Electron. J. Differential Equations* **83** (2014), 1–10.

[367] R. Węgrzyk, Fixed point theorems for multifunctions and their applications to functional equations, *Dissertationes Math. (Rozprawy Mat.)* **201** (1982), 28 pp.

[368] J. West, *Nature's Patterns and the Fractional Calculus*, De Gruyter, 2017.

[369] D. Xue, *Fractional-Order Control Systems, Fundamentals and Numerical Implementations*, De Gruyter, 2017.

[370] X.J. Yang, *General Fractional Derivatives: Theory, Methods and Applications*, CRC Press, Boca Raton, 2019.

[371] X.J. Yang, D. Baleanu, and H.M. Srivastava, *Local Fractional Integral Transforms and Their Applications*, Elsevier, 2016.

[372] H. Ye, J. Gao, and Y. Ding, A generalized Gronwall inequality and its application to a fractional differential equation, *J. Math. Anal. Appl.* **328** (2007), 1075–1081.

[373] K. Yosida, *Functional Analysis,* 6th edn., Springer-Verlag, Berlin, 1980.

[374] D.P. Zielinski and V.R. Voller, A random walk solution for fractional diffusion equations, *Inter. J. Numerical Meth. Heat Fluid Flow* **23** (2013), 7–22.

[375] S. Zhai, X. Feng, and Z. Weng, New high-order compact adi algorithms for 3D nonlinear time-fractional convection–diffusion equation, *Math. Prob. Eng* **2013** (2013), 11 pp.

[376] S. Zhang, Positive solutions for boundary-value problems of nonlinear fractional diffrential equations, *Electron. J. Differential Equations* **36** (2006), 1–12.

[377] X. Zhang, J. Liu, L. Wei, and C. Ma, Finite element method for Gronwald-Letnikov time-fractional partial differential equation, *Appl. Anal.* **92** (10) (2013), 1–12.

[378] L. Zhang and G. Wang, Existence of solutions for nonlinear fractional differential equations with impulses and anti-Periodic boundary conditions, *Electron. J. Qual. Theory Differ. Equ.* **7** (2011), 1–11.

[379] S. Zhang and J. Sun, Existence of mild solutions for the impulsive semilinear nonlocal problem with random effects, *Adv. Difference Equ.* **19** (2014), 1–11.

[380] Y. Zhou, *Fractional Evolution Equations and Inclusions: Analysis and Control*, Elsevier Science, 2016.

[381] Y. Zhou, *Basic Theory of Fractional Differential Equations*, World Scientific Publishing, Singapore, 2014.

Index